Molecules:
Nucleation, Aggregation and Crystallization

Beyond Medical and Other Implications

Molecules: Nucleation, Aggregation and Crystallization

Beyond Medical and Other Implications

edited by

Jan Sedzik
Karolinska Institutet and Royal Institute of Technology, Sweden

Paolo Riccio
Potenza University, Italy

World Scientific

NEW JERSEY · LONDON · SINGAPORE · BEIJING · SHANGHAI · HONG KONG · TAIPEI · CHENNAI

Published by

World Scientific Publishing Co. Pte. Ltd.
5 Toh Tuck Link, Singapore 596224
USA office: 27 Warren Street, Suite 401-402, Hackensack, NJ 07601
UK office: 57 Shelton Street, Covent Garden, London WC2H 9HE

Library of Congress Cataloging-in-Publication Data
Molecules : nucleation, aggregation, and crystallization : beyond
medical and other implications / editors, Jan Sedzik, Paolo Riccio.
p. ; cm.
Includes bibliographical references and index.
ISBN-13: 978-981-283-264-1 (hardcover : alk. paper)
ISBN-10: 981-283-264-5 (hardcover : alk. paper)
ISBN-13: 978-981-283-265-8 (pbk. : alk. paper)
ISBN-10: 981-283-265-3 (pbk. : alk. paper)
1. Macromolecules. 2. Proteins. 3. Nucleation. 4. Aggregation (Chemistry)
5. Crystallization. I. Sedzik, Jan, 1946– II. Riccio, Paolo, 1942–
III. Karolinska institutet. IV. MARIE Network.
[DNLM: 1. Molecular Structure. 2. Protein Conformation. 3. Cell
Aggregation. 4. Crystallization. QU 55.9 M718 2009]
QD381M625 2009
572--dc22

2009009682

British Library Cataloguing-in-Publication Data
A catalogue record for this book is available from the British Library.

Typeset by Stallion Press
Email: enquiries@stallionpress.com

Printed in Singapore by Mainland Press Pte Ltd

Preface

This volume, entitled *Nucleation, Aggregation and Crystallization — Beyond Medical and Other Implications,* is a compilation of articles written based on the lectures presented by leading experts during their participation in the homonymous "International Course"[a] held at the Karolinska Institutet (2007) and in the scientific meetings of the MARIE Network of the European Science Foundation (2004–2006) on "Myelin structure and its Role in Autoimmunity."

The word "nucleation," derived from "nuclear family," is based on the concept of a progenitor, or the mother and father of any family. As a physico-chemical process, *nucleation* is followed by two poorly understood phenomenon: *aggregation and crystallization. Aggregation* underlies disorders like

[a] The following organizations supported/sponsored the course "Nucleation, Aggregation and Crystallization — Beyond Medical and Other Implications," (in alphabetical order): BioVectis Sp.z.o.o www.biovectis.com, CHEMILIA AB www.chemilia.com,
GE Healthcare Life Sciences www.gehealthcare.com/lifesciences,
HAMPTON Research www.hamptonresearch.com,
International Organization for Biological Crystallization www.iobcr.org,
Japanese Society for the Promotion of Science, Stockholm Office www.jsps-sto.com,
Karolinska Institutet Education AB www.karolinskaeducation.ki.se,
ESF MARIE Network, Molecular Dimensions Ltd. www.moleculardimensions.com,
Oxford Diffraction Ltd. www.oxford-diffraction.com,
PCF-CRYSTALLOMIC™ www.crystallomic.eu,
Protein Wave Corp. www.pro-wave.co.jp, RIGAKU www.rigaku.com,
Tekno Optik AB www.teknooptik.se,
Wenner-Gren Foundation www.wennergren.org,
Åke Wiberg Foundation www.ake-wiberg.com.

Alzheimer's and "mad-cow" disease (aggregation of amyloid plaque), and cardiovascular diseases (deposition in coronary vessels of cholesterol and lipids). *Crystallization* refers to the appearance of crystals under physiological conditions (gout, silicoses, and liver or kidney stones).

Crystallization is a physicochemical process appearing in nature or artificially in test tubes, leading to formation of solid crystals from a uniform concentrate of slightly supersaturated solutions. In biological crystallization, crystals are usually needed for the determination of the three-dimensional structure of proteins and other biological macromolecules by X-ray crystallography. The structure is needed for a better understanding of protein function and in order to design effective drugs. If the structure of a protein is precisely solved, the designed drug is supposed to be more effective.

The selection of 20 chapters in this book covers important topics on the principles and methods of the crystallization procedures in one-, two- and three-dimensions, including those regarding organic fine chemicals and pharmaceutical compounds. Moreover, the book describes the use of oils and magnetic fields to grow better quality protein crystals. Other chapters deal with the relevance of bioinformatics and mass spectrometry in crystallomic, and discuss both the role of molecular mimicry and the structure of amyloid proteins and myelin proteins in diseases such as polyneuropaties, Alzheimer's disease and multiple sclerosis.

In recent years, the MARIE Network, funded by the European Science Foundation, re-introduced the concept that recognition of biological structures such as the myelin sheath, is fundamental for the understanding of specific diseases and the design of better therapies. Through MARIE, structural studies on myelin and multiple sclerosis are now enjoying a renaissance after a period of low interest. This can be ascribed to new theoretical or technological developments and to the progress in several fields, including bioinformatics, X-ray crystallography, and proteomics, but also to the MARIE idea to encourage the integration of different kinds of expertise in the fields of biophysics, biochemistry, molecular biology, neurology, neuroimmunology, and bioinformatics. This cooperation between experts in structural studies and specialists in cell biology and neuroimmunology has enabled deeper probing into the underlying structure of the myelin sheath and to understand the processes by which various components involved in immune response, such as cytokines, antibodies, proteinases, and free oxygen radicals, may degrade the myelin over time, causing multiple sclerosis.

MARIE has helped to revive multiple sclerosis research at the structural level. Some aspects of this renaissance, in particular those regarding crystallization, form the fundamentals of this book, which provides a unique perspective of the physical and chemical sciences on one hand, and the biological and medical sciences on the other, and should be of considerable value to scientists, physicians, students.

One of our teachers, Nobel Laureate Dr. Gajdusek, passed away on December 12, 2008, few weeks after the 2nd proofs reading. The 3rd proofs were planned in January 2009. We have decided to treat his chapter, entitled, "From Kuru to Nucleation, Aggregation, Polymerization and Crystallization in Biology and Medicine" as his "scientific testament," and to leave a few very minor mistakes in the text. Chapter I of this book is probably the last scientific writing of Dr. Gajdusek.

I (JS), met Dr. Gajdusek for the first time in Chengdu, Sichuan, China, June 25–28, 2006, during the "First International Conference of Nanobiomedical Technology and Structural Biology." He gave a lecture "Nucleation, Medical Diseases and Beyond." It was a beautiful and informative lecture. He told us why he went to study kuru among the Fore people of New Guinea. New Guinea is the place he loved so much. His research on kuru was never supported financially by any organization, except his own funds. He never wrote any grant application for the whole of his life. From China (2006), at the age 82, he left for Tibet to continue his research. He was an old fashion scientist; he never used a computer for word processing, and he never had a secretary. He wrote this chapter by hand. I will remember him as a very energetic person, full of new ideas and new concepts, very friendly and supportive of teaching. We will miss him in other courses we will organize in the years to come.

We would like also to thank the publisher World Scientific Publishing, in Singapore, especially Ms Sook-Cheng Lim, for their support, encouragement and to take on the "board" publication of our book. I would like to thank Prof. Gösta Winberg (KI, MIT) for introducing me to Dr. Gajdusek and for his support for the course; Prof. Zhang Shuguang (MIT) for inviting me to China, where I met for the fist time, Dr. Gajdusek.

This book is dedicated to all interested in the subject of nucleation, aggregation and crystallization of proteins, particularly to potential course leaders who may use this book during their teaching.

Jan Sedzik
Editor
Associate Professor
Protein Crystallization Facility
Department of Neurobiology
Care Sciences and Society
NOVUM
Karolinska Institutet, Stockholm
Sweden

and

Associate Professor
Protein Crystallization Facility
Royal Institute of Technology
Department of Chemical
Engineering and Technology
KTH, Teknikrigen 28
SE 100 44 Stockholm, Sweden

Board Member, ESF MARIE Network
Chair, JSPS Alumni Club Association
Stockholm, Sweden

Paolo Riccio
Co-editor
Professor
Universitá degli Studi della Basillicata
Department of Biology D.B.A.F.
Campus Macchia Romana
85100 Potenza, Italy
Chair of the ESF MARIE Network

Stockholm, Potenza,
February 12, 2009

Contents

Chapter 1

From Kuru to Nucleation, Aggregation, Polymerization and Crystallization in Biology and Medicine

*D. Carleton Gajdusek**,†

Nucleation is a word derived from nuclear family and refers to the concept of progenitor, the mother and the father of any family, which has been at the root of human life for many thousands of years. From it has emerged the concept of a breeding line of humans. Only in the last few centuries of civilization have physicists borrowed the word, and later biologists for Schwann's cell theory. Very recently it has passed to atomic theory, spectroscopy, radioactivity and to atomic bombs, fission and fusion. In physics any change in the free energy state of matter involves a nucleation or ordering of atoms into a new pattern as in any change to gas, liquid or solid or in the packing of atoms in carbon black, graphite, or diamond. Thus the word *nucleation* is not derived from atomic physics or cell biology. To be so deluded makes it difficult to understand the simple matter of pattern setting in any change of state.

Keywords: Crystallization; molecular casting; twinning of minerals; nucleation; amyloidoses; amyloid enhancing factor; nucleant; infectious nucleant; carbon; diamond; montmorillonite; epitactic.

Molecular Casting

Infectious amyloid nucleants

We have repeated confirmation of resistance of a portion of infectivity of scrapie to temperatures as high as 600°C (Brown *et al.*, 2000). The

*C.N.R.S. Institut de Neurobiologie Alfred Fessard, Gif-sur-Yvette, France, and Medical Biology, University of Tromsø, Norway.
†1923–2008.

enormous resistance to dry heat of a small fraction (about one part in 10^6) of infectious activity may represent a molecular casting, fingerprints of the nucleants.

Infectious nucleant or prion activity is the result of very close three-dimensional matching. Any particle which can sufficiently mimic the ability of the molecule to be nucleated to crystallize, to fibrilize, or to form a two-dimensional molecular sheet can trigger the process. Matching must surely be at atomic distances, close enough to evoke van der Waals forces and Coulombic forces, even H-bonding.

Can we preserve biological specificity of antibodies, antigens, pheronemes, receptors, transmitters, ion channels and enzymes in organic molecular casts or atomic moulds?

The answer is yes, already for the first six items, and if we allow for synthetic hydrocarbon polymers for molecular casting, it may be so for enzymes.

Dermatoglyphic preserving of biological specificity

One of my adopted New Guinea sons has pointed out to me that a fingerprint using battery ink (MnO_2) is an example of such preservation with no atom of carbon and of biological specificity, more individual than the DNA sequence of identical twins.

Fossils show accurate speciation in paleobotany and paleozoology

Another of my adopted New Guinea sons has pointed out that fossil footprints allow for accurate classification and yet are not a source of DNA for speciation by polymer chain reactions. Nucleoli can be counted in inorganic fossils in cells extinct for millions of years.

Osmium shading in electron microscopy reveals details of molecular structure at nanometric distances

There is no contribution from carbon atoms to the image of an osmium or platinum shaded freeze-fracture electron microscopy photograph of individual molecules.

Mineral nucleants for crystal growth in outer space

McPherson and Shlichta (1988) sent a series of proteins into outer space to avoid the convection of fluids by the gravity of the Earth. They nucleated them with a selection of ground minerals. A subset of minerals usually initiated crystal formation in one or another of the proteins, a different set for each protein and differing forms of epitactic crystal growth for a given protein with each mineral.

What makes a diamond hold together?

Two polished ancient Chinese copper mirrors applied face-to-face stick together and require considerable force to separate them. A stack of newly opened clean microscopic cover slips slide one upon the other, yet it requires considerable force to separate them.

A diamond, the hardest of minerals, can scratch steel, yet it is still only made of carbon, the same as carbon black, coal or graphite.

Sulfur may be pure yet malleable, ductile, fragile or clay-like and of many colors depending how the S atoms are packed. Such is the nature of the van der Waals' forces that are brought into play at atomic distances.

Twinning of minerals

There are at least 200 twinning possibilities for quartz (SiO_2). Most possible forms have been found in the over 30 000 years of searching for them. When one new twin "form," "strain" or "species" is found, it is usually named for the region where it has been found. It is common to find other examples of this particular "strain" of twinned quartz in mining shafts for many hundreds of kilometers around the first finding, yet nowhere else on Earth. Much the same is true for diamonds, emeralds and rubies, and for other examples of twinning in mineralogy.

Industrial viruses and "ice nine"

Kurt Vonnegut, Jr wrote of "ice nine" in his book *Cat's Cradle* (1963): a fictional approach to the problem of nucleation based on a sound understanding of the phenomenon. His brother, William, was a major meteorologist fully familiar with non-DNA or non-RNA containing viruses and the World War II ethylene diamine tartrate problem of the industrial viruses, which nucleated the slow appearance of bubbles of large crystals of the compound made for optical purposes. Kurt Vonnegut, Jr. got the idea right.

Amyloid enhancing factors are scrapie infectious amyloid nucleants

For some 35 years, I have been aware of the work of amylidologists in their attempts to accelerate the appearance on AA amyloid deposits in animals primed with inoculation of $AgNO_3$ or heterologous casein. Their discovery of amyloid enhancing factors, which were active in high dilutions and difficult to purify, remind me of our problems with the infectious agents of scrapie or kuru. I suggested that amyloid enhancing factors were scrapie-like agents (Gajdusek, 1977, 1988, 1991, 1994a, 1994b; Niewold *et al.*, 1987).

Any β-pleated polymeric assembly as a two-dimensional sheet or as a fibril may act as a heteronucleant for different amyloidogenic proteins

Amyloid deposits in man or animals are always found to be contaminated with other proteins similarly polymerized into fibrils — even copolymerized. These are all the proteoglycans and glycosoaminoglycans as well as

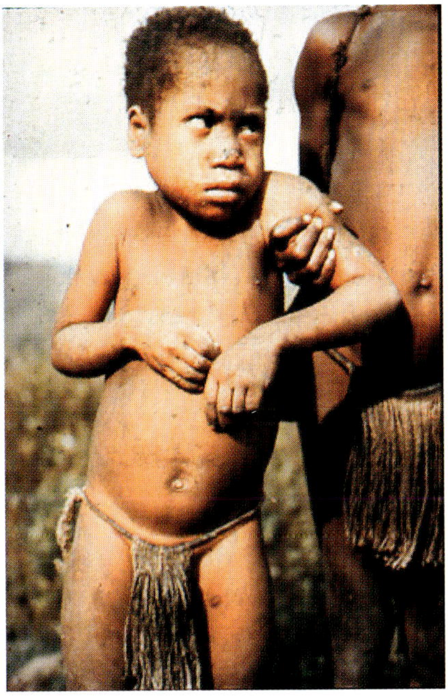

Fig. 1. A prepubertal Gimi boy of about ten with kuru who requires aid to remain standing, He shows the spastic strabismus which most children develop in kuru. He died a few months later in 1957 (DCG-57 NG-1150).

plasma P-protein, chymotrypsin, ubiquitin, light chains of gamma-globins, and other amyloidogenic proteins.

Tropocollagen is nucleated to fibrilize not only by submicroscopic fibrils of tropocollagen but also by dimers and polymers of glycosoaminoglycans or proteoglycans, and not by heparin which is a single-bonded dimer (Obrink, 1973).

Synthesis of prion-like infectious nucleants

Katarzyna Johan (Lundmark) and Per Westermark (1998) succeeded in getting *in vivo* heterologous nucleation of β-fibrillary protein polymerization into amyloid fibrils with synthetic amyloid enhancing factors. Such heterologous nucleants are synthetic peptides from the highly fibrilogenic section of the amyloid precursor protein or both transthyretin and insulin associated amyloid. They serve to nucleate the fibrilization of the AA amyloid precursor protein when polymerized into small fibrils, but not as unpolymerized peptides. Labeling with I^{131} has served to locate AA amyloid fibrils that have been nucleated by these heterologous amyloid-enhancing factors. Thus, if these replicating systems are thought of as being alive, they have already synthesized "life" and published their findings.

Per Westermark and his colleagues have demonstrated the induction of AA amyloidosis by nucleation with heteronucleants such as silk and spider webs and by oral *paté de fois* in transgenic mice (Johan *et al.*, 1998; Solomon *et al.*, 2007).

Biological macromolecules all interact strongly with SiO_2, the most common solid mineral on the surface of Earth. Montmorillonite clay deposits cause delayed neurodegenerative diseases

Iler (1977) and Weiss (1981) have shown how silicon and oxygen in the form of SiO_2 can interact and bond to biological macromolecules or polymers in long series of strong attractions, whether they are carbohydrates, proteins, nucleic acids or fats. Silicon and oxygen are the two most common elements on the surface of Earth. The role of silicon is fully discussed in the Nobel Foundation's *The Biochemistry of Silicon and Related Problems*, in which Iler's article appeared. Thus binding to solids is the most likely origin of life, not a primordial oceanic liquid.

The high incidence foci of two very different diseases, Guamanian amyotrophic lateral sclerosis *(lytico)* and Parkinsonism dementia *(bodig)*, of the Chamoro people on Guam, also occurred in the few remote inland villages of Honshu Island in Japan and among the Auyu and Jakai people around Bade and Kepi in southern West New Guinea (Gajdusek, Salazar, 1987). It has virtually disappeared from all of these places with the introduction of civilization. These three foci were restricted to remote communities in which a depletion of calcium produced a chronic severe deficiency of calcium in the diet, such that calcium sparing resulted in soft tissue deposition of calcium-aluminum-silicon, or montmorillonite clay deposits within brain cells — along with heavy elements as the diet provided. These lay dormant for decades until triggered later in life causing specific neuronal damage that led to either *lytico* or *bodig*.

Nucleation is speleology

In exploring caves speleologists are familiar with nearly identical formations surrounding a fallen and shattered stalagmite or stalactite — very much like the rings of small mushrooms around an old large mushroom on the forest floor or the ring of young sequoia saplings around an old dead tree trunk. They often resemble even the odd idiosyncrasies in the parent formation.

At times a cavern of brown "toadstools" or one of the pink "phallic" organ-pipe cactuses is filled with dozens or hundreds of uncannily similar replicas. Then, several galleries below the dark brown "toadstools" in a gallery of pink "phalluses" stands a dark brown toadstool from nucleant which has tumbled down millions of years ago to the lower gallery from the "toadstools" above.

Nucleation in extragalactic space

The odd patterns of distant galaxies of billions of stars are well-known to all amateur astronomers who have viewed in awe the thousands of photographs we now have of them. They are by no means random patterns of stars, but lend themselves rather easily to classification according to similarities in appearance, as do most patterns in geophysics. These similar patterns appear in groups or in a very small fraction of the 2π sterradions of space around us. The nucleation of such patterns across distances of huge numbers of light years is certainly cause for wonder.

References

Brown P, Rau EH, Johnson BK, Bacote AE, Gibbs CJ Jr, Gajdusek DC. (2000) New studies on the heat resistance of hamster-adapted scrapie agent: Threshold survival after ashing at 600 degrees C suggests an inorganic template of replication. *Proc Nat Acad Sci USA* **97**: 3418–3421.

Gajdusek DC. (1977) Unconventional viruses and the origin and disappearance of kuru. *Science* **197**: 943–960.

Gajdusek DC. (1988) Transmissible and non-transmissible amyloidoses: Autocatalytic post-translational conversion of host precursor proteins to beta-pleated sheet configurations. *J Neuroimmunol* **20**: 95–110.

Gajdusek DC. (1991) The transmissible amyloidoses — Genetical control of spontaneous generation of infectious amyloid proteins by nucleation of configurational change in host precursors: Kuru-CJD-GSS-scrapie-BSE. *Eur J Epidemiol* **7**: 567–577.

Gajdusek DC. (1994a) Spontaneous generation of infectious nucleating amyloids in the transmissible and nontransmissible cerebral amyloidoses. *Mol Neurobiol* **8**: 1–13.

Gajdusek DC. (1994b) Nucleation of amyloidigenesis in infectious and non-infectious amyloidoses of brain. *Ann NY Acad Sci* **724**: 173–190.

Gajdusek DC, Salazar A. (1987) Amyotropic lateral sclerosis and Parkinsonian syndromes in high incidence among the Awyu and Jakai people around Bade and Kepi in southern West New Guinea. *Neurology* **32**: 107–126.

Iler RK. (1977) Hydrogen-bonded complexes of silica with organic compounds. In Bendz G, Lindquist I (eds). *Biochemistry of Silicon and Related Problems*, pp. 53–76. Plenum, Press, London, N.Y.

Johan K, Westermark G, Engstrom U, Gustavsson A, Hultman P, Westermark P. (1998) Acceleration of amyloid protein A amyloidosis by amyloid-like synthetic fibrils. *Proc Nat Acad Sci USA* **95**: 2558–2563.

McPherson A, Shlichta P. (1988) Heterogeneous and epitaxial nucleation of protein crystals on mineral surfaces. *Science* **239**: 385–387.

Niewold TA, Hol PR, van Andel AC, Lutz ET, Gruys E. (1987) Enhancement of amyloid induction by amyloid fibril fragments in hamster. *Lab Invest* **56**: 544–549.

Obrink B. (1973) The influence of glycosaminoglycans on the formation of fibers from monomeric tropocollagen *in vitro*. *Eur J Biochem* **34**: 129–137.

Solomon A, Richey T, Murphy CL, Weiss DT, Wall JS, Westermark GT, Westermark P. (2007) Amyloidogenic potential of foie gras. *Proc Nat Acad Sci USA* **104**: 10998–11001.

Vonnegut K. (1963) *Cat's Cradle*, Holt, Rinehart and Winston, USA.

Weiss A. (1981) Replication and evolution in inorganic systems. *Angew Chem* **20**: 850–860.

Chapter 2

Gels Mimicking Antibodies in Their Selective Recognition of Proteins and Its Potential Use for Protein Crystallization*

***Jan Sedzik*†, *Nasim Ghasemzadeh*‡, *Fred Nyberg*‡ and *Stellan Hjertén*§**

Using a unique molecular-imprinting method we show in this article that human growth hormone, ribonuclease and myoglobin from horse, lysozyme, hemoglobin and albumin can be adsorbed selectively, indicating that the method may be universal or at least applicable to a great number of proteins. A gel with specific adsorption of three model proteins was synthesized in order to demonstrate that the beds can be employed to remove (traces of) several proteins contaminating a sample ("negative purification"). The degree of selective recognition is high, judging from the fact that myoglobin from horse, but not that from whale, was adsorbed onto a column designed to bind specifically the former protein. This selectivity is noteworthy since these two proteins have similar amino acid sequences and 3-D structures. The method for the synthesis of the specific gels

*This paper is reproduced from Hjertén *et al.* (1997) *Chromatrographia* **44**: 227–234 and, with permission from the publisher, with small changes and with references to other articles in this field.

†Protein Crystallization Facility, Department of Neurobiology, Care Sciences and Society, NOVUM, Karolinska Institutet, Stockholm, Sweden, and Department of Chemical Engineering and Technology, Protein Crystallization Facility, KTH, Teknikringen 28, 100 44 Stockholm, Sweden. E-mail: sedzik@swipnet.se.

‡Uppsala University, Department of Pharmaceutical Biosciences, P.O. Box 591, 75124 Uppsala, Sweden. E-mails: fred.nyberg@farmbio.uu.se, nasim.g@farmbio.uu.se.

§Corresponding author, Department of Biochemistry and Organic Chemistry, Uppsala University, Biomedical Center, P.O. Box 576, 75123 Uppsala, Sweden. E-mail: stellan.hjerten@biorg.uu.se.

involves polymerization of appropriate monomers (for instance, acrylamide and its derivatives) in the presence of the protein to be adsorbed specifically, granulation of the gel formed, packing a column with the gel particles, washing the column to remove the protein, and finally application of the sample for selective adsorption of the protein present during the polymerization of the monomers. The approach resembles that used for entrapment (immobilization) of proteins for affinity chromatography and somewhat like that for molecular imprinting of small molecules, with the distinct difference that the monomer composition is quite different and thereby the binding mechanism. This mechanism is discussed, for instance, in terms of (i) a new classification system for chromatographic beds based on the number of bonds between the solute and the matrix and the strength of each bond, and (ii) "non-specific bonds" (these bonds are often harmful in conventional chromatography, but we have used them to our advantage). In this classification system, the selective recognition is characterized by a large number of weak bonds. Therefore, so-called functional monomers are not used for the preparation of the gels because they are often charged and, accordingly, give rise to strong electrostatic interactions, i.e. the beds behave to some extent as ion exchangers or matrices for hydrophobic interaction chromatography. In most experiments we have used a polyacrylamide gel with large pores to facilitate diffusion of proteins into and out of the gel granules. When used in chromatography, these soft gels (which can be used repeatedly) allow only rather low flow rates. This problem can be overcome by a new approach to preparing the granules. Potential applications of the selective beds are discussed, as well as future improvements. These beds can be synthesized for selective adsorption also of bio-particles, for instance viruses and bacteria, and in the form of monoliths (continuous beds).

Keywords: Affinity chromatography; artificial gel antibodies; entrapment; molecular imprinting; recognition of proteins; selectivity.

Introduction

More than a decade ago we described briefly a technique for selective adsorption of a protein (Liao *et al.*, 1996). The method is based on the preparation of a gel (for instance, crosslinked polyacrylamide) in the presence of the protein of interest, granulation of the gel, packing a chromatographic column with the granules, and finally removal of the protein. Upon application of a sample containing this protein and other proteins, only the protein present during the polymerization will become adsorbed. In this paper we present further studies of selective beds and discuss the mechanism of selective recognition.

Experimental Details

Materials

Ribonuclease, lysozyme, cytochrome *c*, and myoglobin from horse and whale were obtained from Sigma (St. Louis, MD, USA). We obtained growth hormone as a gift from Professor Paul Roos, Department of Biochemistry, Uppsala University and human serum albumin from Dr Lars-Olov Andersson, KABI (Stockholm, Sweden). Hemoglobin was prepared from human blood. Acrylamide, N,N′-methylenebisacrylamide, ammonium persulfate, N,N,N′,N′-tetramethylethylenediamine (TEMED), sodium dodecyl sulfate (SDS), and a cation exchanger (CB-S, i.d. 7 mm, length 40 mm) were from Bio-Rad Laboratories (Hercules, CA, USA).

Preparation of a ribonuclease-specific gel

Acrylamide (5.82 mg), N,N′-methylenebisacrylamide (1.8 mg) and ribonuclease (3 mg) were dissolved in 1 mL of 0.01 M sodium phosphate, pH 7.0. Following the addition of 20 μL of a 10% (w/v) solution of ammonium persulfate and deaeration, 20 μL of a 5% (v/v) TEMED solution was added. The polymerization proceeded for 30 minutes, producing a gel with the total concentration (T) = 6% (w/v) and the crosslinking concentration (C) = 3% (w/w), [for definitions of T and C see Hjertén (1962)]. The formed gel was then pressed through a 60-mesh net to produce granules, which were packed into a Pasteur pipette with glass wool in the constriction as support for the gel particles. The Pasteur pipette had an i.d. of 5 mm and the bed height was 4.5 cm. The granules were washed with 0.8 mL of a solution of Savinase (a proteinase obtained from Novo Nordisk A/S, Denmark) to remove ribonuclease and equilibrated with

3 mL of 0.01 M sodium phosphate, pH 7.0. About 50 μL of a sample solution of hemoglobin (10 mg/mL) and ribonuclease (3 mg/mL) was applied. The column was then washed with 0.01 M sodium phosphate, pH 7.0, and the eluate was collected and analyzed by cation-exchange chromatography at 220 nm.

The results are shown in the chromatogram traces of Fig. 1. Figure l(a) is a chromatogram of the sample itself prior to passage through any gel column ("Hb" designating hemoglobin and "R" designating ribonuclease). Figure l(b) is a chromatogram of the fraction collected from a blank column (prepared in the absence of ribonuclease). Figure l(c) is a chromatogram of the fraction collected from the column prepared in the presence of ribonuclease. The ribonuclease peak is present in Figs. l(a) and l(b), but absent in Fig. l(c), indicating that ribonuclease was adsorbed only by the column prepared in the presence of ribonuclease.

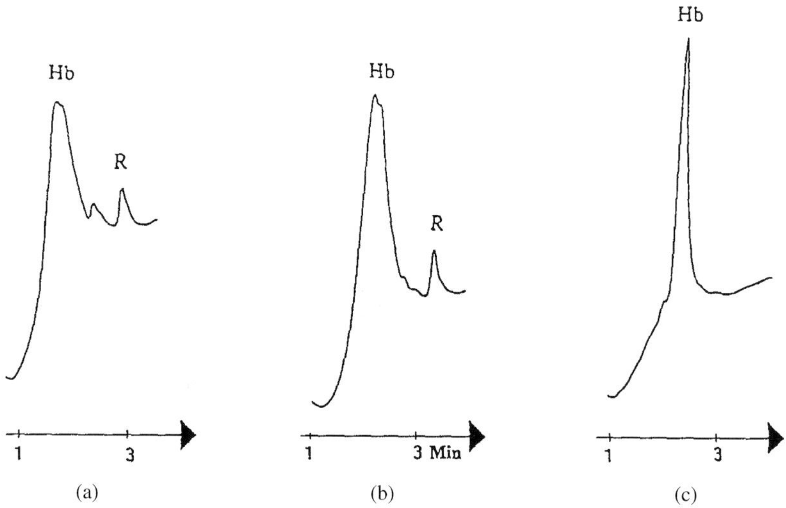

Fig. 1. Selective recognition of ribonuclease. Analysis in a cation exchanger of (a) the sample itself [consisting of hemoglobin (Hb) and ribonuclease (R)] prior to passage through any gel column, (b) the fraction collected from a blank column (prepared in the absence of ribonuclease) and (c) the fraction collected from the column prepared in the presence of ribonuclease. The ribonuclease peak is present in (a) and (b), but absent in (c), indicating that ribonuclease had been specifically adsorbed only onto the gel column prepared in the presence of ribonuclease.

Preparation of a gel for specific adsorption of human growth hormone

Human growth hormone (3 mg) was dissolved in 1 mL of 0.1 M sodium phosphate, pH 6.2. Acrylamide (5.8 mg), N,N′-methylenebisacrylamide (2.4 mg) and ammonium persulfate [10 μL of a 10% (w/v) aqueous solution] were added. The deaerated solution was supplemented with 10 μL of a 5% (v/v) aqueous solution of TEMED. About 30 minutes following polymerization of the gel (which had the composition T6, C4; Hjertén, 1962) it was granulated by pressing it through a 100-mesh net and packed into a Pasteur pipette. To desorb the growth hormone, two column volumes of sodium phosphate (0.1 M, pH 6.2) were passed through the column followed by five column volumes of 10% (v/v) acetic acid containing 10% (w/v) SDS.

The column was washed with the buffer until the pH of the effluent reached 6.2 and no precipitate of dodecyl sulfate could be detected upon addition of potassium chloride (potassium dodecyl sulfate is only slightly soluble in water). A 10 μL volume of the sample, a mixture of growth hormone (5 mg/mL) and human serum albumin (2 mg/mL) in the phosphate buffer, was added. The eluate was analyzed in an anion exchanger using a 5 min linear concentration gradient created by the equilibration buffer and the same buffer containing 0.5 M sodium chloride (Fig. 2(c)). The eluate from a blank column (prepared in the absence of growth hormone) was studied in the same ion exchanger (Fig. 2(b)), as was the original sample (Fig. 2(a)). The chromatograms show that the growth hormone, but not albumin, had adsorbed to the column prepared in the presence of the hormone.

For artificial gel antibodies against other proteins and for gel antibodies against viruses and bacteria (see Hjertén, Liao (1998) and Takátsy *et al.* (2006a; 2006b; 2007)). For other proteins, see Liao *et al.* (1996), Hjertén *et al.* (1997) and Tong *et al.* (2001).

Can a bed be designed with selectivity for more than one protein?

Acrylamide (171 mg), N,N′-methylenebisacrylamide (9 mg), hemoglobin [60 μL of a 1% (w/v) aqueous solution], cytochrome *c* (5 mg) and

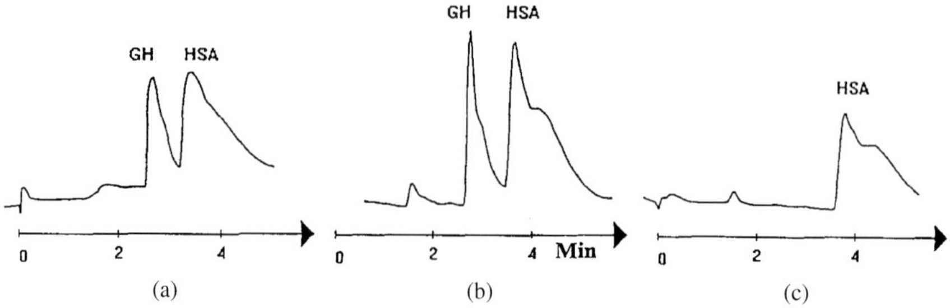

Fig. 2. Selective recognition of human growth hormone. The sample, consisting of human growth hormone (GH) and human serum albumin (HSA), was applied (a) directly on an anion exchanger for analysis, (b) on a blank column prepared in the absence of growth hormone and the eluated protein fraction was analyzed on the same ion-exchanger, and (c) on a column prepared in the presence of growth hormone and the eluated protein fraction was analyzed on the ion-exchanger. A comparison of the three chromatograms shows that the growth hormone was adsorbed only onto the gel synthesized for specific recognition of this protein (i.e. the gel prepared in the presence of the growth hormone).

lysozyme (5 mg) were dissolved in 2.94 mL of water. Following the addition of 30 μL of a 10% (w/v) aqueous solution of ammonium persulfate and deaeration by aspiration, the mixture was supplemented with 30 μL of a 5% (v/v) aqueous solution of TEMED. The polymerization was allowed to proceed for three hours. The gel, which had the composition T6, C5 (Hjertén, 1962), was granulated by pressing it through a 100-mesh net and packed into a 5 μL plastic Eppendorf syringe (with the piston removed) that had been fitted at the outlet with a piece of a cut nylon stocking. The column was washed for three hours with 10 mL of a 10% (w/v) SDS solution in 10% (v/v) acetic acid to desorb the three proteins, but was still somewhat colored, indicating that hemoglobin and cytochrome *c* (and therefore probably also lysozyme) could not be released completely. The column was equilibrated with 12 mL of 100 mM sodium phosphate, pH 6.2, for one hour and was then stored for two days before it was tested for adsorption of hemoglobin, cytochrome *c* and lysozyme by applying 35 μL of these proteins dissolved in 0.1 M sodium phosphate, pH 6.2, at a final concentration of 0.020%, 0.085% and 0.085% (w/v), respectively. The column was washed with 100 mM

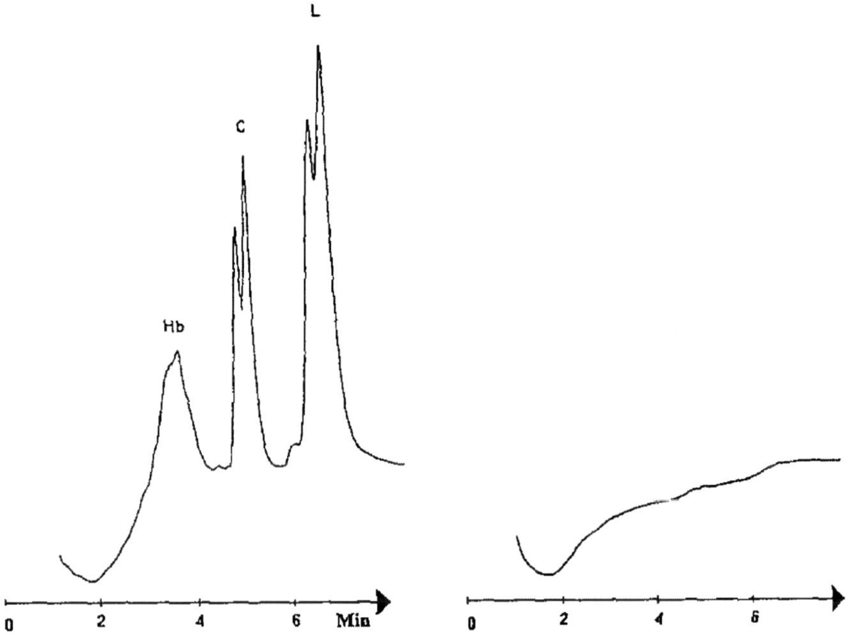

(a) (b)

Fig. 3. Selective recognition of three proteins: hemoglobin (Hb), cytochrome *c* (C) and lysozyme (L). Analysis in a cation exchanger of (a) the fraction collected from a blank column (prepared in the absence of any protein) and (b) the fraction collected from the column prepared in the presence of the sample. A comparison of the two chromatograms indicates that hemoglobin, cytochrome *c* and lysozyme were adsorbed onto the column prepared in the presence of these three proteins.

sodium phosphate, pH 6.2, and the eluate was analyzed at 220 nm by cation-exchange chromatography (Fig. 3(b)). Evidently, all three proteins had been adsorbed.

A similar experiment was performed on a blank column, i.e. a column prepared in the absence of the three proteins (Fig. 3(a)). These figures show that more than one protein can be adsorbed selectively on the same column. In a pre-experiment on the same adsorbent we used glass wool to support the bed. Because glass wool appeared to interact with hemoglobin, we exchanged it for a piece of a nylon stocking in the runs described.

The degree of selectivity of the adsorption

Horse myoglobin (3 mg) was dissolved in 1 mL of 10 mM sodium phosphate, pH 6.2. Acrylamide (57 mg), N,N′-methylenebisacrylamide (2.5 mg) and 10 μL of a 10% aqueous solution of ammonium persulfate were added. Following deaeration for one minute the solution was supplemented with 5 μL of a 5% aqueous solution of TEMED. The polymerization proceeded overnight. The gel formed was pressed through a 60-mesh net and packed into a Pasteur pipette (i.d. = 5 mm; bed height = 4 cm) with glass wool at the outlet. The column was washed overnight with a 10% (v/v) solution of acetic acid containing 10% (w/v) sodium dodecyl sulfate to remove the horse myoglobin and was then equilibrated with 10 mM sodium phosphate, pH 6.2. This column is denoted "horse myoglobin column." Another column was prepared in the same way, but in the absence of horse myoglobin ("blank column"). Horse myoglobin, whale myoglobin, ribonuclease and cytochrome *c* were dissolved in 10 mM sodium phosphate, pH 6.2. The final concentration of each protein was 1 mg/mL. A 10 μL volume of this sample was applied on each of the two columns. Non-adsorbed proteins were eluted with the phosphate buffer used for equilibration of the columns and analyzed in a cation exchanger. Detection was done at 220 nm. The chromatogram in Fig. 4(a) was obtained from the "blank column" and that in Fig. 4(b) from the "horse myoglobin column." Horse myoglobin was adsorbed only on the latter column, and whale myoglobin on neither of the two columns. Figure 5 shows that the 3-D structures of these proteins are similar.

Selective bed with higher flow rate

The polyacrylamide beds described above had the composition T6, C5. The beds described previously had a similar composition (Liao *et al.*, 1996). From molecular-sieve chromatography experiments it is known that such gels give a relatively low flow rate (Hjertén, 1962). Since this can be improved by using denser, i.e. more rigid, gels, we prepared a bed with the composition T20, C3. However, the selective adsorption of proteins

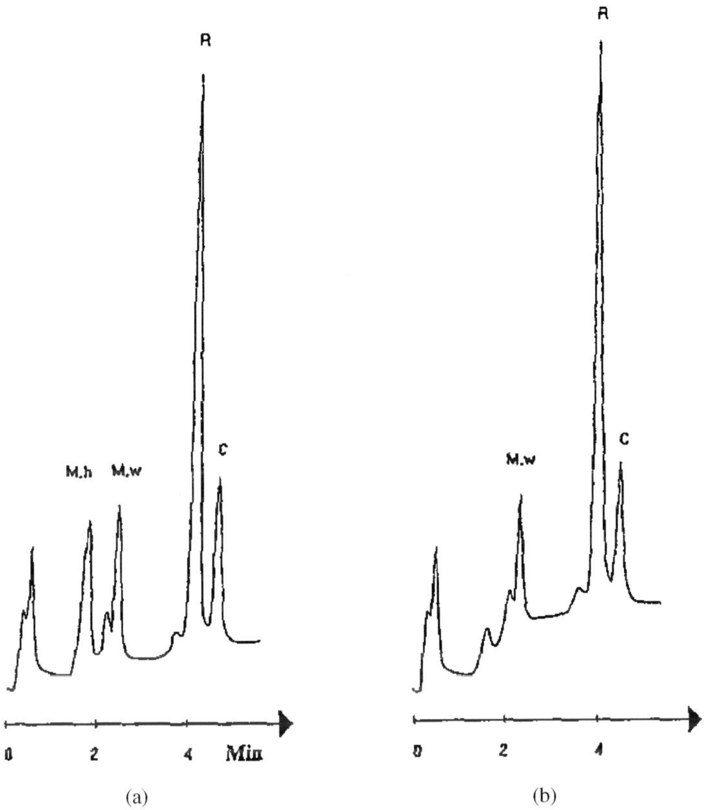

Fig. 4. Demonstration of the high degree of selective recognition of the adsorbent for myoglobin from horse (M,h), myoglobin from whale (M,w), ribonuclease (R) and cytochrome *c* (C). Analysis in a cation exchanger of (a) the fraction collected from a blank column prepared in the absence of any protein, and (b) the fraction collected from a column prepared in the presence of myoglobin from horse. A comparison of the two chromatogram traces shows that myoglobin from horse, but not that from whale, was adsorbed onto the column prepared in the presence of the former protein, indicating a high degree of selective recognition.

was low, which could be due to the smaller pore size of these gels with the attendant difficulty of removing the protein to be specifically adsorbed. Therefore, we studied another alternative for increasing the flow rate: embedding grains of a soft gel with specific adsorption of a certain protein into grains of a rigid gel (Hjertén *et al.*, 1997).

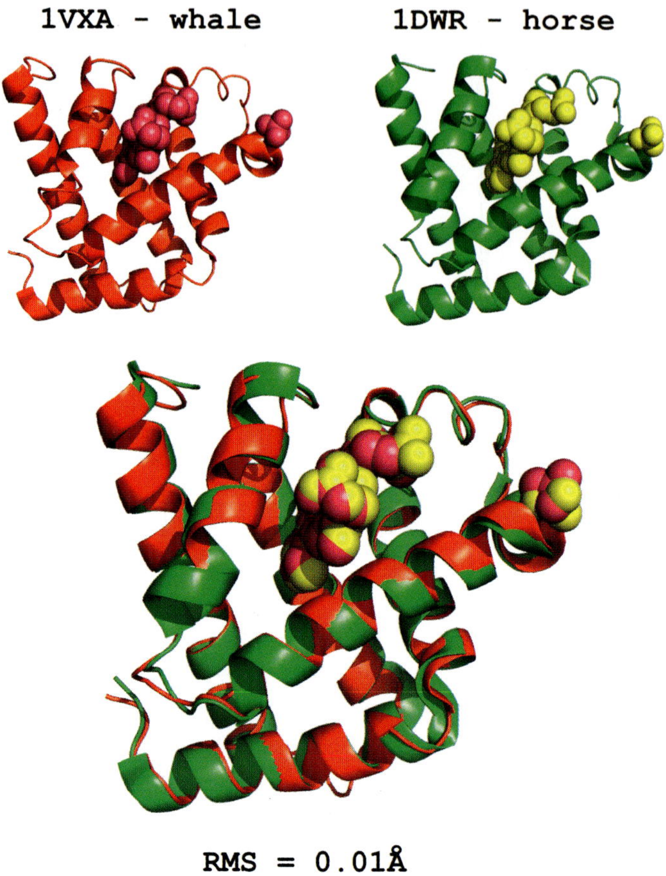

Fig. 5. The structural model of native form of whale sperm myoglobin (1VXA) has been colored red (polypeptide chain) and purple (ligands–heme and sulphate ion). The structural polypeptide chain of horse myoglobin (1DWR) has been colored green and the ligands (heme and sulphate ions) have been colored yellow. Both proteins can be fitted with the RMS = 0.01 Å (carbons alpha only), including that both molecules are structurally almost identical. The amino acids sequences of both proteins are 94% identical. Despite the great structural similarities only horse myoglobin, but not whale myoglobin, was adsorbed onto the column prepared specifically for horse myoglobin (Fig. 4). The column was thus highly selective.

The detailed procedure was as follows. 100-mesh granules of a T6, C5 gel specific for adsorption of hemoglobin were prepared in 1 mL of 0.01 M sodium phosphate pH 7.0, essentially as described for the above "growth hormone column" (the hemoglobin concentration in the monomer solution was 1.2 mg/mL). The column was emptied and the gel particles were pressed three times through a 100-mesh net and mixed with a monomer solution of the composition T20, C3 in the volume ratio 1:2. The same catalyst was used as in the above experiment with the "ribonuclease column" although the volumes of the ammonium per-sulfate and TEMED solutions were increased 2.5-fold. About 40 minutes following polymerization the gel was disintegrated by forcing it once through a 100-mesh net and then packed into a Pasteur pipette. The column was washed with 6% SDS in 10% acetic acid and equilibrated with 20 mM sodium phosphate, pH 7.0. The sample consisted of 20 μL of a solution of hemoglobin (10 mg/mL), cytochrome *c* (2 mg/mL) and lysozyme (2 mg/mL) in 1.0 mM sodium phosphate, pH 7.0. The flow resistance of this bed was considerably lower than that obtained with the original T6 C5 gel. The selectivity test (Fig. 6), performed as described for the experiment presented in Fig. 4, demonstrates that it is possible to combine high specificity with high flow rate on beds with small gel particles which have the advantage of permitting shorter run times and less zone broadening.

Results and Discussion

The selectivity of the artificial gel antibodies

In an article published more than a decade ago, we described a method for the preparation of gels with the property to permit selective recognition of cytochrome *c* (Liao *et al.*, 1996). From the chromatograms in Figs. 1, 2 and 4, one can conclude that the same method can be employed to adsorb specifically ribonuclease from horse, human growth hormone and myoglobin from horse. From the many experiments we have done since then, we can conclude that, with great probability, similar beds can be designed for all proteins. It is interesting to note that one can synthesize a gel which can recognize several proteins (Fig. 3). Such a gel can be of interest for rapid "negative purification" of a certain protein by suspending the gel particles (selective for the non-desirable proteins) in the protein sample and decanting; see section on "The design of a chromatographic bed for removal of impurities", (page 19).

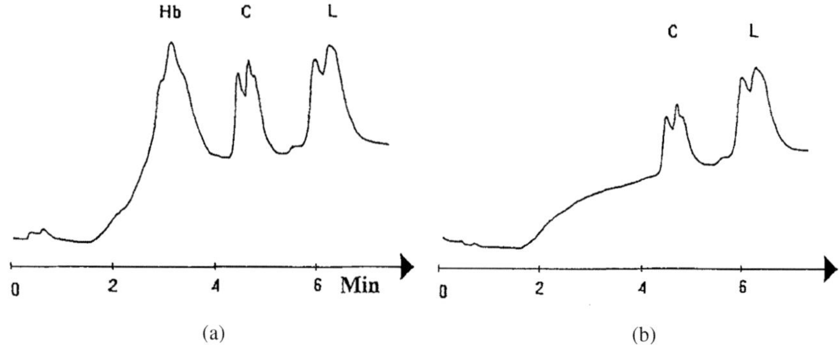

Fig. 6. Preparation of a selective bed with higher flow rate. A polyacrylamide column of the composition T6 C5 and specific for hemoglobin was prepared essentially as described for the "ribonuclease column" (Fig. 1). The column was emptied and the granules were suspended in a monomer solution of the composition T20, C3. Following polymerization the gel was granulated by pressing it through a 100-mesh net and packed in a Pasteur pipette. A blank column was prepared in the same way but in the absence of hemoglobin. A standard solution consisting of hemoglobin (Hb), cytochrome *c* (C) and lysozyme (L) was (a) passed through the blank column and analyzed in a cation exchanger, and (b) passed through the column prepared in the presence of hemoglobin and analyzed in the ion exchanger. The chromatogram traces indicate that the T6, C5/T20, C3 column adsorbed hemoglobin. Accordingly, entrapment of the T6, C5 granules into small, rigid T20, C5 gel grains did not affect the selective recognition of hemoglobin. Owing to the rigidity of these grains the column had the advantage of permitting high flow rates.

Do some protein molecules become attached covalently to the gel matrix?

This could be so because "hemoglobin columns" are somewhat colored even after extensive washing with different buffers and a 10% solution of acetic acid containing 10% SDS. Since monomer radicals form in the polymerization process, one cannot exclude the possibility that these react with hemoglobin molecules, which thus may become linked to the gel network. To investigate whether such a reaction occurs, a "hemoglobin column" was prepared as described by Liao *et al.* (1996) with the exception that many amino acids were present at high concentrations along with hemoglobin during the polymerization. The free radicals should react also with the free amino acids if they react with the hemoglobin molecules, and one should expect a weaker color of this "hemoglobin column" — which, however, was not the case. The experiment indicates that the proteins are not linked covalently to the gel matrix.

What conclusions can be drawn from the finding that (bio-)affinity methods do not always give the selectivity theoretically expected?

Selective interactions between different constituents in the living cell form the basis for its function. This bio-recognition has been utilized also in non-biological systems, for instance, in methods utilizing biosensors (Jakusch *et al.*, 1999) and in bio-affinity chromatography (Porath, 1981). These artificial systems have been employed with success in many investigations in biochemistry and related disciplines. However, the specificity is not always as high as expected, the main reason being that the ligands (for instance, enzyme substrates, antibodies) and the functional monomers used for imprinting of small molecules (for instance, methacrylic acid) are charged and/or non-polar, and therefore create beds that to a greater or lesser extent also behave as conventional media for ion exchange and/or hydrophobic interaction chromatography, i.e. the selectivity is lost because having *many strong* bonds is often not compatible with high selectivity (see below). The situation is of particular concern when electrostatic and hydrophobic interactions occur at the same time since an increase in the ionic strength of the eluent decreases the former type of interaction but increases the latter (Hjertén, 1973). In such cases it may be impossible to find a buffer concentration such that the adsorbed sample can be released. Accordingly, there are reasons to investigate whether methods can be designed in which ligands and functional monomers are omitted, i.e. methods based solely on weak interactions with the (chromatographic) matrix. The experiments described herein should be considered against this background.

Classification of chromatographic techniques in terms of the number of bonds between the solute and the stationary phase and the strength of these bonds

Proteins and other macromolecules with many charged and/or non-polar groups bind strongly to a bed which has a high density of ligands of the same types of groups by virtue of electrostatic and hydrophobic interactions, respectively. However, the adsorption is not selective because such a bed can, via different combinations of the ligands, strongly bind *many* proteins of *varying* structure. This alternative (*several strong* bonds) is used in conventional gradient elution of macromolecules (proteins) in ion exchange and hydrophobic-interaction chromatography. For isocratic elution of proteins these separation methods require *few strong* bonds (Hjertén *et al.*, 1986; Yao, Hjertén, 1987). As for low molecular weight compounds, the strength of the bonds can (or perhaps, must) be *high* since the number of bonds is necessarily *small.* An example of this alternative for small molecules is the molecular imprinting method, which has been employed with great success for the separation of enantiomers and other low molecular weight compounds, often using strong electrostatic bonds created by functional monomers (Wulff *et al.*, 1985; Wulff, Mindrik, 1988; 1990; Wulff, 1995; Shea, 1994; Schweitz *et al.*, 1997). In ideal molecular imprinting, *all* functional monomer molecules in the bed have positions close to the complimentary groups in the solute molecules. However, a gel in which only some fraction of the functional monomer molecules fulfills this requirement also exhibits (a certain degree of) selectivity because the binding caused by the rest of these monomer molecules will be weaker, more or less resembling that occurring in conventional ion exchangers (when charged functional monomers are used) or beds for hydrophobic-interaction chromatography (when non-polar functional monomers are used). The difficulty of synthesizing a gel for ideal molecular imprinting (or the latter semi-ideal approach) increases with the number of binding sites, often equivalent to an increase in molecular weight. Therefore, it is not surprising that no molecular imprinting experiments with high selectivity for proteins have been reported. These considerations indicate that monomers which give rise to very strong bonds should be replaced by those interacting more *weakly.* To attain a strong overall binding the number of bonds must, accordingly, be *large*. Examples of weak interactions are those originating from hydrogen bonds, charge transfer, faint induced dipoles and slightly non-polar groups. Acrylamide and N,N′-methylenebisacrylamide represent appropriate monomers. Gels synthesized from these monomers are widely used in many standard methods for electrophoresis and chromatography. All accumulated experience shows that these gels are very inert to biopolymers, including proteins.

A very close contact between a protein molecule and polymer chains in the gel is, accordingly, a prerequisite for the generation of bonds. This proximity is created when the monomers polymerize around the protein molecule (and may increase the strength of each bond). If the discussed hypothesis for selective recognition is correct, one should expect a protein of a relatively small size (and therefore fewer adsorption sites) to interact only weakly with gels of a composition similar to that employed in this investigation. Therefore, we studied the behavior of insulin (molecular weight: 5700 daltons) on a T6, C5 gel prepared in the presence of insulin. The protein was much less adsorbed than were larger proteins, such as growth hormone and transferrin, but the interaction was nevertheless selective because the elution volume was 40% larger than that obtained on a blank gel synthesized in the absence of insulin (not shown here).

Some comments on the mechanism of selective recognition

In the 1960s it was shown that proteins could be immobilized in a gel for bio-affinity chromatography if they were included in the monomer solution (Bernfeld, Wan, 1963; Kennedy, Cabra, 1983). It was postulated that the immobilization consisted of occlusion of the protein molecules in the polymeric network. In view of the properties of the gels described herein, it is likely that at least part of the immobilization originated from selective adsorption and not by entrapment caused by occlusion. It is known that cross-linking of agarose gels with divinyl sulfone causes some non-specific interaction with proteins (Hjertén *et al.*, 1987). Since our working hypothesis on selective recognition is partly based on non-specific interactions, an agarose gel was prepared in the presence of hemoglobin and at the same time cross-linked with divinyl sulfone (Hjertén *et al.*, 1987). This gel showed specific adsorption of hemoglobin although the loading capacity was low. It should be stressed that cross-linking of agarose only slightly affects the pore size of the gel (Porath *et al.*, 1975).

Artificial gel antibodies for detection of biomarkers

The high selectivity of the gel antibodies can be used to "fish out" a biomarker from a body fluid for diagnosis and prognosis of a particular disease (Ghasemzadeh *et al.*, 2008a; 2008b). From a standard curve one can then rapidly determine the concentration of the biomarker in, for instance, serum or cerebrospinal fluid (CSF).

For the design of the standard curve CBB (Coomassie Brilliant Blue), staining of proteins selectively adsorbed to the artificial gel antibodies has the great advantage that it absorbs light at 588 nm, where the light scattering from polyacrylamide gel granules is negligible (Fig. 7). The measured adsorption value of the biomarker (the protein selectively adsorbed to the gel granules) inserted into the calibration curve (Fig. 7) gives on the x-axis the concentration of the biomarker in the sample solution.

We applied this method to estimate the concentration of albumin in plasma and CSF samples from patients with amyotrophic lateral sclerosis (ALS). Difficulties in detection and treatment of diseases together with a lack of suitable diagnostic and prognostic tools have indicated a need for discovery or identification of relevant proteins or peptides which can serve as biomarkers for a disease. An example is given in Fig. 8, which shows that albumin in cerebrospinal fluid is a biomarker for neurological diseases such as ALS.

The mean levels of albumin in CSF from ALS patients were more than twice as high as in the control subject (Fig. 8(a)). This finding is in agreement with results from earlier studies, which confirms the validity of our analysis technique and suggests that the barrier permeability for

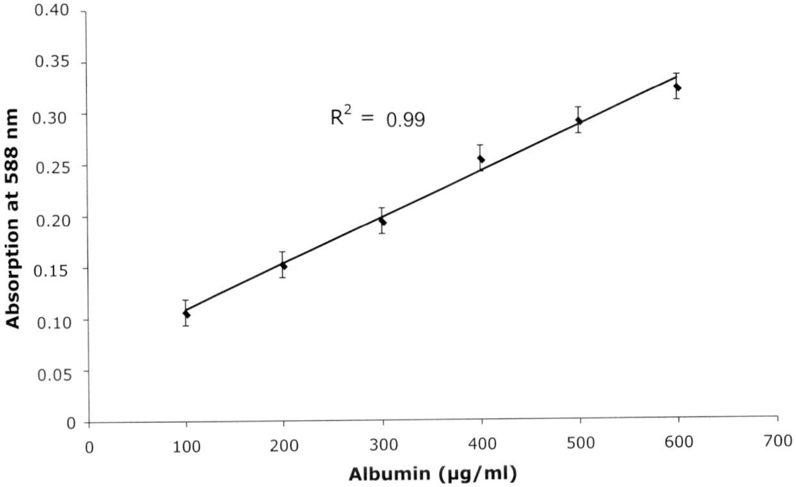

Fig. 7. Standard curve for the determination of the concentration of human albumin in a sample (in this case CSF and plasma). The sample cuvette contained stained, artificial gel antibodies selective for human albumin, prepared as described in Ghazemzadeh *et al.* (2008) The reference cuvette contained a suspension of albumin-selective, stained gel granules depleted of albumin (control gel).

albumin may be affected in ALS and perhaps also for other proteins. In contrast, a low level of albumin was observed in plasma from ALS patients compared to the control plasma samples (Fig. 8(b)). The results further indicate that our approach might also apply well to other biomarkers for the actual neurological disease and other disorders.

Selectivity, protein capacity and imprinting of bioparticles

A comparison between the chromatograms in Figs. 4(a) and 4(b) (obtained from analyses with a cation exchanger of the eluate from the "blank column" and the "horse myoglobin" column, respectively) shows that only horse myoglobin was adsorbed onto the latter column but none of the other proteins, not even whale myoglobin, although the amino acid sequence of the two myoglobins differs in only twenty of the 153 amino acid positions and in such a way that the three-dimensional structure is

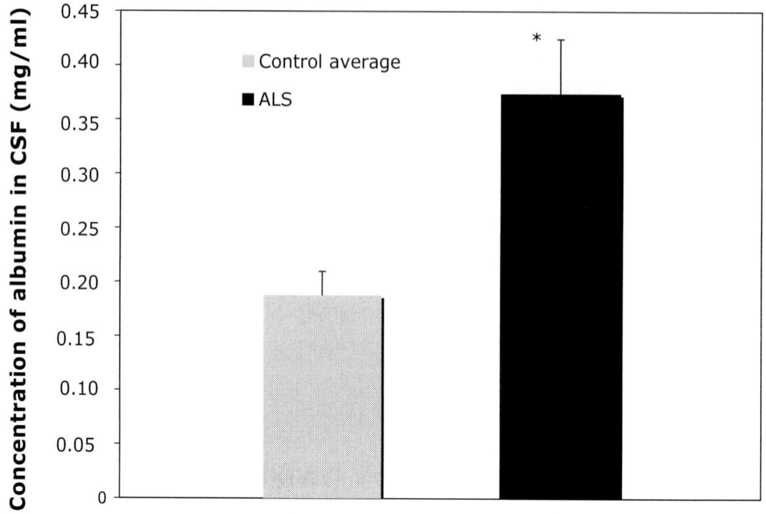

(a)

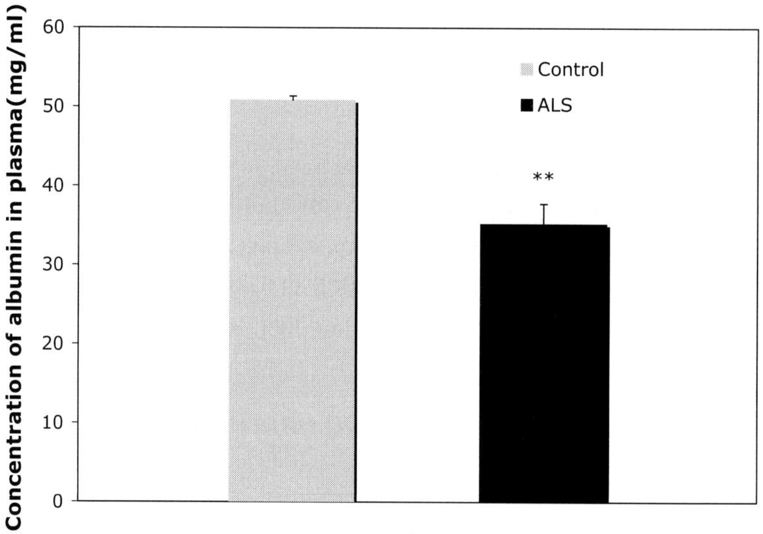

(b)

Fig. 8. Quantitative analysis of albumin in CSF and plasma in control subjects and in ALS patients. For determination of the concentration of albumin the standard curve of Fig. 7 was used. (a) The mean level of albumin in CSF from ALS patients is more than two-fold higher than that from control subjects. (b) The plasma from ALS patients exhibits a large decrease in albumin concentration compared to the albumin concentration in plasma from the control subject.

only slightly affected (Fig. 5). The selectivity of the bed was, accordingly, high. During a test period of one month a "hemoglobin column" was used in 14 experiments without loss of selectivity. We have used the imprinting approach not only to "fish out" a particular protein "biomarker" (for diagnosis and prognosis of diseases) but also a particular virus or bacterium (Takátsy *et al.*, 2006a; Bacskay *et al.*, 2006). Important studies on molecular imprinting of proteins are also found in Guo *et al.* (2005) and Hawkins *et al.* (2006). Interestingly, the artificial gel antibodies can also be synthesized as monoliths, i.e. the granulation and the packing steps are eliminated (Rezeli *et al.*, 2006).

Potential applications

A comparison of the three chromatograms in Fig. 2 shows that the human growth hormone was adsorbed only on the gel prepared in the presence of the hormone. The gel grains thus behave as beads to which antibodies have been attached and can therefore probably be used in RIA (radioimmunoassay) or ELISA (enzyme-linked immunosorbent assay). If so, animal experiments aimed at the generation of antibodies can be avoided not only in these assays but also in other experiments, including receptor studies. The high selectivity of the gels described herein makes it tempting to use them also in the area of biosensors, where the binding can be studied by rapid response from transducers based on different principles. It is relatively easy to investigate whether the gels, in practice, can be utilized in such analytical systems. More time-consuming experiments are required if the gels are to be designed to function as artificial enzymes. Those interested in these and related areas should read an excellent article by Mosbach and Ramström (1996).

We have also used the molecular imprinting approach to synthesize enzyme reactors, particularly because they have a great advantage compared to conventional enzyme reactors. They can be regenerated, following the unavoidable denaturation of the enzyme, by desorbing the enzyme and adding a crude extract of the enzyme because the high selectivity of the artificial gel antibodies (the imprinted gels) is a guarantee that only the enzyme will be adsorbed. The "enzyme reactors" can also be used as the screening element in biosensors (Hjertén *et al.*, 2008)

The design of a chromatographic bed for removal of impurifying proteins ("negative purification")

The impurifying proteins can be found for instance in the mother liquor (crystallization cocktail) at crystallization and in the supernatant at precipitation. Artificial gel antibodies against the proteins can easily be prepared by the addition of acrylamide and bis-acrylamide to these solutions. Following polymerization the gel formed is granulated and the proteins are removed (see, Section "Preparation of a ribonuclease specific gel").

Forthcoming studies

The mechanism of selective recognition is only one of many questions which importunately insist upon an answer. Others are: How does one elute the protein adsorbed specifically in a small volume and in a gentle way? Does a high selectivity require a high rigidity of the solute; if so, should the solute molecules be cross-linked (reversibly)? Analogous electrophoretic methods should also be developed, and the studies of methods based on "negative purification" should be intensified and employed in experiments aimed at crystallization of proteins.

Conclusions

The experiments described in this paper strengthen the hypothesis that it is possible to design gels which resemble chemically and biologically antibodies in the sense that they recognize selectively and bind strongly a particular protein, virus or bacterium (and other bioparticles). A prerequisite seems to be that the number of bonds between the gel and the antigen is large and that each bond is relatively weak. The ease and simplicity by which the gels can be prepared render them potentially attractive for many different techniques based on selective recognition, including the attractive possibility of using them for purification of membrane proteins.

Acnowledgments

This work was partially supported by the Swedish Research Council, grant VR-2007-6890 (J.S). We wish to thank Dr. Jan Pawel Jastrzebski (Poland) for making bioinformatic calculations and plotting Fig. 5.

References

Bacskay I, Takátsy A, Végvári A, Elfwing A, Ballagi-Pordány A, Kilár F, Hjertén S. (2006) Universal method for synthesis of artificial gel antibodies by the imprinting approach combined with a unique electrophoresis technique for detection of minute structural differences of proteins, viruses, and cells (bacteria). III: Gel antibodies against cells (bacteria). *Electrophoresis* **27**: 4682–4687.

Bernfeld P, Wan J. (1963) Antigens and enzymes made insoluble by entraping into lattices of synthetic polymers. *Science* **142**: 678–679.

Ghasemzadeh N, Nyberg F, Hjertén S. (2008a) Highly selective artificial gel antibodies for detection and quantification of biomarkers in clinical samples. I. Spectrophotometric approach to design the calibration curve for the quantification. *J Sep Sci* **31**: 3945–3953.

Ghasemzadeh N, Nyberg F, Hjertén S. (2008b) Highly selective artificial gel antibodies for detection and quantification of biomarkers in clinical samples. II. Albumin in body fluids of patients with neurological disorders. *J Sep Sci* **31**: 3954–3958.

Guo TY, Xia YQ, Hao GJ, Zhang BH, Fu GQ, Yuan Z, He BL, Kennedy JF. (2005) Chemically modified chitosan beads as matrices for adsorptive separation of proteins by molecularly imprinted polymer. *Carbohyd Polym* **62**: 214–221.

Hawkins DM, Trache A, Ellis EA, Stevenson D, Holzenburg A, Meininger GA, Reddy SM. (2006) Quantification and confocal imaging of protein specific molecularly imprinted polymers. *Biomacromolecules* **7**: 2560–2564.

Hjertén M, Rezeli M, Kilár F, Hjertén S. (2008) Renewable enzyme reactors based on beds of artificial gel antibodies. *J Biochem Biophys Meth* **70**: 188–191.

Hjertén S. (1962) "Molecular sieve" chromatography on polyacrylamide gels, prepared according to a simplified method. *Arch Biochem Biophys Suppl* **1:** 147–151.

Hjertén S. (1971) In Niederwieser A, Pataki G (eds), *New Techniques in Amino Acid, Peptide and Protein Analysis*, p. 227. Science Publishers, Ann Arbor, Michigan.

Hjertén S. (1973) Some general aspects of hydrophobic interaction chromatography. *J Chromatog* **87**: 325–331.

Hjertén S, Liao JL. (1998) Chromatography columns with continuous beds formed *in situ* from aqueous solutions. *US Patent* **5814**: 223.

Hjertén S, Liao J-L, Nakazato K, Wang Y, Zamaratskaia G, Zhang H-X. (1997) Gels mimicking antibodies in their selective recognition of proteins. *Chromatographia* **44**: 227–234.

Hjertén S, Yao K, Eriksson K-O, Johansson B. (1986) Gradient and isocratic high-performance hydrophobic interaction chromatography of proteins on agarose columns. *J Chromatog* **359**: 99–109.

Hjertén S, Wu BL, Liao JL. (1987) A high-performance liquid chromatographic matrix based on agarose cross-linked with divinyl sulphone. *J Chromatog* **396**: 101–113.

Jakusch M, Janotta M, Mizaikoff B, Mosbach K, Haupt K. (1999) Molecularly imprinted polymers and infrared evanescent wave spectroscopy. A chemical sensors approach. *Anal Chem* **71**: 4786–4791.

Kennedy JE, Cabra JM. (1983) In Scouten WH (ed), *Immobilized Enzymes*, Solid Phase Biochemistry, p. 253. John Wiley & Sons, New York.

Liao JL, Wang Y, Hjertén S. (1996) A novel support with artificially created recognition for the selective removal of proteins and for affinity chromatography. *Chromatographia* **42**: 259–262.

Mosbach K, Ramström O. (1996) The emerging technique of molecular imprinting and its future impact on biotechnology. *Bio/Technology* **14**: 163–170.

Porath J. (1981) Development of modern bioaffinity chromatography (a review). *J Chromatog* **218**: 241–259.

Porath J, Låås T, Janson JC. (1975) Agar derivatives for chromatography, electrophoresis and gel-bound enzymes. III. Rigid agarose gels cross-linked with divinyl sulphone (DVS). *J Chromatog* **103**: 49–62.

Rezeli M, Kilár F, Hjertén S. (2006) Monolithic beds of artificial gel antibodies. *J Chromatog* A **1109**: 100–102.

Schweitz L, Andersson L, Nilsson S. (1997) Capillary electrochromatography with predetermined selectivity obtained through molecular imprinting. *Anal Chem* **69**: 1179–1183.

Shea KJ. (1994) Molecular imprinting of synthetic network polymers: The *de novo* synthesis of macromolecular binding and catalytic sites. *Trends Polymer Sci* **2**: 166–173.

Takátsy A, Sedzik J, Kilár F, Hjertén S. (2006a) Universal method for synthesis of artificial gel antibodies by the imprinting approach combined with a unique electrophoresis technique for detection of minute structural differences of proteins, viruses, and cells (bacteria): II. Gel antibodies against virus (Semliki Forest Virus). *J Sep Sci* **29:** 2810–2815.

Takátsy A, Kilár A, Kilár F, Hjertén S. (2006b) Universal method for synthesis of artificial gel antibodies by the imprinting approach combined with a unique electrophoresis technique for detection of minute structural differences of proteins, viruses, and cells (bacteria): Ia. Gel antibodies against proteins (transferrins). *J Sep Sci* **29**: 2802–2809.

Takátsy A, Végvári Á, Kilár F, Hjertén S. (2007) Universal method for synthesis of artificial gel antibodies by the imprinting approach combined with a unique electrophoresis technique for detection of minute structural differences of proteins, viruses and cells (bacteria). Ib. Gel antibodies against proteins (hemoglobins). *Electrophoresis* **28**: 2345–2350.

Tong D, Hetényi C, Bikádi Z, Gao JP, Hjertén S. (2001) Some studies of the chromatographic properties of gels ("artificial antibodies/receptors") for selective adsorption of proteins. *Chromatographia* **54**: 7–14.

Wulff G, Mindrik M. (1988) In Zief M, Crane LJ (eds), *Chromatographic Chiral Separations*, p. 15. Marcel Dekker, Inc., New York.

Wulff G, Mindrik M. (1990) Template imprinted polymers for HPLC separation of racemates. *J Liq Chromatog* **13**: 2987–3000.

Wulff G. (1995) Molecular imprinting in cross-linked materials with the aid of molecular templates — A way towards artificial antibodies. *Angew Chem Int Ed Engl* **34**: 1812–1832.

Wulff G, Oberkobusch D, Minarik M. (1985) Enzyme-analogue built polymers. Chiral cavities in polymer layers coated on wide-pore silica. *React Polym* **3**: 261–275.

Yao K, Hjertén S. (1987) Gradient and isocratic high-performance liquid chromatography of proteins on a new agarose-based anion exchanger. *J Chromatog* **385**: 87–98.

Chapter 3

Bioinformatics of Myelin Membrane Proteins

Gunnar von Heijne,† and Jan Sedzik*‡

While tertiary structure cannot in general be predicted for membrane proteins, membrane topology is easier to predict. In this chapter, our focus is on the proteins found in the myelin membrane. The myelin membrane is unique, since it contains almost 80% (w/w) lipids, making it one of the most lipid-rich biomembranes. Not all abundant myelin proteins are membrane proteins; there are also water soluble proteins. The difference between these two types of proteins can easily be detected by studying the amino acid sequence of the proteins. In this chapter, we present predictions of the membrane topology and the presence/absence of signal peptides in the myelin proteins, and discuss the results in light of published experimental data.

Keywords: Crystallogenesis; bioinformatics; membrane; protein; lipid; myelin sheath.

Introduction

The rapid development of bioinformatics in recent years has been driven to a large extent by the accumulation of large databases, compiling amino acid sequences of proteins and sequences of nucleotides in nucleic acids. Obviously, to extract biologically relevant information from such

*Corresponding author.

†Gunnar von Heijne, Department of Biochemistry and Biophysics, Stockholm University, SE-10691 Stockholm, Sweden. Email: gunnar@dbb.su.se.

‡Protein Crystallization Facility, Department of Neurobiology, Care Sciences and Society, NOVUM, Karolinska Institutet, Stockholm, Sweden, and Department of Chemical Engineering and Technology, Protein Crystallization Facility, KTH, Teknikringen 28, 100 44 Stockholm, Sweden. E-mail: sedzik@swipnet.se.

databases, one needs computers and sophisticated software. For proteins, computerized sequence analysis and three-dimensional (3D) molecular dynamics simulations may yield information on the following:

(i) the similarities between (groups of) proteins in terms of primary, secondary and tertiary structure;
(ii) the secondary and tertiary structure of a given protein as predicted from its amino acid sequence;
(iii) prediction of membrane-spanning α-helices;
(iv) prediction of membrane-spanning β-sheet structures;
(v) prediction of antigenic epitopes;
(vi) how membrane lipids influence protein structure and behavior;
(vii) prediction of crystallization conditions from the amino acid sequence.

For tackling some of the above, there are good methods available, e.g. prediction of transmembrane segments in proteins; others are hopelessly difficult, such as prediction of the crystallizability of a protein.

Definition of Bioinformatics

For the purposes of this review, the word "bioinformatics" is defined as follows: "any use of computers to handle biological information" (http://wiki.bioinformatics.org/Bioinformatics, 2008), which includes analysis of all type of sequences stored in biological databases, e.g. amino acid sequences of proteins, nucleotide sequences, structural information on proteins, viruses, plants, and so on. These kinds of data are inputs for computational data analysis as required by bioinformatics.

The Basis of Biological Membranes

Biomembranes enclose or separate cells, forming compartments. In these compartments cells may maintain a chemical or biochemical environment that differs from the environment outside. Membranes are composed of bilayer-forming phospholipids, other kinds of lipids such as cholesterol, and peripheral and integral proteins. Those proteins which are not soluble in water are called membrane proteins (or integral proteins). Membranes are semi-permeable and are involved in a vast array of cellular processes such as cell adhesion, ion channels, and transport of small molecules.

The most important feature of a biomembrane is that it has a selective permeability to ions and molecules. This means that the size, charge and other chemical properties of the ions or molecules attempting to cross the biomembrane will determine whether they succeed to do so. Such selective permeability is essential for effective separation of a cell or organelle from its surroundings. Viruses are too large and are unable to cross the membrane by themselves; in such cases transport into the cell depends on endocytosis.

In general, explaining how the membrane proteins are involved in the organization and function of any membranes requires knowing their molecular structures. The tertiary structures of the P2 basic protein of PNS myelin (bovine, 1PMP; equine, 1YIV), the soluble part of P0 protein (rat, recombinant, 1NEU) and the soluble part of MOG (mouse, recombinant 1PY9) are known. The three-dimensional structures of other (and full sequence) myelin membrane proteins are not yet known.

Biological Membranes — Protein/Lipid Ratio

There are no biological membranes which are made up 100% of lipids or 100% of proteins. For all known membranes the ratio of proteins to lipids (w/w) is in the range 1:3 to 3:1. This is summarized in Table 1.

One of the most lipid-rich membranes is the myelin membrane: 75% by weight are lipids. In contrast, the purple membrane contains only 25% lipids by weight.

Whether such large differences in the lipid to protein ratio can directly affect the structure of membrane proteins is not known, but might be an interesting area of future study.

Basic Bioinformatics

At present, membrane-protein bioinformatics predictors perform quite well on two tasks: identifying membrane-protein encoding genes

Table 1. Summary of Lipid and Protein Composition of some Biological Membranes

Membrane	Lipid, protein compositions	Proteins (abundant)	Lipids
Myelin PNS	25% lipids 75% proteins	P0, P2	Cholesterol, Phospholipids
Myelin CNS	25% protein 75% protein	MBP, PLP	Cholesterol, Phospholipids, Cerebrosides
SFV lipid bilayer	75% protein 25% lipids	E1, E2	Phospholipids, Cerebrorozide, cholesterol (lipid composition depends of host membrane)
Purple membrane (Corcelli *et al.*, 2002) *Halobacterium salinarum*	25% lipids 75% protein (10 lipids per one molecule of retinal)	retinal	Phosphatidylglycerophosphate methyl ester, glycolipid sulfate, phosphatidylglycerol, archaeal glycocardiolipin, squalene, minor amounts of phosphatidylglycerosulfate bisphosphatidylglycerol archaeal cardiolipin vitamin MK8
Thylakoids membrane (Li *et al.*, 1990, Aro *et al.*, 2005)	0.37–0.67 (lipids) 0.13–0.25 chrorophyl 1 proteins	Few proteins MW 10–100 kDa	Monogalactosyl, diglyceryde phosphatidylglycerol
Erythrocyte membrane (Ballas, Krasnov 1980)	50% lipids 50% proteins	Sprectrin, Actin, Band 3, Glycophorin 35–40 enzymes	Choline phospholipids Phosphiolipids, Cholesterol

in genome sequences and predicting the membrane topology of membrane proteins. When it comes to predicting membrane-protein 3D structure, however, there are no automatic methods available, and models have to be built on a case-by-case basis taking as much experimental information as possible into account (Elofsson, von Heijne, 2007).

For α-helical membrane proteins, methods such as TMHMM (a method for prediction of transmembrane helices based on a hidden Markov model; Krogh *et al.*, 2001), prodiv-TMHMM (Viklund, Elofsson, 2004), HMMTOP (Tusnády, Simon, 2001) and MEMSAT (Jones, 2007) are among the best choices for topology predictions. Recently, we have developed an experimentally based algorithm called the "ΔG predictor," which predicts, from the amino acid sequence of a protein, the Gibbs' free energy of membrane insertion for transmembrane α-helices (Hessa *et al.*, 2007). This last predictor is the one used below in the analysis of the myelin membrane proteins.

Some membrane proteins have N-terminal, cleavable signal peptides that guide them to the endoplasmic reticulum membrane. Signal peptides are on the order of 10–20 residues long and have a three-partite structure: a short N-terminal region (n-region) including one or two basic residues, a central hydrophobic region (h-region), and a more polar C-terminal region (c-region) that is 4–7 residues long and is recognized by the signal peptidase enzyme. The signal peptidase cleavage site is defined by small residues in positions −1 and −3 relative to the cleavage site [the so-called (−1, −3)-rule]. Signal peptides can be reliably identified using the SignalP predictor (Dyrløv-Bendtsen, 2004), which scores the typical signal peptide n-, h- and c-regions. SignalP is used below in the analysis of the myelin proteins.

Many other prediction methods based on the primary structure of proteins are available (www.expasy.ch/tools). One basic assumption of these methods is that structurally similar proteins are similar in their function (Aloy *et al.*, 2005, 2006). Prediction of protein–protein interactions based on structural information is still far from perfect, but allows in some cases the production of structural models at exceptionally high resolution (Kiel *et al.*, 2005).

Focus on the Myelin Membrane

The last report analyzing the structure and function of the myelin proteins from primary sequence data was published more than a decade ago

(Inouye, Kirschner, 1991), at a time when the term "bioinformatics" had not even been invented (Boguski, 1998). It is interesting to compare the results then and now.

Water Soluble Versus Integral Myelin Proteins

Myelin basic protein

Human MBP (Uniprot P02686) is 171 residues long. It is a water-soluble protein that lacks both a signal peptide and putative transmembrane helices (Fig. 1). Bovine myelin basic protein is the most studied protein; it is remarkable for its "random coil" conformation in aqueous solutions,

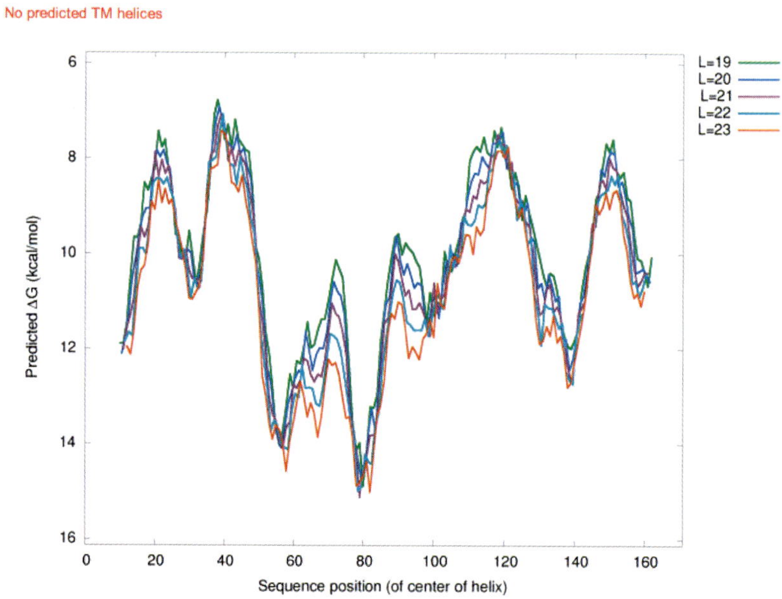

(a)

Fig. 1. Analysis of human MBP by (a) the ΔG_{app} predictor and (b) SignalP 3.0. There is no indication of transmembrane helices [i.e. segments with predicted ΔG (Gibb's free energy for membrane insertion) near or less than zero kcal/mol; note that negative values are upwards. Sequence scans with different lengths are displayed in different colors]. The SignalP predictor finds no indication of a signal peptide (i.e. there is no N-terminal segment with n-, h- and c-region probabilities > 0.5).

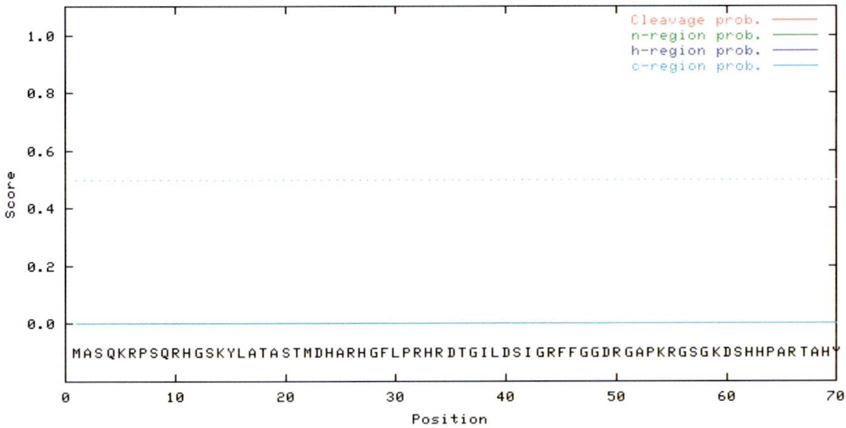

(b)

Fig. 1. (*Continued*)

pH 2–10, however prediction of the secondary structure indicates a large content of β-structure (Stoner 1984; 1990). Myelin basic protein is a water-soluble protein (up to 100 mg/ml), but has so far "stubbornly" resisted attempts at crystallization (Sedzik, Kirschner, 1991). More about the curious nature of myelin basic protein can be found in this book in Chap. 10 by Paolo Riccio.

P2 protein

Human P2 (Uniprot P02689) is 132 residues long and is classified as a peripheral membrane protein. It lacks both a signal peptide and putative transmembrane helices (Fig. 2). It has multiple phosphoserines, consistent with a cytosolic location.

P0 protein

Human P0 (Uniprot P25189) is 248 residues long, including a strongly predicted N-terminal signal peptide (residues 1–29) and a very stable transmembrane α-helix (Fig. 3). Residue 122 is annotated in Uniprot as

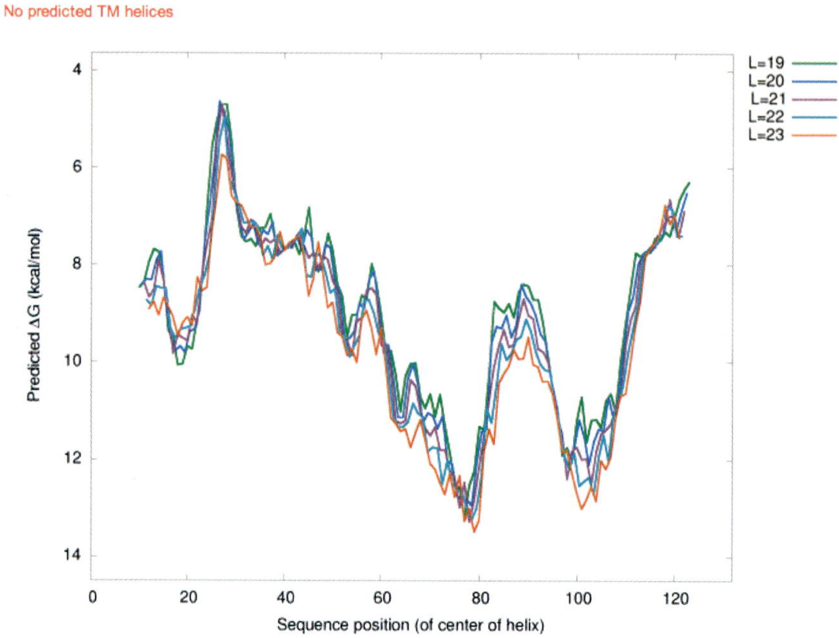

(a)

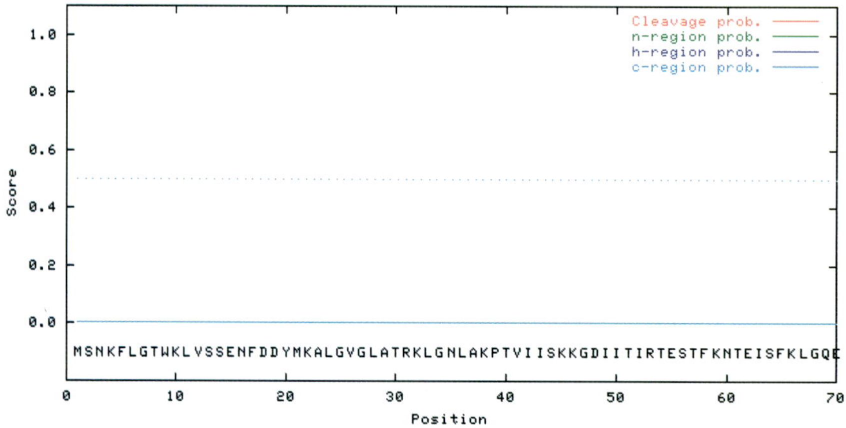

(b)

Fig. 2. Analysis of the human P2 protein by (a) the ΔG_{app} predictor and (b) SignalP 3.0. There is no indication of transmembrane helices or a signal peptide.

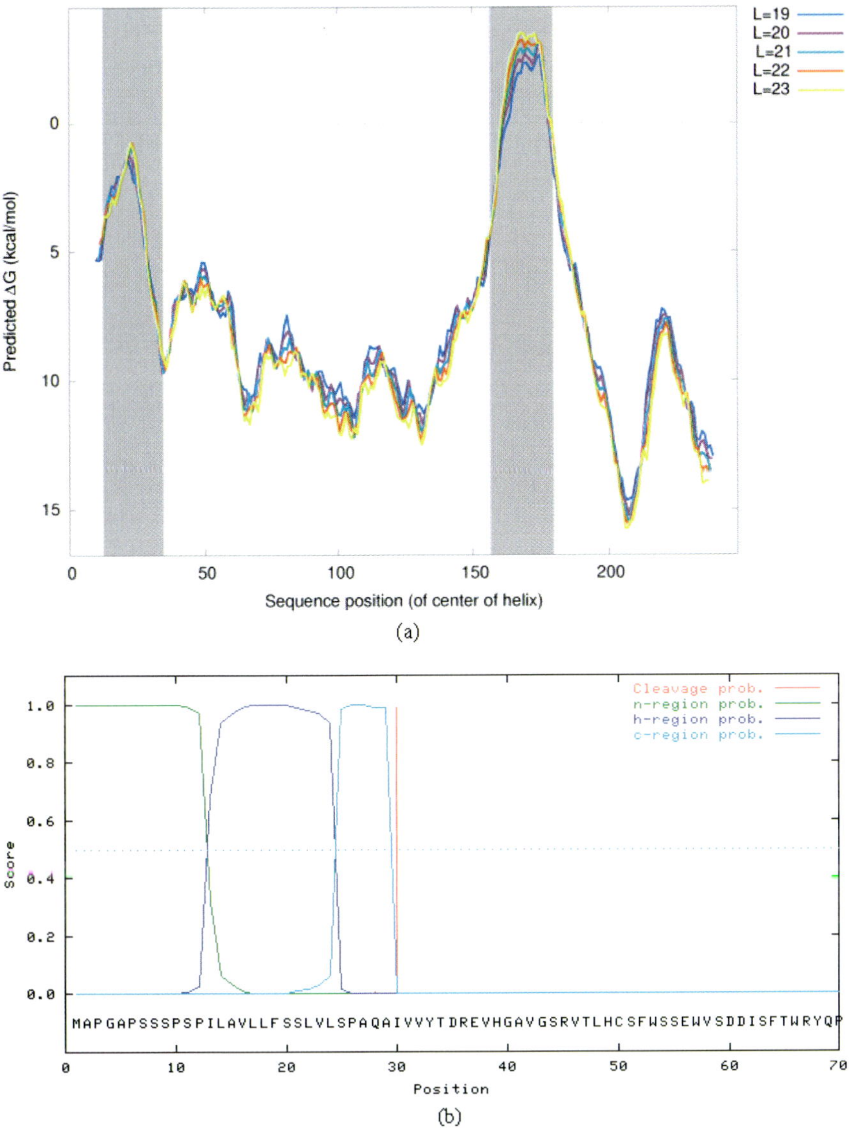

Fig. 3. Analysis of the human P0 by (a) the ΔG_{app} predictor and (b) SignalP 3.0. Putative transmembrane segments are shaded in the ΔG_{app} plot. The predicted N terminus of the mature protein after signal peptidase cleavage is indicated by a red line in the SignalP plot. The n-, h- and c-regions in the predicted signal peptide are shown in green, blue and sky blue respectively.

carrying an N-linked glycan (consistent with an extracellular location), while Ser residues in the region 226–243 are annotated as phosphorylated (consistent with a cytosolic localization). The topology of the mature protein (after removal of the signal peptide) is thus N_{out}-1TM-C_{in}, i.e. the C-terminus is in the cytosol, cf. D'Urso *et al.*, (1990).

PMP22 protein

Human PMP22 (Uniprot Q01453) is 160 residues long. Like PLP1, it has four clear hydrophobic regions that are all predicted to form transmembrane α-helices (Fig. 4). Residue 41 is annotated in Uniprot as carrying an N-linked glycan (consistent with an extracellular location) and both the N- and C-termini have been localized to the cytosolic side of the membrane by antibody mapping (D'Urso, 1997). Despite the high hydrophobicity of putative TM2 and TM3, antibody mapping of a construct with an HA-tag inserted between residues 93 and 94 has localized this short loop to the extracellular side of the membrane (Taylor, 2000), suggesting that PMP22 may only have two bonafide TM α-helices (corresponding to the most N- and C-terminal hydrophobic segments) and an N_{in}-C_{in} orientation. Considering the rather high hydrophobicity of the second and third hydrophobic segments, this is a rather surprising result that warrants further study. The most peculiar nature of this protein is its lack of stainability on SDS-PAGE by the silver method, and measured by circular dichroism large content of β-secondary structure in the presence of SDS (Sedzik *et al.*, 2001).

PLP protein

Human PLP1 (Uniprot P60201) is 272 residues long. It has four clear hydrophobic regions that are all predicted to form transmembrane α-helices (Fig. 5). Antibody mapping studies have located the C-terminus and a region around residues 103–116 to the cytosolic side of the plasma

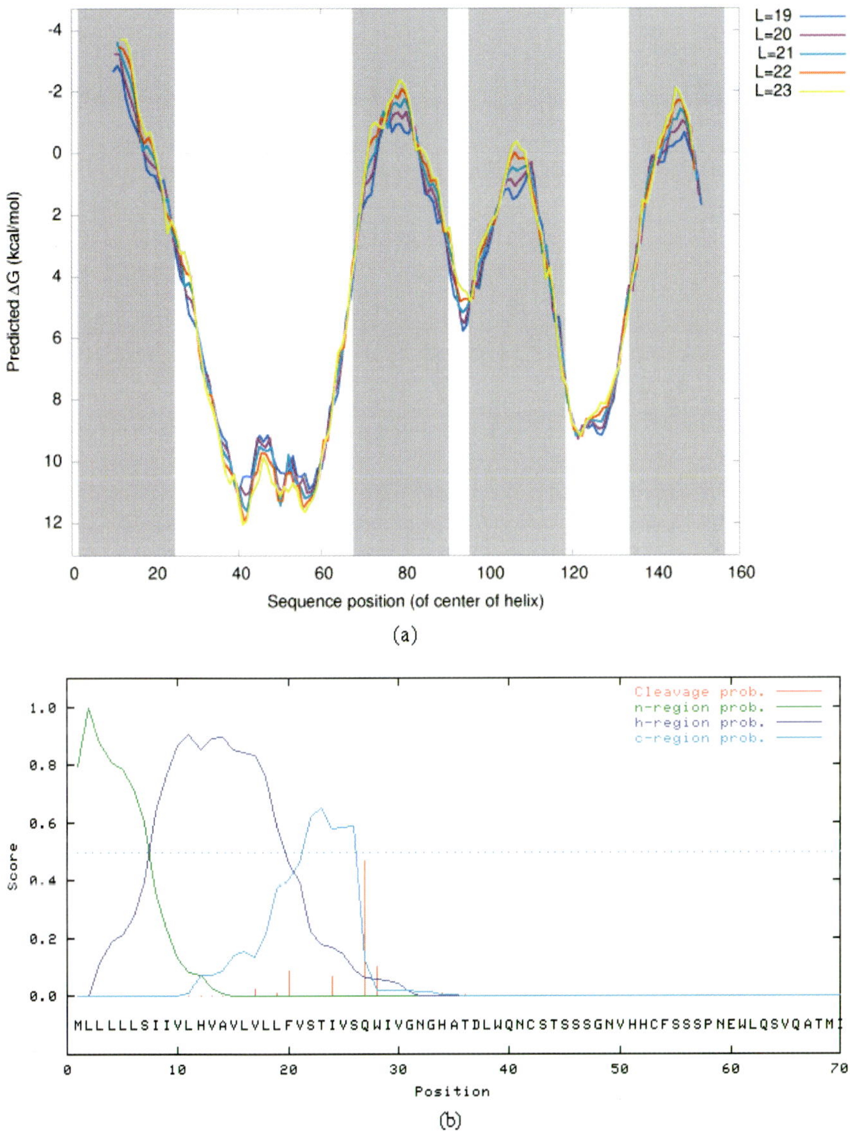

Fig. 4. Analysis of human PMP22 by (a) the ΔG_{app} predictor and (b) SignalP 3.0. Putative transmembrane segments are shaded in the ΔG_{app} plot. The signal peptide is most likely not cleaved but remains the most N-terminal transmembrane helix.

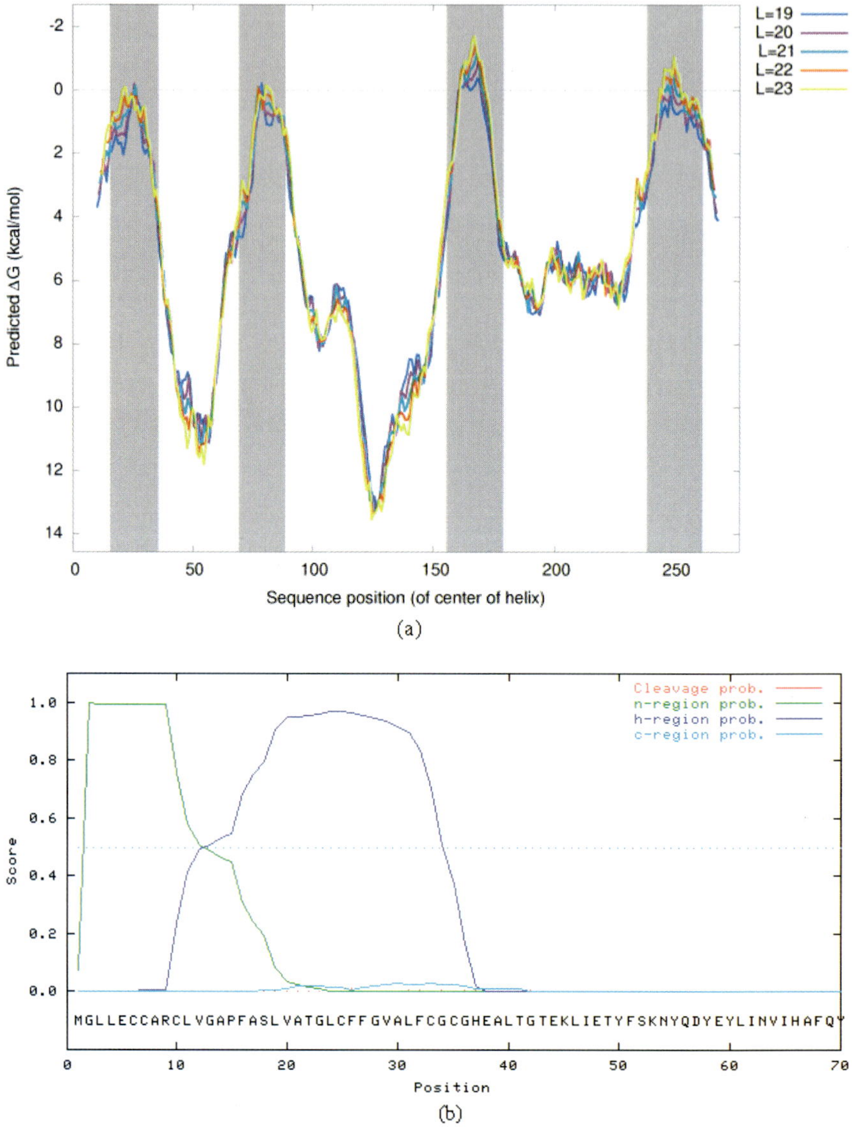

Fig. 5. Analysis of human PLP by (a) the ΔG_{app} predictor and (b) SignalP 3.0. Putative transmembrane segments are shaded in the ΔG_{app} plot. The signal peptide is most likely not cleaved but remains the most N-terminal transmembrane helix.

membrane, and regions around residues 40–59, 178–191 and 215–232 to the extracellular side. Disulfide bonds are thought to form between residues 184–228 and 201–220, suggesting an extracellular location for these residues. Taken together, the data strongly suggests an N_{in}-4TM-C_{in} topology with both termini in the cytosol, cf. Wahle, Stoffel (1998). The DM-20 isoform lacks residues 117–151 in the central cytosolic loop. Proteolipid is implicated in Pelizaeus–Merzbacher disease, a neurological condition that affects myelin (insulation surrounding the axons) in the brain and spinal cord. The symptoms are first noticed shortly after birth or in infancy (Garbern *et al.*, 1999). There is no known cure for this disease.

MAG

Human MAG (Uniprot P20916) is 626 residues long, including a strongly predicted N-terminal signal peptide (residues 1–19) and a very stable transmembrane α-helix (Fig. 6). It carries multiple N-linked glycans and disulfide bonds in the first 500 residues (consistent with an extracellular location of this region). The topology of the mature protein (after removal of the signal peptide) is hence N_{out}-1TM-C_{in}.

MOG

Human MOG (Uniprot Q16653) is 247 residues long, including a strongly predicted N-terminal signal peptide (residues 1–29) and two C-terminal hydrophobic segments (Fig. 7). It carries a potential N-linked glycan at residue 60 and a potential disulfide between residues 53 and 127 (consistent with an extracellular location of this region). The most C-terminal hydrophobic domain is lacking in a number of isoforms of the protein. Available biochemical data suggest that only the first of the two hydrophobic domains forms a transmembrane α-helix, and that the C-terminus is cytosolic (della Gaspera, 1998). The topology of the mature protein (after removal of the signal peptide) is hence N_{out}-1TM-C_{in}. Given

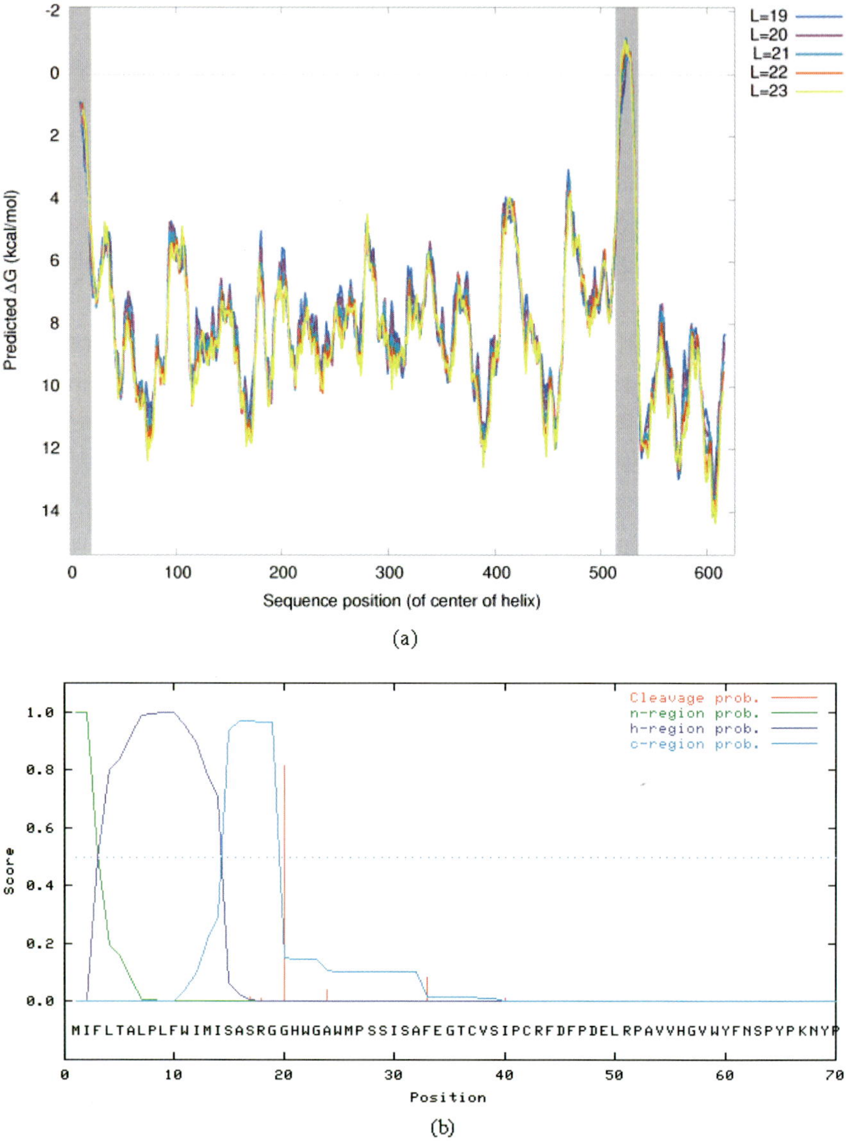

Fig. 6. Analysis of human MAG by (a) the ΔG_{app} predictor and (b) SignalP 3.0. Putative transmembrane segments are shaded in the ΔG_{app} plot. The predicted N-terminus of the mature protein after signal peptidase cleavage is indicated by a red line in the SignalP plot.

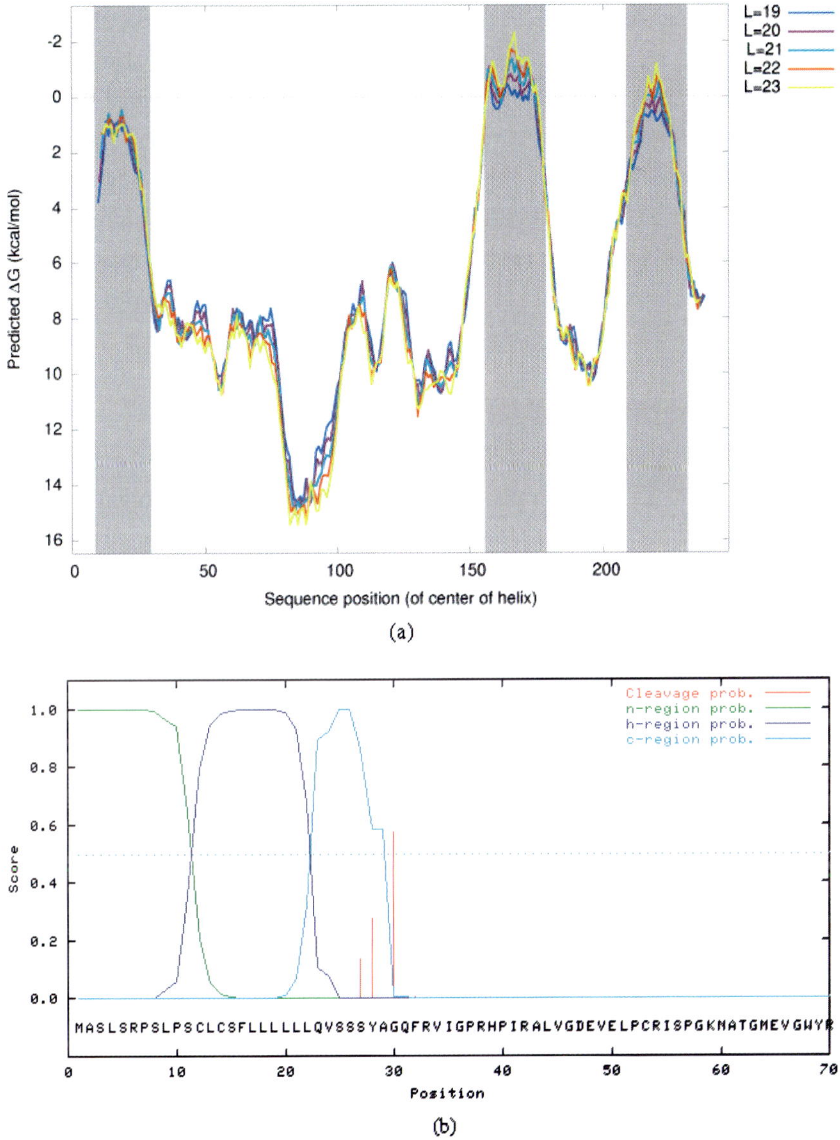

Fig. 7. Analysis of human MOG by (a) the ΔG_{app} predictor and (b) SignalP 3.0. Putative transmembrane segments are shaded in the ΔG_{app} plot. The predicted N-terminus of the mature protein after signal peptidase cleavage is indicated by a red line in the SignalP plot.

the hydrophobic nature of the second hydrophobic domain, the presence of two proline residues near the middle of the hydrophobic stretch, and positively charged residues on both flanks, it is possible that this segment forms a short "helical hairpin" that penetrates only partly across the membrane (Monné *et al.*, 1999; Sääf *et al.*, 2000). The segment is protected from protease attack and cannot be extracted from membranes with chaotropic agents (della Gaspera, 1998). So far, MOG is the only candidate autoantigen responsible for causing multiple sclerosis (Bernard *et al.*, 1997).

Summary

Explaining how the myelin proteins are involved in the organization and function of the myelin sheath requires knowing their molecular structures. From the bioinformatics analysis presented above, along with published experimental data, we conclude that MBP and P2 are both cytosolic and either soluble or peripherally bound to the myelin membrane. P0 and MAG both have a cleavable signal peptide and a C-terminal transmembrane anchor. The same conclusion was made by Kirschner and Inouye (1991): their comparison of the corresponding myelin basic proteins (MBP) and P0 glycoproteins for rodent and shark showed that the conserved residues included most of the amino acids which were predicted to form the alpha or beta conformations, while the altered residues were mainly in the hydrophilic and turn or coil regions. In both rodent and shark the putative extracellular domain of P0 glycoprotein displayed consecutive peaks of beta propensity similar to that for the immunoglobulins, while the cytoplasmic domain showed alpha-beta-alpha folding. The flat β-sheets of P0 are orientated parallel to the membrane surface to facilitate their homotypic interaction in the extracellular space (Inouye, Kirschner, 1991).

PLP has four transmembrane helices, and PMP22 has either two or four transmembrane helices, and both proteins have their N- and C-termini orientated towards the cytosol. MOG, finally, has a signal peptide and one transmembrane helix. A second, markedly hydrophobic segment in the

C-terminal part of the protein is apparently not transmembrane but may form a re-entrant loop penetrating part of the way across the membrane.

Neither MBP, PLP, PMP22, MAG nor full-length P0 or MOG membrane glycoproteins have been crystallized for structural determination and structurally based drug design. Bioinformatics and protein crystal growers will be very busy in the years to come.

Acknowledgment

Jan Sedzik acknowledges the partial support of the Swedish Research Council, grant 2007–6890.

References

Aloy P, Pichaud M, Russell RB. (2005) Protein complexes: Structure prediction challenges for the 21st century. *Curr Opin Struct Biol* **1**: 15–22.

Aloy P, Russell RB. (2006) Structural systems biology: Modelling protein interactions. *Nat Rev Mol Cell Biol* **3**: 188–197.

Aro E-M, Suorsa M, Rokka A, Allahverdiyeva Y, Paakkarinen V, Saleem A, Battchikova N, Rintamäki E. (2005) Dynamics of photosystem II: A proteomic approach to thylakoid protein complexes. *J Exp Botany* **56**: 347–356.

Ballas SK, Krasnov SH. (1980) Structure of erythrocyte membrane and its transport functions. *Ann Clin Lab Sci* **10**: 209–219.

Bernard CC, Johns TG, Slavin A, Ichikawa M, Ewing C, Liu J, Bettadapura J. (1997) Myelin oligodendrocyte glycoprotein: A novel candidate autoantigen in multiple sclerosis. *J Mol Med* **75**: 77–88.

Boguski MS. (1998) Bioinformatics — A new era. *Trends Biotech* **16**(S1): 1–3.

Corcelli A, Lattanzio VMT, Mascolo G, Papadia P, Francesco F. (2002) Lipid-protein stoichiometries in a crystalline biological membrane: NMR quantitative analysis of he lipid extract of the purple membrane. *J Lipid Res* **43**: 132–140.

della Gaspera B, Pham-Dinh D, Roussel G, Nussbaum JL, Dautigny A. (1998) Membrane topology of the myelin/oligodendrocyte glycoprotein. *Eur J Biochem* **258**: 478–484.

D'Urso D, Brophy PJ, Staugaitis SM, Gillespie CS, Frey AF, Stempakshort JG, Colman DR. (1990) Protein zero of peripheral nerve myelin: Biosynthesis, membrane insertion, and evidence for homotypic interaction. *Neuron* **4**: 449–460.

D'Urso D, Müller HW. (1997) Ins and outs of peripheral myelin protein-22: Mapping transmembrane topology and intracellular sorting. *J Neurosci Res* **49**: 551–562.

Dyrløv-Bendtsen J, Nielsen H, von Heijne G, Brunak S. (2004) Improved prediction of signal peptides — SignalP 3.0. *J Mol Biol* **340**: 783–795.

Elofsson A, von Heijne G. (2007) Membrane protein structure: Prediction vs. reality. *Ann Rev Biochem* **76**: 125–140.

Garbern J, Cambi F, Shy M, Kamholz J. (1999) The molecular pathogenesis of Pelizaeus–Merzbacher disease. *Arch Neurology* **56**: 1210–1214.

Gaspera BG, Pham-Dinh D, Roussel G, Nussbaum JL, Dautigny A. (1998) Membrane topology of the myelin/oligodendrocyte glycoprotein. *Eur J Biochem* **258**: 478–484.

Hessa T, Meindl-Beinker NM, Bernsel A, Kim H, Sato Y, Lerch-Bader M, Nilsson IM, White SH, von Heijne G. (2007) Molecular code for transmembrane-helix recognition by the Sec61 translocon. *Nature* **450**: 1026–1030.

Inouye H, Kirschner DA. (1991) Folding and function of the myelin proteins from primary sequence data. *J Neurosci Res* **28**: 1–17.

Jones DT. (2007) Improving the accuracy of transmembrane protein topology prediction using evolutionary information. *Bioinformatics* **23**: 538–544.

Kiel C, Wohlgemuth S, Rousseau F, Schymkowitz J, Ferkinghoff-Borg J, Wittinghofer F, Serrano L. (2005) Recognizing and defining true Ras binding domains II: *In silico* prediction based on homology modelling and energy calculations. *J Mol Biol* **348**: 759–775.

Krogh A, Larsson B, von Heijne G, Sonnhammer ELL. (2001) Predicting transmembrane protein topology with a hidden Markov model: Application to compete genomes. *J Mol Biol* **305**: 567–580.

Li G, Knowles PF, Murphy DJ, Marshal D. (1990) Lipid-protein interactions in thylakoid membranes of chilling-resistant and -sensitive plants stufied by spin label, electron spin label electron spin resonance spectroscopy. *J Biol Chem* **165**(28): 16867–16872.

Monné M, Nilsson IM, Elofsson A, von Heijne G. (1999) Turns in transmembrane helices: Determination of the minimal length of a "helical hairpin" and derivation of a fine-grained turn propensity scale. *J Mol Biol* **293**: 807–814.

Sedzik J, Kirschner DA. (1992) Is myelin basic protein crystallizable? *Neurochem Res* **17**: 157–166.

Sedzik J, Kotake Y, Uyemura K. (1998) Purification of PASII/PMP22 — An extremely hydrophobic glycoprotein of PNS myelin membrane. *Neuroreport* **9**: 1595–1600.

Stoner GL. (1984) Predicted folding of beta-structure in myelin basic protein. *J Neurochem* **43**: 433–437.

Stoner GL. (1990) Conservation throughout vertebrate evolution of the predicted beta-strands in myelin basic protein. *J Neurochem* **55**: 1404–1411.

Sääf A, Hermansson M, von Heijne G. (2000) Formation of cytoplasmic turns between two closely spaced transmembrane helices during membrane protein integration into the ER membrane. *J Mol Biol* **301**: 191–197.

Taylor V, Zgraggen C, Naef R, Suter U. (2000) Membrane topology of peripheral myelin protein 22. *J Neurosci Res* **62**: 15–27.

Tusnády GE, Simon I. (2001) The HMMTOP transmembrane topology prediction server. *Bioinformatics* **17**: 849–850.

Viklund H, Elofsson A. (2004) Best alpha-helical transmembrane protein topology predictions are achieved using hidden Markov models and evolutionary information. *Protein Sci* **13**: 1908–1907.

Wahle S, Stoffel W. (1998) Cotranslational integration of myelin proteolipid protein (PLP) into the membrane of endoplasmic reticulum: Analysis of topology by glycosylation scanning and protease domain protection assay. *Glia* **24**: 226–235.

Chapter 4

Biomolecular Mass Spectrometry: Applications to Proteins and Peptides

Leopold L. Ilag and Gianluca Maddalo*†

Mass spectrometry is an analytical technique which, as its name implies, "weighs" molecules, and thereby can define molecular identity, interactions and reactions. With this technique, molecules from a source are ionized (i.e. made to carry a formal charge, either negative or positive) and subsequently enter the mass analyzer where ions are separated according to their mass/charge ratio (*m*/*z*), and then a spectrum is recorded as ions reach the detector. Initially applications were limited to small molecules but nowadays even large macromolecular complexes are amenable for analysis.

Keywords: Mass spectrometry; biomolecules; electrospray ionization (ESI); matrix-assisted laser desorption ionization (MALDI); collision induced dissociation/activation (CID/CAD); high molecular weight mass spectrometry; molecular machines; proteomics; tandem mass spectrometry; imaging.

Introduction

The advent of soft ionization methods, namely, electrospray ionization (ESI) and matrix-assisted laser desorption ionization (MALDI), revolutionized the field of mass spectrometry (MS) particularly extending its domain well into the life sciences. This made possible the ionization of biological macromolecules, which are species too fragile for previous

*Corresponding author.
Department of Analytical Chemistry, Stockholm University, Stockholm, Sweden. E-mail: leopold.ilag@anchem.su.se.
†Department of Analytical Chemistry, Stockholm University, Stockholm, Sweden.

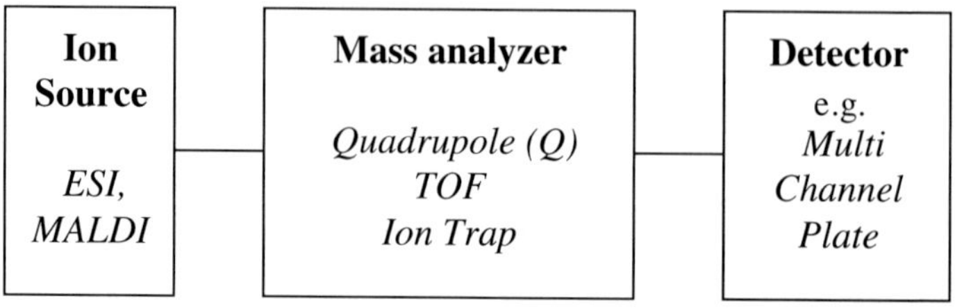

Fig. 1. Basic modular structure of a mass spectrometer.

ionization methods. This ground-breaking advance was recognized in 2002 with a Nobel Prize in Chemistry.

Jargon in MS can be daunting, but much can be simplified when one bears in mind the modular structure of the instrument, i.e. ion source, mass analyzer, detector (Fig. 1). Examples of ion sources as mentioned above are ESI and MALDI, whereas types of mass analyzers include quadrupoles (Q), ion traps (IT), and time-of-flight (ToF), among others. The combination of ion source and mass analyzer(s) gives rise to the name of the configuration of a particular instrument. A MALDI-ToF is a time of flight instrument with a MALDI source whereas an ESI-Q-ToF (Fig. 2) is a tandem mass spectrometer made up of two analyzers: quadrupole and ToF with an electrospray ion source. Technical descriptions of these analyzers are beyond the scope of this chapter, thus the reader is referred to an excellent introductory book (Siuzdak, 1996).

Briefly, a ToF analyzer measures the mass as a function of the time it takes for the ions to get from the source to the detector. Heavier ions will take a longer time than lighter ions. The important thing to remember with quadrupoles is that this can be used in transmission mode, where one can let all ions through, or as a mass filter, wherein one can select ions of a particular mass (m/z) and subject these to collisional activated dissociation (CAD) in order to be fragmented (or dissociated) and the products analyzed, giving structural information. This technique using CAD is referred to as tandem mass spectrometry, MS/MS or MS^2. Finally, ion traps like quadrupoles can perform MS/MS but can do so in more than one cycle, which gives rise to MS^n capabilities resulting in greater structural information. A detector then

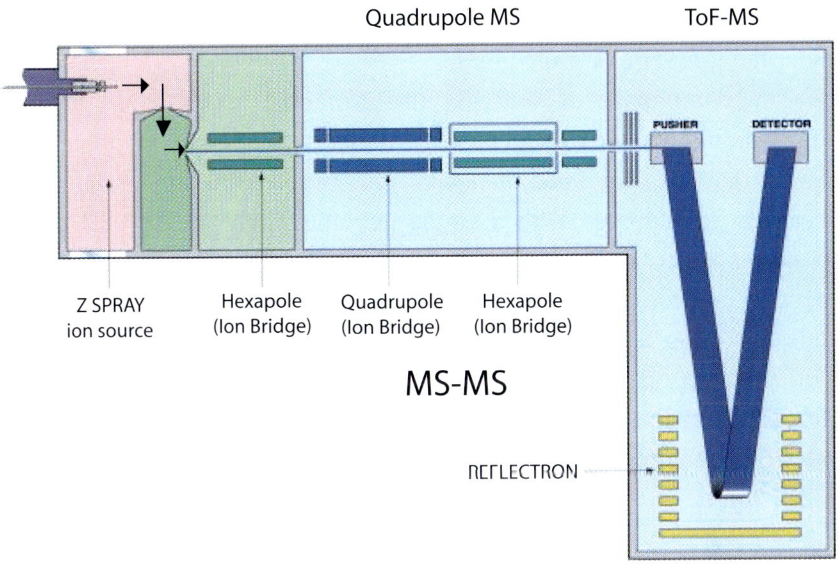

Fig. 2. The QToF is one of the most versatile configurations for biological mass spectrometry. It combines a quadrupole mass filter and a hexapole collision cell with a ToF analyzer, which allows MS/MS for *de novo* sequencing and dissociation experiments on non-covalent complexes. Arrows indicate the trajectory of the ions into the spectrometer.

detects the ions that are separated in the mass analyzer. One of the most common detectors is the multi-channel plate detector, which is closely related to electron multipliers. When ions hit the detector, the secondary ions produced are accelerated into the continous dynode electron multiplier, resulting in an electron cascade that gives rise to a measurable current at the end of the electron multiplier (de Hoffmann, Stroobant, 2002).

Proteomics

The most prominent role of MS in biological research today is in the post-genomic arena. The genome is a mere blueprint of a vastly complex network of interactions orchestrated principally by proteins. The protein complement of the genome is what we refer to as the *proteome* and the variety of disciplines aiming to survey and understand the proteome falls under the umbrella of proteomics.

The complexity of studying the proteome arises from the fact that proteins encoded by genes can undergo a number of processing events that are regulated in space and time. Although a particular organism is endowed with a single genome, there are several corresponding proteomes for each cell population at a certain time and under specific environmental influences. At the molecular level, a single gene encoding a protein yields a product that is then processed and modified (e.g. glycosylation or phosphorylation), amplifying the variants. Thus the estimated 25 000 genes in humans, for example, get amplified to more than a million protein species (Cho, 2007; Kosak *et al.*, 2004).

Techniques and strategies

Given the above numbers, it seems almost impossible to make sense of the proteome. Fortunately the methods used to address issues in proteomics simplify the existing variation into more manageable magnitudes because the techniques offer the possibility for efficient profiling. There is no immediate need to account for all minor differences allowing a survey, for example, of relevant proteins that may fluctuate as a result of a certain disease.

The techniques involved in proteomics include those which can be used for protein separation, identification and quantification. More advanced applications allow for structural analysis, identifying/mapping modifications, defining complex interactions and even tracking protein regulation in the cellular context. One of the most common strategies include one-dimensional (1D) and two-dimensional (2D) gel electrophoresis, various types of chromatography, and mass spectrometry (Link, 2002). The latter is powerful as a stand-alone technique in proteomics research but maximal effectiveness is achieved when it is coupled to the above-mentioned separation methods.

It is amazing how much information one can get by merely "weighing things." However, when one realizes that MS offers enough resolution to distinguish a mass difference in the range of a single proton to half a millidalton using Fourier transform ion cyclotron mass spectrometry (FT-ICR-MS) (Marshall *et al.*, 1998), then it is not so surprising why this technique is so powerful.

Most of the variations we try to track as indicators of abnormal physiology can be related to a mass change. When two molecules interact or are modified, there is a corresponding change in mass (except rearrangement reactions). When peptides or proteins are modified chemically, there would also be a change in mass seen as an increase or decrease attributed to the moiety added or removed, respectively. Because MS allows directed fragmentation, even isomers can be distinguished based on the different fragments generated as a function of structure; hence, products of different masses are obtained giving insight into chemical structure.

Analytes or samples to be studied by MS can be solid, liquid or gas as starting materials as long as charged analytes can be generated from them, because MS depends on gas-phase mass separation and detection of ions. Early methods of ionization were limited to analysis of small molecules. With the introduction of fast atom bombardment (FAB) (Morris *et al.*, 1981), it became possible to ionize intact peptides. However, it was not until soft-ionization methods, namely, ESI (Fenn *et al.*, 1989, 1990) and MALDI (Karas *et al.*, 1988; Chaurand *et al.*, 1999) were developed, that rapid and facile applications to large biomolecules were made routinely possible.

Peptide mass fingerprinting

MALDI coupled to the time-of-flight (ToF) analyzer has been a workhorse for the strategy called peptide mass fingerprinting (PMF). This is based on the fact that proteins having unique sequences have distinct patterns of cleavage sites for a given enzyme. The cleaved protein therefore yields peptides that reflect this unique site distribution (distinct sequence) allowing the matching of the experimental digest with a theoretical digest stored in databases, thus aiding the identification of a protein (Yates, 1998).

De novo sequencing

Mass fingerprinting is not a very high-resolution method and often fails when the sample is not pure, protein of interest is not in the database, or when you need to distinguish related proteins or isoforms. Therefore, it is still important to get actual sequence information through *de novo* sequencing, which can be achieved when powerful mass analyzers, for example, the quadrupole time of flight (Q-ToF), are used (Fig. 2) (Wood *et al.*, 2002). The Q-ToF is endowed with the capability to perform tandem MS. As described previously, this refers to the possibility of using the quadrupole as a mass filter allowing the isolation of ions of a desired mass/charge ratio (*m*/*z*) and subsequently subjecting these to collisions with inert gas molecules to cause fragmentation [collision induced dissociation (CID) or collision activated dissociation (CAD)] (Sleno *et al.*, 2004). The accelerated ions will impact against inert gas molecules (usually argon) resulting in an increase in internal energy and inducing the fragmentation of the ions. The weakest bonds are those along the peptide backbone. Statistically, the peptide is going to break frequently at the amide bond, and hence the most abundant fragments that are going to be produced are the *b* and *y* series (Fig. 3), according to the retention of the charge at the amino or carboxy terminal of the fragments, respectively. Fragments vary by a discernable mass difference corresponding to known amino acids from which the sequence can be pieced together.

Post-translational and Chemical Modifications

MS also allows determination of post-translational and chemical modifications when mass differences not corresponding to any of the essential amino acids are detected. Typical modifications may include a mass increase of 80 Da for phosphorylation or addition of 16 Da for oxidation. There

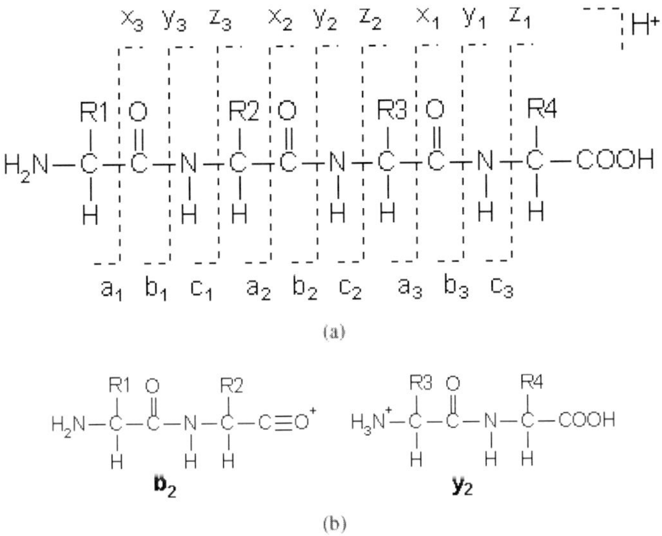

Fig. 3. (a) Fragmentation of a peptide along the backbone with corresponding nomenclature of fragments generated. (b) *b* and *y* ions are obtained when the charge is retained on the amino or carboxy terminus fragment, respectively. (Figure used with permission from Matrix Science Ltd., www.matrixscience.com.)

are around 300 known post-translational modifications (PTMs) found in proteins, some of which are listed in the Table 1. For more detailed lists, the reader is referred to other resources (see footnote in Table 1).

MALDI Imaging

In recent times, the application of MS has moved into the realm of MS-imaging. This is typically based on MALDI-ToF or MALDI-ToF/ToF instrumentation wherein the laser is used to raster over a whole tissue sample (Caprioli *et al.*, 1997; Chaurand *et al.*, 2006; Andersson *et al.*, 2008). Each spectrum collected over time not only encodes the chemical entities present but also the spatial distribution of those entities over the tissue sample. Through a simple algorithm, it becomes possible to display the distribution of specific molecules of a particular *m*/*z*, e.g. peptides, thus monitoring fluctuations that may be correlated with disease or any

Table 1. Examples of Post-translational and Chemical Modifications Observed in Peptides/Proteins

Modification	Monoisotopic Mass	Average Mass	Chemical Species +/–
Acetylation	+42.0106	+42.0373	$+COCH_3$
Formylation	+27.9949	+28.0104	$+CHO$
Methylation	+14.0157	+14.0269	$+CH_3$
Phoshorylation	+79.9663	+79.9799	$+HPO_3$
Citrullination	+0.9840	+0.9847	(Arg)-$CH_4N_3 \rightarrow CH_3N_2O$
Palmitoylation	+238.2297	+238.4136	$+C_{16}H_{31}O_1$ (palmitic acid moiety) and –H (from modified amino acid)
Sulfation	+79.9568	+80.0642	$+SO_3$
O-GlcNAc	+203.0794	+203.1950	$+C_8H_{13}O_5N$
Oxidation of Met	+15.9949	+15.9994	+O
Homoserine formed from Met by CNBr treatment	–29.9928	–30.0935	C_5H_9NOS (Met) $\rightarrow C_4H_7NO_2$ (homoSer)
Pyroglutamic acid formed from Gln	–17.0265	–17.0306	$-NH_3$
Disulfide bond formation	–2.0156	–2.0159	$-H_2$
Deamidation of Asn	+0.9840	+0.9847	$C_4H_6N_2O_2$ (Asn) $\rightarrow C_4H_5NO_3$ (isoaspartic acid)
Cholesterol	+368.3443	+368.6400	Cholesterol glycine ester

http://www.expasy.ch/tools/findmod/findmod_masses.html
http://www.abrf.org/index.cfm/dm.home

abnormal physiology. This is an excellent approach for obtaining global difference maps without previous knowledge of what exact biomolecular species is of interest. More information on this development was reviewed recently (McDonnell, Heeren, 2007).

MS in structural biology

MS has also become an important tool in structural biology. Because of the extension in mass range for tandem MS and a phenomenon known as collisional cooling, which allows efficient focusing of macromolecular ions in the mass spectrometer, very large non-covalent complexes can be maintained and analyzed (Sobott *et al.*, 2002; Chernushevich *et al.*, 2004). It turns out to be an excellent complementary tool for X-ray crystallography.

When the molecules exceed the size of typical peptides, such as proteins and their complexes, the energies absorbed by ions in the collision cell are not sufficient to cause fragmentation but only simple dissociation (of complexes). Although this cannot give sequence information, it provides useful data in terms of structural organization and stoichiometry. Given a macromolecular complex, through dissociation it is possible to detect heterogeneous composition (van den Heuvel *et al.*, 2004; Ilag *et al.*, 2005; Hernandez *et al.*, 2006). Finally, it could also set the stage for comparing relative binding strengths of complexes opening avenues, for example, in screening drug binding (Tjenberg *et al.*, 2004; Ilag *et al.*, 2004). One should bear in mind, however, that in any MS analysis the interactions being studied are gas-phase interactions. Though interactions existing in solution may still be reflected in the gas phase, one should not assume that one is still looking at exactly native solution structures or interactions.

Fibroblast growth factor (FGF) and fibroblast growth factor receptor (FGFR)

The fibroblast growth factor (FGF) is a heparin binding protein which requires this interaction for its signal transduction. Crystal structures of FGF complexed with its receptor (FGFR) and heparin have been determined. However, two markedly different structures (Pellegrini *et al.*, 2000; Schlessinger *et al.*, 2000) have been elucidated, reflecting two different stoichiometries: one with two molecules of heparin bound and another with only one heparin bound. The question is: which stoichiometry (and therefore structure) is most likely to occur in solution? This is a straightforward matter for MS to resolve because the difference in the number of bound heparin molecules would be clearly reflected in the mass of the complex. In this particular case, it was shown that the measured mass is consistent with 2:2:1 stoichiometry (Harmer *et al.*, 2004) (Fig. 4). Further

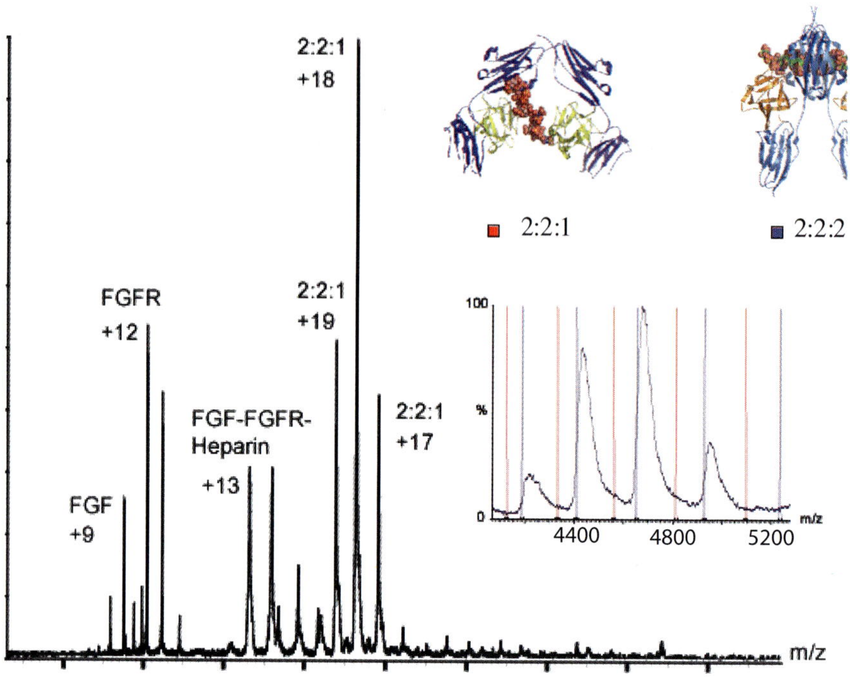

Fig. 4. The FGF:FGFR:heparin complex exhibits a 2:2:1 stoichiometry. The MS spectrum clearly demonstrates that the stoichiometry of this complex is the 2:2:1 with simulations (sharp red and blue peaks in inset) clearly showing the correspondence of the 2:2:1 (blue) theoretical spectrum with the onset of the measured peaks. Broad peaks are typically observed for non-covalent complexes because buffer molecules and other adducts remain bound to the complex. Charge states are indicated for the different molecular species.

confirmation of this was established by employing tandem MS experiments where ions corresponding to the intact complex were isolated and dissociated in the collision cell of the spectrometer and the products analyzed. The dissociation products indeed show only one bound heparin molecule.

RNA polymerase

It should be noted that broadening of peaks is characteristic of non-covalent complexes because buffer molecules and some adducts are often retained.

However, there exists a simple but efficient strategy whereby very narrow peaks for complexes can be obtained. In the work on bacterial RNA polymerase (~400 kDa) this strategy has led to the discovery of a previously undetected complex which would otherwise have been buried by the typical broad peaks of the complex (Ilag *et al.*, 2004). This allowed the identification of the regulatory factor Rsd as interacting with the RNA polymerase holoenzyme and binding to sigma factor 70 displacing it from the holoenzyme (Fig. 5). This is a strong demonstration of the power of MS, not only in characterizing complexes but also in detecting dynamic reorganizations that may have otherwise been overlooked. It is important to note that such reorganizations may have important regulatory implications.

Ribosome

From analysis of small to medium sized complexes, MS has also tackled huge assemblies (MDa range) such as viruses (Bothner, Siuzdak, 2004) and the ribosome (Videler *et al.*, 2005). The latter is a bit more challenging because this molecular machine is an asymmetric and heterogeneous complex. It is made up of three RNA chains and over 50 associated proteins. Two subunits called 50S and 30S make up the intact 70S ribosome. The crystal structure of the ribosome has been solved (Schuwirt *et al.*, 2005; Wimberly *et al.*, 2000; Harms *et al.*, 2001; Schluenzen *et al.*, 2000; Ban *et al.*, 2000). However, the stalk proteins on the 50S subunit have rather poor electron density because they are rather dynamic structures. Here again we see the complementary nature of MS to X-ray crystallography because the stalk proteins seem to be the most accessible for MS and successful measurements of intact 30S, 50S and 70S ribosomes have been made (Ilag *et al.*, 2005; McKay *et al.*, 2006; Videler *et al.*, 2005). Using tandem MS on the 50S ribosomal subunit established new heights for MS. For the first time, ions above 20000 *m/z* were isolated and dissociated (Fig. 6), revealing that a population of stalk proteins are apparently phosphorylated (Ilag *et al.*, 2005). This has never been observed in *E. coli* ribosomes (Hanson *et al.*, 2003) but is a known

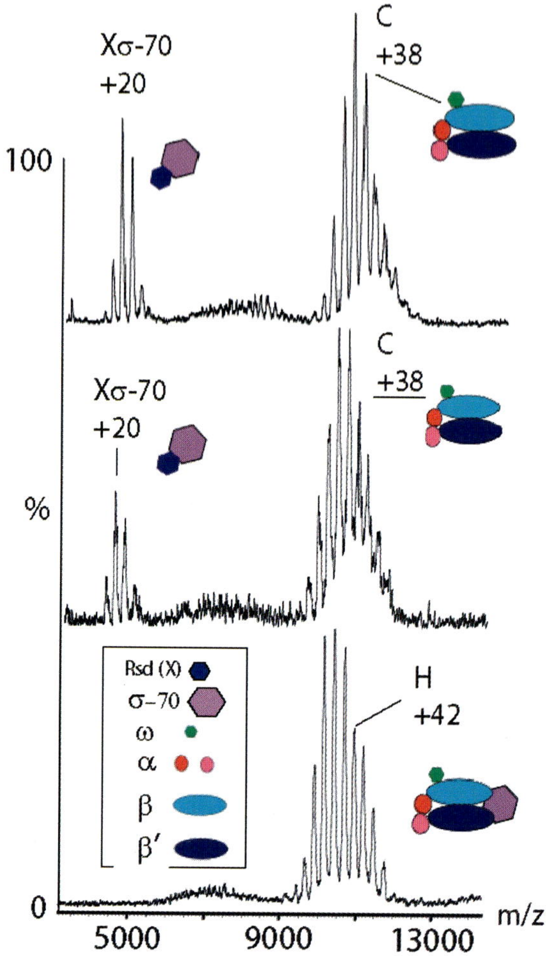

Fig. 5. Addition of Rsd (indicated as X) to the RNA polymerase holoenzyme (H). The holoenzyme alone (bottom spectrum) compared to addition of equimolar amount of Rsd (middle spectrum), and with two molar excess of Rsd (top spectrum) shows the appearance of a species which corresponds in mass to a 1:1 complex of Rsd-σ^{70}, indicating a reorganization of the holoenzyme. Inset shows Rsd and the components of the holoenzyme.

modification on the yeast ribosomal stalk proteins (Hanson *et al.*, 2004). Furthermore, by assessing dissociated products from the samples, a complex with a mass not attributable to any known ribosomal protein(s) was observed. By employing further tandem MS on this complex, it turns out

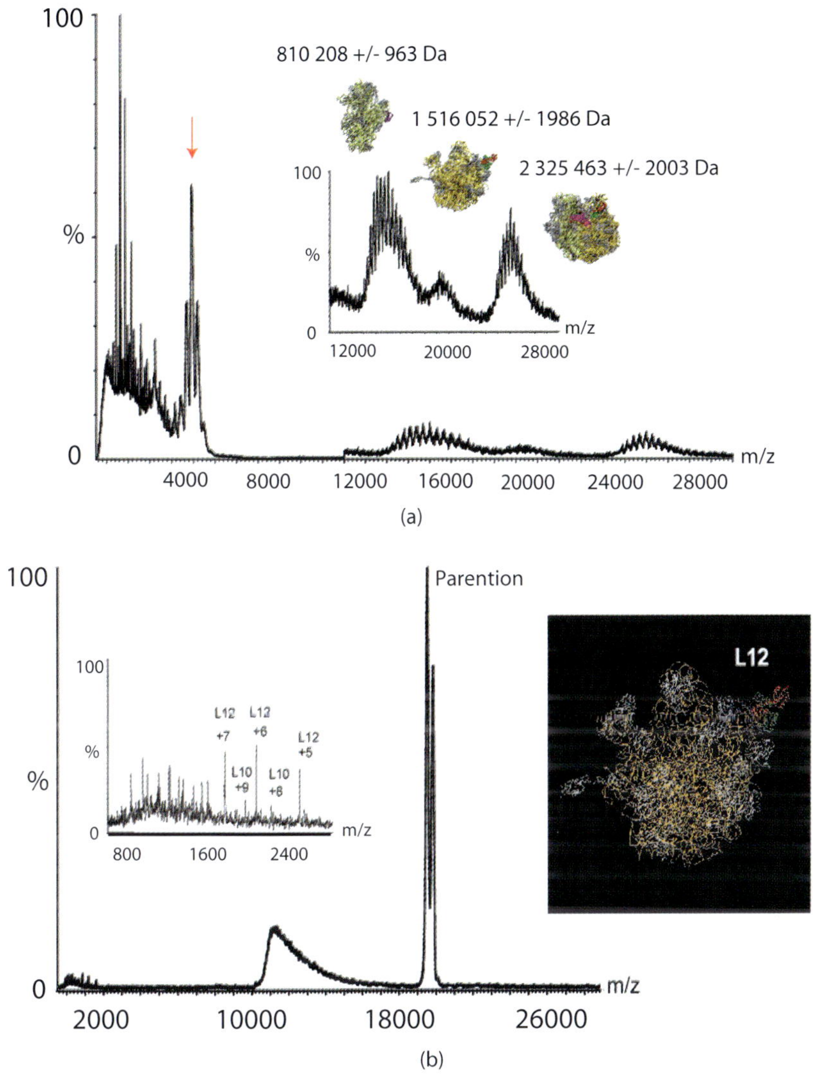

Fig. 6. (a) MS of intact *Thermus thermophilus* ribosomes is accompanied by release of some individual protein components. An unknown component of ~96 kDa (red arrow) not corresponding to any known proteins or subcomplexes of the ribosomes was later determined to be a non-canonical trimer of the L7/L12 stalk proteins. (b) Tandem MS of 50S ribosomal subunit allows detection of stalk proteins (inset) without loss of the structural information as to their origin (i.e. bound to the 50S as opposed to free proteins in solution). Further analysis of the data indicated the presence of phosphorylated L7/L12 proteins. Models are based on the Protein Data Bank structures 1GIX and 1GIY.

to be a trimer of the usual dimer of stalk proteins. This non-canonical stalk structure seemed to be a peculiarity of the thermophilic ribosomes studied (Fig. 6).

Protein-lipid interactions

An important aspect of structural mass spectrometry is in identifying lipid ligands of proteins not readily discernible by other structural methods. Lipids are structural components of biomembranes but they also play significant roles in signal transduction and energy storage. As regulatory molecules some have been found to be ligands of orphan nuclear receptors (so called because their endogenous ligands were unknown).

In the case of retinoid-X-receptor or RXR, which is a nuclear receptor acting as a ligand activated transcription factor, the identification of the ligand was done by MS (de Urquiza *et al.*, 2000). Through a combination of biochemical work-up and bioassays of brain conditioned-media, the isolation of an active component was achieved. Analysis of an intense ion peak associated with active fractions as well as some minor ones made it possible to deduce the molecular formula as $C_{22}H_{32}O_2$. This was later confirmed by identical fragmentation patterns from MS/MS experiments of the intense ion peak (from active fractions) with that of the standard cis-4,7,10,13,16,19-docosahexaenoic acid (DHA). The true ligand binding activity of DHA was further confirmed by its activation being very sensitive to mutations that alter the ligand-binding specificity of RXR.

Another example is the case of horse myelin P2, which is an essential myelin (insulating membrane of neuronal axons) associated protein belonging to the fatty acid binding protein family (FABP). It was possible not only to identify the ligands in the binding pocket but also preserve the complex such that a higher than expected molecular weight was accounted for by the presence of lauryl dimethylamine-N-oxide and HEPES, both components of the crystallization buffer for this membrane associated protein (Hunter *et al.*, 2005). FABPs have been shown to have as natural ligands, palmitic acid, oleic acid among others (Balendiran *et al.*, 2000).

Conclusion

Mass spectrometry is indeed a powerful technique that is not limited to mere mass measurements for small molecules but could also be powerfully employed for the identification, characterization and structural analysis of any biomolecule, particularly proteins and their interactions. So far no technique can parallel the accuracy, rapidity, sensitivity and reliability of MS for proteomics applications. Bearing in mind that most biological pathways are governed by proteins and their different temporal isoforms, it is easy to appreciate the paramount role to be further played by MS in elucidating the dynamics of various metabolic pathways in the cell. By weighing molecules and their fragments, we can identify and characterize different biologically significant entities. Moreover, this can lead to the search for potential biomarkers related to diseases and aid further unravelling of the fundamental mechanics of molecular and cellular networks. Furthermore, recent improvements in MS instrumentation has allowed MS to open an important new avenue for addressing real biological samples. One major pitfall of the technique is that molecules to be analyzed must form ions sufficiently stable under typical MS conditions. For other practical considerations, the reader is referred to a review by Lubec and Afjehi-Sadat (2007). Overall, MS is not only important for proteomics but has come of age as a significant tool for molecular, cellular and structural biology.

Acknowledgments

We would like to thank Nicholas Harmer (Cambridge University, UK) for the FGF/FGFR/heparin model structures and Andrew Carter (Medical Research Council, Cambridge, UK) for the ribosomal structure models. We also wish to acknowledge helpful comments from Frank Sobott (Oxford University, UK), Ulrika Nilsson, Therese Redeby, Mohammadreza Shariatgorji and Gunnar Thorsen (Stockholm University, Sweden).

References

Andersson M, Groseclose MR, Deutch AY, Caprioli RM. (2008) Imaging mass spectrometry of proteins and peptides: 3D volume reconstruction. *Nat Methods* **5**: 101–108.

Balendiran GK, Schnutgen F, Scapin G, Borchers T, Xhong N, Lim K, Godbout R, Spener F, Sacchettini JC. (2000) Crystal structure and thermodynamic analysis of human brain fatty acid-binding protein. *J Biol Chem* **275**: 27045–27054.

Ban N, Nissen P, Hansen J, Moore PB, Steitz TA. (2000) The complete atomic structure of the large ribosomal subunit at 2.4 Å resolution. *Science* **289**: 905–920.

Bothner B, Siuzdak G. (2004) Electrospray ionization of a whole virus: Analyzing mass, structure, and viability. *Chem BioChem* **5**: 258–260.

Caprioli R, Farmer TB, Gile J. (1997) Molecular imaging of biological samples: Localization of peptides and proteins using MALDI-TOF MS. *Anal Chem* **69**: 4751–4760.

Chaurand P, Luetzenkirchen F, Spengler B. (1999) Peptide and protein identification by matrix-assisted laser desorption ionization (MALDI) and MALDI-post-source decay time-of-flight mass spectrometry. *J Am Soc Mass Spectrom* **10**: 91–103.

Chaurand P, Norris JL, Cornett DS, Mobley JA, Caprioli RM. (2006) New developments in profiling and imaging of proteins from tissue sections by MALDI mass spectrometry. *J Proteome Res* **5**: 2889–2900.

Chernushevich IV, Thomson BA. (2004) Collisional cooling of large ions in electrospray mass spectrometry. *Anal. Chem* **76**: 1754–1760.

Cho WCS. (2007) Proteomics technologies and challenges. *Genom Proteom Bioinform* **5**: 77–85.

de Hoffmann E, Stroobant V. (2002) *Mass Spectrometry: Principles and Applications*, 2nd edn. John Wiley & Sons, New York, pp. 125–126.

de Urquiza AM, Liu S, Sjöberg M, Zetterström RH, Griffiths W, Sjövall J, Perlmann T. (2000) Docosahexaenoic acid, a ligand for the retinoid X receptor in mouse brain. *Science* **290**: 2140–2144.

Fenn JB, Mann M, Meng CK, Wong SF, Whitehouse CM. (1989) Electrospray ionization for mass spectrometry of large biomolecules. *Science* **246**: 64–71.

Fenn JB, Mann M, Meng CK, Wong SF, Whitehouse CM. (1990) Electrospray ionization — Principles and practice. *Mass Spectrom Rev* **9**: 37–70.

Hanson CL, Fucini P, Ilag LL, Nierhaus KH, Robinson CV. (2003) Dissociation of intact *Escherichia coli* ribosomes in a mass spectrometer. Evidence for conformational change in a ribosome elongation factor G complex. *J Biol Chem* **278**: 1259–1267.

Hanson CL, Videler H, Santos C, Ballesta JP, Robinson CV. (2004) Mass spectrometry of ribosomes from *Saccharomyces cerevisiae*: Implications for assembly of the stalk complex. *J Biol Chem* **279**: 42750–42757.

Harmer NJ, Ilag LL, Mulloy B, Pellegrini L, Robinson CV, Blundell TL. (2004) Towards a resolution of the stoichiometry of the fibroblast growth factor (FGF)-FGF receptor-heparin complex. *J Mol Biol* **339**: 821–334.

Harms J, Schluenzen F, Zarivach R, Bashan A, Gat S, Agmon I, Bartels H, Franceschi F, Yonath A. (2001) High resolution structure of the large ribosomal subunit from a mesophilic eubacterium. *Cell* **107**: 679–688.

Hernández H, Dziembowski A, Taverner T, Séraphin B, Robinson CV. (2006) Subunit architecture of multimeric complexes isolated directly from cells. *EMBO Rep* **7**: 605–610.

Hunter DJ, Macmaster R, Roszak AW, Riboldi-Tunnicliffe A, Griffiths IR, Freer AA. (2005) Structure of myelin P2 protein from equine spinal cord. *Acta Crystallogr D Biol Crystallogr* **61**: 1067–1071.

Ilag LL, Westblade LF, Deshayes C, Kolb A, Busby SJ, Robinson CV. (2004) Mass spectrometry of *Escherichia coli* RNA polymerase: Interactions of the core enzyme with Sigma 70 and Rsd protein. *Structure* **12**: 269–275.

Ilag LL, Ubarretxena-Belandia I, Tate CG, Robinson CV. (2004) Drug binding revealed by tandem mass spectrometry of a protein-micelle complex. *J Am Chem Soc* **126**: 14362–14363.

Ilag LL, Videler H, McKay AR, Sobott F, Fucini P, Nierhaus KH, Robinson CV. (2005) Heptameric (L12)6/L10 rather than canonical pentameric complexes are found by tandem MS of intact ribosomes from thermophilic bacteria. *Proc Nat Acad Sci USA* **102**: 8192–8197.

Karas M, Hillenkamp F. (1988) Laser desorption ionization of proteins with molecular masses exceeding 10,000 daltons. *Anal Chem* **60**: 2299–2301.

Kosak ST, Groudine M. (2004) Gene order and dynamic domains. *Science* **306**: 644–647.

Link AJ. (2002) Multidimensional peptide separations in proteomics. *Trends Biotechnol* **20**: 8–13.

Lubec G, Afjehi-Sadat L. (2007) Limitations and pitfalls in protein identification by mass spectrometry. *Chem Rev* **107**: 3568–3584.

Marshall AG, Hendrickson CL, Jackson GS. (1998) Fourier transform ion cyclotron resonance mass spectrometry: A primer. *Mass Spectrom Rev* **17**: 1–35.

McDonnell CA, Heevan KMA. (2007) Imaging mass spectrometry. *Mass Spectrom Rev* **26**: 606–643.

McKay AR, Ruotolo BT, Ilag LL, Robinson CV. (2006) Mass measurements of increased accuracy resolve heterogeneous populations of intact ribosomes. *J Am Chem Soc* **128**: 11433–11442.

Morris HR, Panico M, Barber M, Bordoli RS, Sedgwick RD, Tyler A. (1981) Fast atom bombardment: A new mass spectrometric method for peptide sequence analysis. *Biochem Biophys Res Commun* **101**: 623–631.

Pellegrini L, Burke DF, von Delft F, Mulloy B, Blundell TL. (2000) Crystal structure of fibroblast growth factor receptor ectodomain bound to ligand and heparin. *Nature* **407**: 1029–1034.

Schlessinger J, Plotnikov AN, Ibrahimi OA, Eliseenkova AV, Yeh BK, Yayon A, Linhardt RJ, Mohammadi M. (2000) Crystal structure of a ternary FGF-FGFR-heparin complex reveals a dual role for heparin in FGFR binding and dimerization. *Mol Cell* **6**: 743–750.

Schluenzen F, Tocilj A, Zarivach R, Harms J, Gluehmann M, Janell D, Bashan A, Bartels H, Agmon I, Franceschi F, Yonath A. (2000)

Structure of functionally activated small ribosomal subunit at 3.3 Ångstroms resolution. *Cell* **102**: 615–623.

Schuwirth BS, Borovinskaya MA, Hau CW, Zhang W, Vila-Sanjurjo A, Holton JM, Cate JH. (2005) Structures of the bacterial ribosome at 3.5 Å resolution. *Science* **310**: 827–834.

Siuzdak G. (1996) *Mass Spectrometry for Biotechnology*. Academic Press, San Diego, CA, pp. 4–40.

Sleno L, Volmer DA. (2004) Ion activation methods for tandem mass spectrometry. *J Mass Spectrom* **39**: 1091–1112.

Sobott F, Hernández H, McCammon MG, Tito MA, Robinson CV. (2002) A tandem mass spectrometer for improved transmission and analysis of large macromolecular assemblies. *Anal Chem* **74**: 1402–1407.

Tito MA, Tarsk K, Valegard K, Hajdu J, Robinson CV. (2000) Electrospray time-of-flight mass spectrometry of the intact MS2 virus capsid. *J Am Chem Soc* **122**: 3550–3551

Tjenberg A, Carnö S, Oliv F, Benkestock K, Edlund PO, Griffiths WJ, Hallen D. (2004) Determination of dissociation constants for protein-ligand complexes by electrospray ionization mass spectrometry. *Anal Chem* **76**: 4325–4331.

van den Heuvel RH, Heck AJ. (2004) Native protein mass spectrometry: From intact oligomers to functional machineries. *Curr Opin Chem Biol* **8**: 519–526.

Videler H, Ilag LL, McKay AR, Hanson CL, Robinson CV. (2005) Mass spectrometry of intact ribosomes. *FEBS Lett* **579**: 943–947.

Wimberly BT, Brodersen DE, Clemons WM Jr., Morgan-Warren RJ, Carter AP, Vonrhein C, Hartsch T, Ramakrishnan V. (2000) Structure of the 30S ribosomal subunit. *Nature* **407**: 327–339.

Wood DD, She YM, Freer AD, Harauz G, Moscharello MA. (2002) Primary structure of equine myelin basic protein by mass spectrometry. *Arch Biochem Biophys* **405**: 137–146.

Yates JR III. (1998) Database searching using mass spectrometry data. *Electrophoresis* **19**: 893–900.

Chapter 5

Myelin: A One-Dimensional Biological "Crystal" for X-Ray and Neutron Scattering

*Hideyo Inouye and Daniel A. Kirschner**

The periodic nature of the myelin sheath makes it well-suited for examination of its molecular organization by diffraction techniques. Diffraction provides a means not only of monitoring the separation between membranes, but also of analyzing the forces and interactions between them. The diffraction technique is non-perturbing, and is uniquely suited to analyzing myelin structure and stability in physiologically intact, unfixed tissue. In principle, changes in structure can be followed in real time during physiological events or experimental treatments. The correlation of results from diffraction with results from electron microscopy and chemical analysis has led to a description of the distribution of lipid, protein, and water in the membrane array and to the localization of specific proteins and lipids within the membrane. Such studies are providing the foundation for understanding the molecular roles of particular myelin lipids and of specific myelin proteins — both native and mutated — in membrane–membrane adhesion and myelin stability.

Keywords: Myelin; electron density; paracrystalline; neutron scattering amplitude; electron microscopy; Fourier transform; lipid; protein; membranes.

Introduction

The nerve myelin sheath, which is responsible for rapid, saltatory conduction in the vertebrate nervous system, is a closely-packed, concentric

*Biology Department, Boston College, Chestnut Hill, MA 02467-3811, USA. E-mail: inouyeh@bc.edu and kirschnd@bc.edu.

We dedicate this article to the memories of our colleagues, myelin structural biologists, Dr Allen E. Blaurock (1940–2007) and Dr Leonardo Mateu (1939–2008).

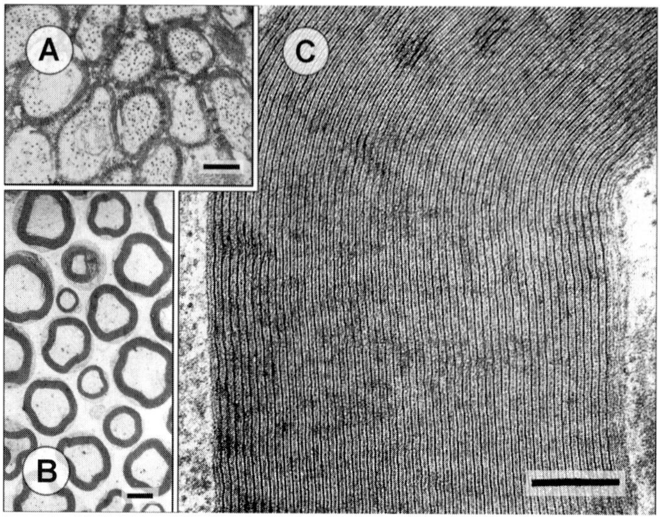

Fig. 1. The ultrastructure of nerve myelin is revealed by electron micrographs. Tissues of mice were prepared by chemical stabilization using glutaraldehyde, then dehydrated, and embedded in plastic (Kirschner, Hollingshead, 1980). The plastic blocks containing tissue were cut into thin-sections ~800 Å thick, stained by heavy metals (lead or uranyl salts), and observed at high magnification in the electron microscope. The images are from tissue that had been sectioned at right angles to the long axis of the nerve, and therefore the myelin appears at low magnification as darkly-stained rings. (A) Optic nerve. Note that the different myelinated fibers abut one another. The dense bands that frequently traverse the width of the myelin sheaths are specialized junctions that are unique to central nervous system myelin (Kosaras, Kirschner, 1990). Scale bar: 1 μm. (B) Sciatic nerve. The individual myelinated fibers are separated from one another by collagen fibers. Scale bar: 3 μm. (C) High magnification view of a portion of one myelin sheath from sciatic nerve. The densely-staining interface between cytoplasmic faces of the membranes ("major dense line") are clearly distinguished from the more lightly-staining surfaces of the extracellular faces of the membranes ("intraperiod line"). The most electron lucent feature in the periodic array of membrane pairs is the center of the membrane bilayers. Scale bar: 0.2 μm. (The micrographs were obtained by Dr Béla Kosaras, Dr Allen L Ganser, and Ms Carol J Hollingshead, respectively, in the laboratory of Dr DA Kirschner.)

multilamellar structure between nodes of Ranvier, as shown by the complementary structural techniques of electron microscopy (Fig. 1) and X-ray diffraction (Fig. 2) (Schmitt *et al.*, 1941; Robertson, 1958). The sheath results from the spiral wrapping of the plasma membranes of Schwann

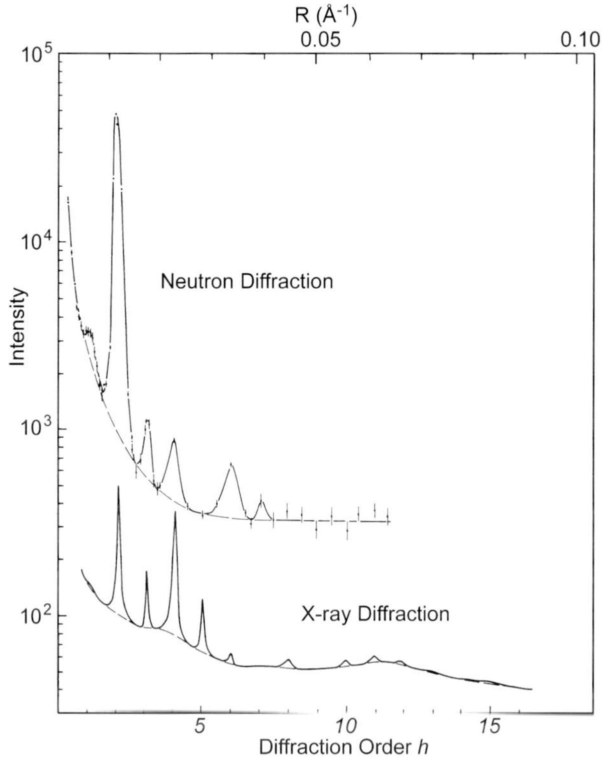

Fig. 2. Diffraction patterns for rabbit sciatic nerves, with the intensity on a log scale plotted against diffraction angle $R(=2\sin\theta/\lambda)$. The neutron diffraction data (upper) was obtained from nerve in D_2O-Ringer's solution, using a neutron beam having a wavelength of 4.06 Å. The X-ray diffraction data (lower) was from nerve in Ringer's solution, and the wavelength was 1.54 Å. Note that the relative intensities are very different between the two spectra, reflecting the fact that neutrons and X-rays interact very differently with matter — neutrons interact with atomic nuclei whereas X-rays interact with atomic electrons. [Modified from Fig. 11 of Kirschner and Caspar (1977), and used with kind permission of Springer Science and Business Media.]

cells in the peripheral nervous system and of oligodendrocytes in the central nervous system. The myelination occurs in such a way that the extracellular membrane leaflets become apposed to one another as do the cytoplasmic leaflets (Geren, 1954). In the internode of the sheath, this geometry results in formation of a one-dimensional (1D) lattice of membrane pairs

(Kirschner, Blaurock, 1992). The electron density profile for this 1D lattice, which can be calculated from the X-ray diffraction data, confirms that the myelin consists of lipid bilayers, and provides a measure of the widths of the extracellular and cytoplasmic spaces (Finean, Burge, 1963; Moody, 1963; Blaurock *et al.*, 1971; Caspar, Kirschner, 1971; McIntosh, Worthington, 1974; Inouye *et al.*, 1989; Luzzati, Mateu, 1990; Mateu *et al.*, 1990).

The disruption of myelinated tissue in the CNS and PNS results in nerve conduction abnormalities including transmission failure, and correlates with neurological disorders including multiple sclerosis, peroxisomal dysfunction, and peripheral neuropathies. Mutations in genes that encode myelin or myelin-related proteins or lipids can cause serious neurological disorders via amyelination or dysmyelination — i.e. the failure of formation or abnormal formation of myelin — during development, or eventual demyelination (breakdown of myelin). For example, the severe CNS leukodystrophy known as Pelizaeus–Merzbacher disease is caused by mutations in myelin proteolipid protein; and PNS demyelinating neuropathies having varying phenotypes (or observable characteristics at the clinical, morphological, or biochemical level) such as Charcot–Marie–Tooth disease and Dejerine–Sottas syndrome are caused by mutations in the genes for myelin protein zero (P0 glycoprotein), PMP22, and others (Lazzarini, 2004). Many of these correlate with myelin packing defects and lattice instability (Avila *et al.*, 2005; Wrabetz *et al.*, 2006).

The compact extracellular apposition in PNS myelin under native conditions (i.e. physiological pH and ionic strength) is stabilized by contacts between P0 molecules, most likely involving ionic interactions of histidine residues (Inouye, Kirschner, 1988). The packing of myelin membranes to form the multilamellar sheath (Fig. 1), however, is not static but rather dynamic (Uzman, Hedley-Whyte, 1968). The inter-membrane separation at the extracellular apposition is highly labile, and the swelling and compaction here as a function of pH and ionic strength can be accounted for by the DLVO (Derjaguin, Landau, Verwey and Overbeek) theory of colloid stability (Verwey, Overbeek, 1948). By contrast, the cytoplasmic apposition is invariant over a wide range of pH and ionic strength. What underlies this relative stability has been enigmatic; however, recent studies on model systems consisting of zwitterionic phosphatidylcholine (PC), negatively charged phosphatidylserine (PS), and the cytoplasmic domains of positively charged human and zebrafish P0's show that the apposition is likely stabilized by the hydrogen-bonding between β-sheets of apposed P0 molecules (Luo *et al.*, 2007a, 2007b).

To calculate the number of charges on the membrane surfaces and the electrostatic repulsion force between the apposed membranes as a function of pH and ionic strength requires having a plausible chemical model (Inouye, Kirschner, 1988; Luo *et al.*, 2008). Such a model should be consistent with the observed electron and neutron scattering length density distributions. In this brief review, we first show how one may apply elementary diffraction theory to derive an absolute density distribution for the 1D myelin lattice of the internode (Fig. 2), and then interpret this density distribution in terms of the available chemical composition data.

Electron Density Profile on an Absolute Scale

The electron density distribution for the myelin sheath on an absolute scale — i.e. electrons per Å^3 — is not readily calculated from the observed intensity because of the missing intensity at the origin of the pattern and the unknown scale factor that relates the observed intensity to the structure amplitude. Sucrose and glycerol, for which electron densities can be calculated, have been used to derive an absolute scale between the myelin membrane and the medium (Blaurock, 1971; Blaurock, Caspar, 1971; McIntosh, Worthington, 1974; Inouye *et al.*, 1999). To illustrate, we consider the case where the fluid layer is at the extracellular space. Here, r is in the radial direction of the cylindrically symmetric multilamellar structure of period d. The exclusion thickness of the membrane r_0 is defined as that part of the structure where the extracellular medium is not accessible. The "minus" fluid electron density of the unit structure $\Delta\rho(r)$ is written as $\Delta\rho(r) = \rho(r) - f$ at $-r_0/2 \le r \le r_0/2$, and $\Delta\rho(r) = 0$ at $d/2 > r > r_0/2$ and $-r_0/2 < r < r_0/2$, where f is the electron density of the fluid. For example, the fluid electron densities (in electrons/Å^3) are 0.334 for distilled water, 0.336 for 154 mM NaCl, 0.335 for 60 mM NaCl, 0.343 for 0.24 M sucrose, and 0.347 for 10% glycerol.

Because of the apposition of like surfaces to one another (see above and Fig. 1C), the electron densities $\rho(r)$ and $\Delta\rho(r)$ are centrosymmetric, and $\rho(r) = \rho(-r)$ and $\Delta\rho(r) = \Delta\rho(-r)$. Assuming that a cylindrically symmetric structure is equivalent to the fully rotated flat structure, Fourier transforms of $\rho(r)$ and $\Delta\rho(r)$ are given by

$$F(R) = 2\int_0^{d/2} \rho(r)\cos(2\pi rR)dr,$$
$$\Delta F(R) = 2\int_0^{d/2} \Delta\rho(r)\cos(2\pi rR)dr = F(R) - fd\,\mathrm{sinc}(\pi dR). \qquad (1)$$

Here, R is the reciprocal coordinate in the same direction as real coordinate r with $|R| = 2\sin\theta/\lambda$, where 2θ is the scattering angle and λ is wavelength (1.542 Å for CuK_α). The structure factor contains only the real term as the phase is either 0 or π. With the 1D periodic lattice of period d (corresponding to the myelin repeat of membrane pairs), $F(R)$ gives values at discrete reciprocal $R = h/d$, where h is an integer. With slit collimation or using a line-focused X-ray beam, the observed intensity $I_{obs}(h/d)$ (e.g. Fig. 2) can be measured as the average intensity over the height of the beam as measured by a position-sensitive detector (Gabriel, 1977; Boulin *et al.*, 1988) or by digitizing the exposed X-ray film. These methods integrate the scattered intensity due to disorientation of the myelinated nerve fibers. For the line-collimated beam, the Lorentz type correction factor is $C(h) = h$, whereas for the point-focused beam it is h^2. The structure factor $F(h/d)$ is related to the observed intensity by the scale factor K and Lorentz correction $C(h)$ according to

$$F\left(\frac{h}{d}\right) = \Delta F\left(\frac{h}{d}\right) = KF_{\mathrm{obs}}\left(\frac{h}{d}\right),$$

where

$$F_{\mathrm{obs}}\left(\frac{h}{d}\right) = \pm\sqrt{C(h)I_{\mathrm{obs}}\left(\frac{h}{d}\right)}. \qquad (2)$$

From the observed structure amplitudes, the electron density on an absolute scale is, therefore, given by

$$\rho(r) = \left(\frac{1}{d}\right)\sum_{-\infty}^{\infty} F\left(\frac{h}{d}\right)\exp\left(-\frac{i\,2\pi rh}{d}\right)$$
$$\approx \frac{F(0)}{d} + \left(\frac{2}{d}\right)\sum_{h=1}^{h_{\max}} KF_{\text{obs}}\left(\frac{h}{d}\right)\cos\left(\frac{2\pi rh}{d}\right) \quad (3)$$

and

$$F(0) = \Delta F(0) + fd. \quad (4)$$

The values for *F(0)*, *K,* and the phase of the structure factors should be known in order to calculate the absolute electron density. *F*(*R*), which is the continuous Fourier transform of $\rho(r)$, can be derived from Eqs. (1) and (3) as

$$F(R) = \sum_{-\infty}^{\infty} F\left(\frac{h}{d}\right)\text{sinc}(\pi dR - \pi h)$$

and

$$\Delta F(R) = F(R) - fd\,\text{sinc}(\pi dR)$$
$$= \sum_{-\infty}^{\infty} \Delta F\left(\frac{h}{d}\right)\text{sinc}(\pi dR - \pi h). \quad (5)$$

Internodal myelin swells and compacts under different conditions of pH and ionic strength, giving various myelin periods. To compare the observed intensities with different myelin periods, the following scaling is needed. Using discretely observed structure amplitudes and the scale factor *K*, the origin of the Patterson function of $\rho(r)$ at $r = 0$ is given by

$$\int \rho(r)^2 dr = \int \left|F(R)\right|^2 dR \approx \left(\frac{1}{d}\right)\left|F(0)\right|^2 + \left(\frac{2}{d}\right)K^2\sum\left|F_{\text{obs}}\left(\frac{h}{d}\right)\right|^2.$$

When the fluid electron density is similar to that of the membrane, then $f \approx \langle\rho(r)\rangle_{r_0}$ and $\Delta F(0) \approx 0$. The observed structure amplitudes can subsequently be scaled by assuming a constant $\sum|F_{obs}(h/d)|^2/d$ for the different myelin periods. These properly scaled structure factors sample the continuous structure factor $F(R)$ given by Eq. (5).

It is also possible to derive the electron density on an absolute scale by using two fluids having different densities f_1 and f_2. The Fourier transform of $\Delta\rho(r)$ gives

$$\Delta F(R) = F_i(R) - r_0 f \operatorname{sinc}(\pi r_0 R), \tag{6}$$

where $F_i(R)$ is the Fourier transform of the bounded $\rho(r)$ within $|r|<r_0/2$, and is independent of the fluid density f. For myelin lattices having two different fluid densities, the structure factors $\Delta F_1(R)$ and $\Delta F_2(R)$ are expressed as

$$\Delta F_1(R) - \Delta F_2(R) = r_0 \operatorname{sinc}(\pi r_0 R)(f_2 - f_1). \tag{7}$$

When the structure factors are measured at $R = h/d$, it follows that

$$K_1 F_{1_{obs}}\left(\frac{h}{d}\right) - K_2 F_{2_{obs}}\left(\frac{h}{d}\right) = r_0 \operatorname{sinc}\left(\frac{\pi r_0 h}{d}\right)(f_2 - f_1), \tag{8}$$

where K_1, and K_2 are scale factors for the observed structure factors $F_{1_{obs}}$ and $F_{2_{obs}}$. Because these structure amplitudes, and d, f_1 and f_2 are known in Eq. (8), the unknown phases of the structure factors, and K_1, K_2 and r_0 may be obtained by searching for the values satisfying Eq. (8) (Fig. 3). The electron density $\rho(r)$ and $F(R)$ on an absolute scale are then derived according to Eqs. (3) and (5) (Fig. 3).

Chemical Interpretation of the Electron Density and Neutron Scattering Length Density Profiles

Method

The electron density distribution and the neutron scattering length density are compared with the chemical dispositions of phospholipids, cholesterol,

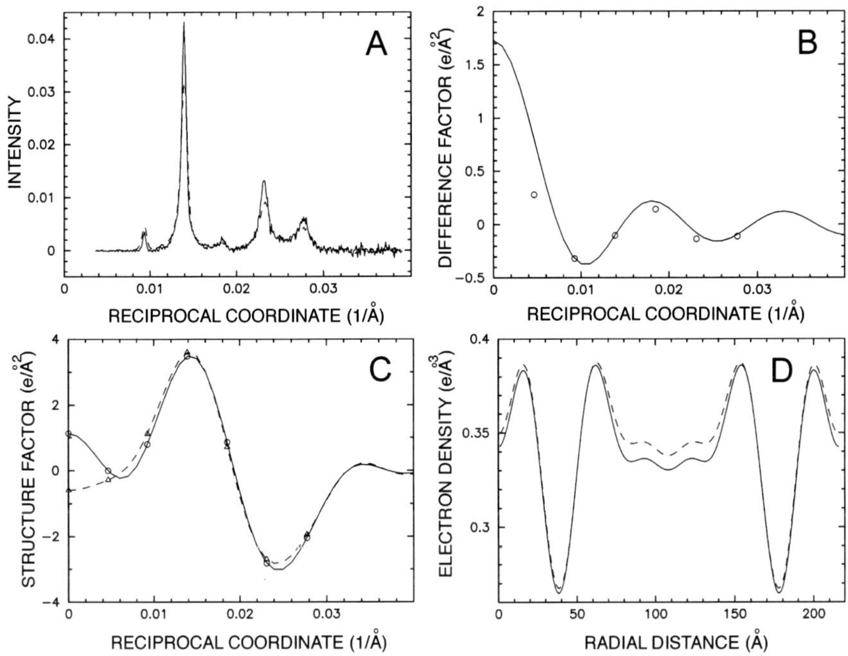

Fig. 3. (A) Densitometer tracings of X-ray diffraction patterns after background subtraction for mouse sciatic nerve treated in buffered medium with fluid electron density of 0.3347 e/Å^3 at ionic strength 0.06 and pH 6.0 (solid line) and for the nerve treated in the same buffered medium containing 10% glycerol with fluid electron density 0.3474 e/Å^3 (dashed line). The myelin period for mouse sciatic nerve in the two different media is 216 Å. (B) Difference Fourier transform (control structure factors minus those from glycerol treated myelin) between the structure factors with different fluid electron densities. The continuous curve is calculated and the circles are from the observed structure factors. Using 2.0 for the scale parameter and 136 Å for the exclusion length give 43% as the best goodness-of-fit. (C) The continuous Fourier transform and the structure amplitudes on an absolute scale (e/Å^2) in control medium (solid line and circle) and in 10% glycerol solution (dashed line and triangle). (D) Absolute electron density profiles of the membrane pair for the control (solid line) and the glycerol-treated nerve (dashed line). The difference in the two density profiles is at the extracellular space. The average electron density within the exclusion length is 0.343 e/Å^3. The cytoplasmic separation (32 Å), the distance between the polar head groups (46 Å), and the extracellular separation (92 Å) are measured from the profile.

water distribution and protein. The neutron scattering lengths in femtometers (fm or 10^{-15} m; in parenthesis) for various isotopes, i.e. H (–3.74), ^{2}H (6.67), C (6.65), N (9.36), O (5.80), P (5.13), S (2.85) are tabulated in *www.ncnr.nist.gov/resources/n-lengths*. The neutron scattering length density in fm/Å^3 for water having different volume fraction of D_2O is calculated as 0.636 for 100% D_2O, 0.394 for 65% D_2O, 0.082 for 20% D_2O, and –0.056 for 100% H_2O by using the density 1.104 g/ml for D_2O and 0.997 g/ml for H_2O, and molecular masses 20.03 and 18.02, respectively (Eisenberg, 1981).

To correlate the observed X-ray and neutron data with the chemical data, it is necessary to calculate the electron density and neutron scattering length density from the composition and disposition of molecular constituents, and the molecular structures. The increment of the projection Δr and surface area S are specified. The atomic positions satisfying $r_j < r < r_j + \Delta r$ are determined from the known atomic coordinates of the phospholipids and cholesterol. When $n(i,r_j)$ is the number of the i-type of atom at coordinate r_j, and e_i is the number of electrons or the neutron scattering length of the ith atom, the net number of electrons or neutron scattering length at r_j is calculated from $\sum n(i, r_j)e_i N_k$, where N_k is the number of k-type of molecule (phospholipid, cholesterol, and protein). The density is given by using the surface area $D(r_j,k) = [\sum n(i, r_j)e_i]N_k/(S\Delta r)$.

For internodal myelin, the water distribution $w(r)$ is obtained from the reported neutron scattering data for rabbit sciatic nerve (Kirschner *et al.*, 1976). The relative scale $w'(r)$ is derived from $w'(r) = \sum[F_D(h) - F'_D(h)]\cos(2\pi rh/d)$, where F_D and F'_D are the observed structure factors for rabbit sciatic nerve myelin with period d = 180 Å in 100% D_2O and in 65% D_2O (Kirschner *et al.*, 1976). Bragg orders h = 1–6 are used. At coordinate r_j in the scattering density profile, the number of water molecule n_j is

$$n_j = K\,\Delta r S w'(r_j) = \frac{N_{\text{water}} w'(r_j)}{\sum w'(r_j)} \quad \text{and} \quad \sum n_j = N_{\text{water}},$$

where N_{water} is the total number of water molecules. Here, the scale factor K is calculated from $K\Delta rS = N_{water}/\Sigma w'(r_j)$. Thus, the number of electrons at r_j is qn_j, where q is 10 electrons per water molecule, or for neutron scattering −1.678 fm (10^{-15} m) for 100% H_2O, 2.486 for 20% D_2O, 11.856 for 65% D_2O, and 19.145 for 100% D_2O. The net number of electrons or neutron scattering length $N_e(r_j)$ is, therefore, calculated from $N_e(r_j) = \Sigma_k\Sigma_i n(i, r_j)e_iN_k + qN_{water}w'(r_j)/\Sigma w'(r_j)$. Finally, the net electron density or neutron scattering length density is $D(r_j) = N_e(r_j)/(S\Delta r)$.

To obtain the neutron scattering density distribution with comparable resolution to that of the electron density profile, $D(r_j)$ is first Fourier transformed and then the structure factor is sampled using a comparable period. If the continuous function $D(r)$ is the density of the asymmetric unit, its Fourier transform is given as $a(R) + ib(R)$, where a and b are the real and imaginary terms. Since the unit density distribution for internodal myelin is centrosymmetric, the structure factor $F(R)$ can be written as $F(R) = 2a(R)\cos(2\pi uR) - 2b(R)\sin(2\pi uR)$, where u is the position of the asymmetric unit from the origin. Here, the value for u is chosen such that $\Sigma|b(R)|^2/\Sigma|a(R)|^2$ gives the minimum value. Thus, the density profile at a resolution of h_{max}/d is

$$\rho'(r) = \frac{F(0)}{d} + \left(\frac{2}{d}\right)\sum_{h=1}^{h_{max}} F\left(\frac{h}{d}\right)\cos\left(\frac{2\pi rh}{d}\right).$$

Model calculation

The chemical composition, i.e. N_k and N_{water} for the lipids and water, are from the published chemical data summarized in Table V of Inouye and Kirschner (1988). The crystal data for phosphatidylethanolamine (Elder *et al.*, 1977) and cholesteryl dihydrogen phosphate (Pascher, Sundell, 1982) were used for the myelin phospholipids and cholesterol in myelin. The atomic coordinates of the phosphate group in cholesteryl dihydrogen phosphate were excluded. As the hydrogen coordinates in the crystal are not known, they were assigned to the nearest carbon atoms.

Positions of the molecules were determined in the following manner. The origin of the membrane is taken at the center of the cytoplasmic apposition. The absolute electron density profile from swollen mouse sciatic nerve (Fig. 3) shows the first density peak at 16 Å from the center of the cytoplasmic space, and the second at 62 Å. The phosphate group of the phosphatidylethanolamine (PE) (Elder *et al.*, 1977) is assigned to those peaks (Fig. 4(A)). The cholesterol molecule is positioned so that the oxygen of cholesterol is 4 Å away from the phosphate (Franks, Lieb 1979). The two cholesterol oxygens at the cytoplasmic and extracellular leaflets, therefore, are localized 20 Å and 58 Å from the origin (Fig. 4(B)). The water distribution is derived from the neutron diffraction according to the above method, and the electron density is shown in Fig. 4(C).

The chain length of the reported crystallographic data for PE is 12 carbon atoms (Elder *et al.*, 1977), while 18 is the mean chain length of lipid hydrocarbon in myelin (O'Brien, Sampson, 1965). The electron densities for PE with different lengths of hydrocarbon chain were considered. The chains in all *trans* configurations is extended along the long axis of the fatty acids. To account for the electron density, the 18-carbon hydrocarbon chains should interdigitate at the center of the lipid bilayer. Chains having fewer than 16 carbons do not interdigitate, and give significantly lower electron density than observed (Fig. 4(D)).

The disposition of cholesterol in the leaflets of the PNS myelin is not certain. Cholesterol can be asymmetrically placed at a 1:2 molar ratio between the cytoplasmic and extracellular leaflets (Fig. 5(A)) (Caspar, Kirschner, 1971), or can be placed at 1:1 (Fig. 5(B)) (Scott *et al.*, 1980; Allt *et al.*, 1985). The net electron density profiles are asymmetric largely due to an asymmetric distribution of water, i.e. more water molecules are present in the cytoplasmic half of the leaflet than in the extracellular half. The position of protein is estimated from the difference profile between the observed and calculated electron densities (Fig. 5(C)). The two prominent peaks at the cytoplasmic and extracellular space likely come from myelin basic protein (MBP) (Inouye *et al.*, 1985) and P0-glycoprotein. The electron density at the center of the membrane bilayer arises from the transmembrane domain of P0. The net average electron density of the myelin

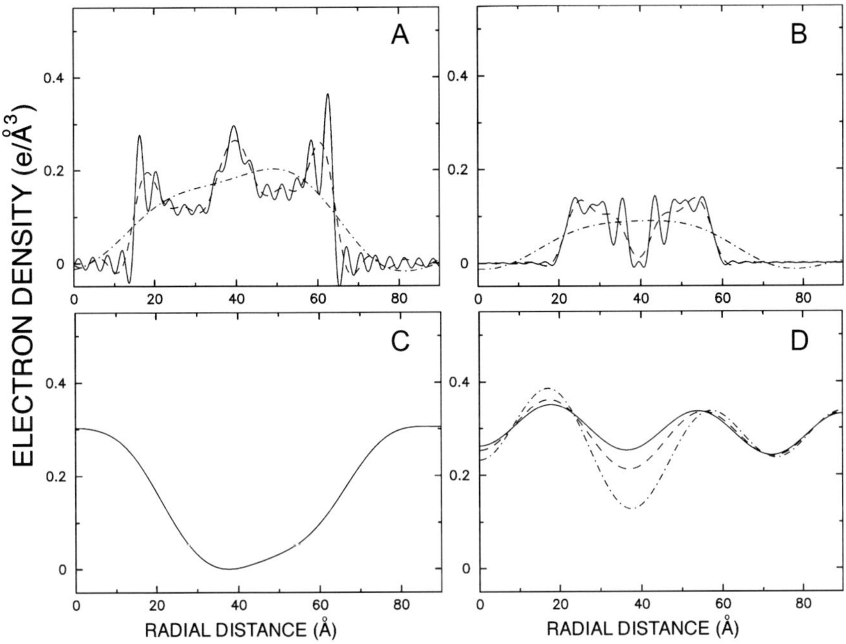

Fig. 4. Electron density projection calculated for phospholipid (A), cholesterol (B), water (C), and PNS myelin (D). The number of molecules at the cytoplasmic and extracellular side of the PNS myelin membrane in the surface area $S = 3724$ Å^2 are given by Inouye and Kirschner (1988). The number of phospholipid molecules at the cytoplasmic and extracellular sides are 34 and 45, and for cholesterol molecules 30(20) and 31(41), where the values in parenthesis are for the 1:2 molar distribution of cholesterol. (A) The electron density projection of phosphatidylethanolamine (PE) molecules having 18-carbon chain lengths. The electron density projection was first calculated at every 0.2 Å from the atomic coordinates of the molecule. Its Fourier transform was then sampled using either 50, 20, or 5 orders for a 180 Å period. The electron density profile at a resolution of 180 Å/50 (solid line) was derived by reverse Fourier transform of the sampled data. Similar electron density distributions are shown at lower resolution (180 Å/20, dashed line, and 180 Å/5, dash-dot-dash). (B) The electron density projection of symmetrically-distributed cholesterol molecules at a resolution of 180 Å/50 (solid line), 180 Å/20 (dashed), and 180 Å/5 (dash-dot-dash). Panel (C) shows 5337 water molecules per $S = 3724$ Å^2 at a resolution of 180 Å/6. (D) The modeled myelin without any protein contribution, i.e. total electron density of PE, cholesterol, and water per $S = 3724$ Å^2 with PE containing different chain lengths (18:0 carbons, solid line; 16:0 carbons, dashed; and 12:0 carbons, dash-dot-dash). The cholesterol molecules are shown symmetrically disposed at the cytoplasmic and extracellular sides of the membrane. The resolution is at 180 Å/5, which is comparable to that of the observed electron density.

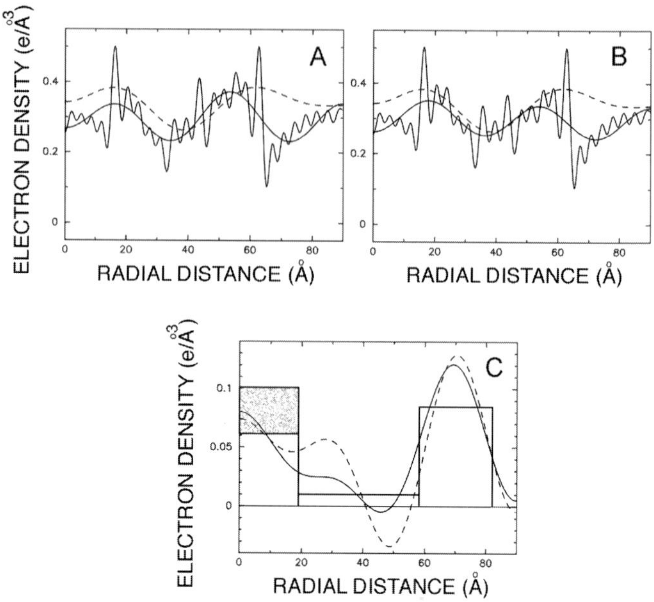

Fig. 5. (A) Modeled PNS myelin with the asymmetric cholesterol disposition (1:2 molar ratio at the cytoplasmic and extracellular sides) at a resolution of 180 Å/50 (solid line), 180 Å/5 (thick solid line), and mouse sciatic nerve from X-ray diffraction data at 216 Å/6 resolution (dashed line). The number of carbons for the chain length of PE is 18:0. Note that the water distribution is at a resolution of 180 Å/6. (B) Modeled myelin with the symmetric cholesterol disposition at the same resolution as in (A). (C) The difference between the observed and calculated electron densities with the symmetric cholesterol disposition (solid line) and with the asymmetric cholesterol disposition (dashes). Note that because the calculated curve does not include the electron density of protein, then the difference here is accounted for by the disposition of protein. The box plot shows the estimated electron density level from the primary sequences of the mouse P0 (without sugar moiety) and myelin basic protein (14 kDa and 18 kDa isoforms; shaded box).

membrane calculated from the chemical composition is 0.347 e/Å^3, which is similar to the measured value of 0.343 e/Å^3 from the electron density on an absolute scale.

The neutron scattering lengths (in fm/Å^3) are calculated for PE with 18-carbon chains (Fig. 6(A)), symmetrically-distributed cholesterol (Fig. 6(B)), and water for different D_2O volume fractions (Fig. 6(C)). The net density for the mixture of PE, cholesterol and water is compared with

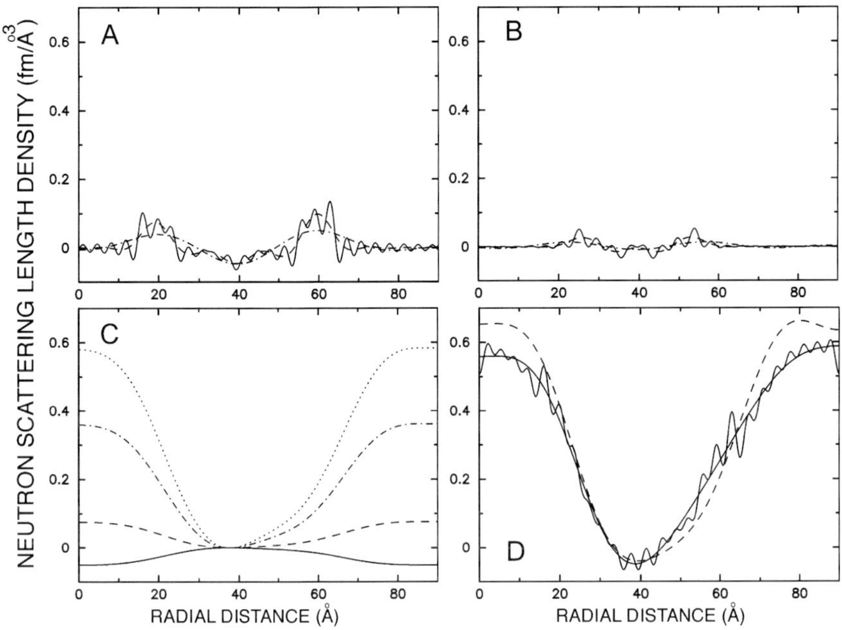

Fig. 6. Neutron scattering length density (fm/Å^3) calculated for phospholipid (A), cholesterol (B), water (C), and PNS myelin (D). The number of molecules at the cytoplasmic and extracellular side of the PNS myelin membrane in the surface area $S = 3724$ Å^2 are given by Inouye and Kirschner (1988). (A) The neutron scattering length density for PE molecules (hydrocarbon chain length, 18 carbons). The density projection was first calculated at every 0.2 Å from the atomic coordinates of the molecule. Its Fourier transform was then sampled by either 50, 20 or 6 orders with a 180 Å period. The density profile at a resolution of 180 Å/50 (solid line) was derived by reverse Fourier transform of the sampled data. Similar density distributions are shown at lower resolution of 180 Å/20 (dashed line) and 180 Å/6 (dash-dot-dash). (B) The density projection for symmetrically-distributed cholesterol molecules at a resolution of 180 Å/50 (solid line), 180 Å /20 (dashed line) and 180 Å/6 (dash-dot-dash). Panel (C) shows 5337 water molecules per $S = 3724$ Å^2 at a resolution of 180 Å/6 for different D_2O volume fractions, i.e. H_2O (solid line), 20% D_2O (dashed line), 65% D_2O (dash-dot-dash), and 100% D_2O (dotted). (D) The modeled myelin without any protein contribution, i.e. total density of 18:0 PE, symmetrically-distributed cholesterol molecules, and 100% D_2O per $S = 3724$ Å^2. The density from the model are at resolutions of 180 Å/6 (smoother solid line) and 180 Å/50 (solid line). The observed neutron scattering length density (dashed line) was calculated from the reported structure factors for rabbit sciatic nerve myelin in D_2O (Kirschner, 1974; Kirschner *et al.*, 1976). The absolute scale was derived by assuming that the density at the center of the lipid bilayer as –0.04 fm/Å^3 and the density at the center of the extracellular space as 0.636 fm/Å^3 (100% D_2O).

the observed profile (Fig. 6(D)). The model calculation shows that the densities of PE and cholesterol are much smaller than that of D_2O. The difference between the observed and calculated neutron scattering lengths (Fig. 6(D)) shows the localization of protein at the cytoplasmic and extracellular spaces.

The above calculation shows that the chemical contents of protein, lipids, and water can account for the density profile of myelin as determined from the observed X-ray and neutron diffraction. This analysis also shows that a large portion of the transmembrane P0 is localized at the cytoplasmic and extracellular spaces, and the hydrocarbon chains are likely interdigitated at the center of the myelin membrane.

Conclusion

Elucidating from X-ray and neutron diffraction the molecular organization of lipids, proteins, and water in the internode, where the myelin shows 1D packing, will help to illuminate the dynamic physical/chemical processes that not only underlie the ensheathment of the axon and stabilize the myelin, but can also lead to demyelination in a variety of CNS and PNS disorders.

Acknowledgments

This work was supported by Institutional Research Support Funds from Boston College (DAK) and by a research grant from La Fondation de Recherche ELA (DAK and RJ Wanders).

References

Allt G, Blanchard CE, MacKenzie ML, Sikri K. (1985) Distribution of filipin-sterol complexes in the myelinated nerve fiber. *J Ultrastruct Res* **91**: 104–111.

Avila RL, Inouye H, Baek RC, Yin X, Trapp BD, Feltri ML, Wrabetz L, Kirschner DA. (2005) Structure and stability of internodal myelin in mouse models of hereditary neuropathy. *J Neuropathol Exp Neurol* **64**: 976–990.

Blaurock AE. (1971) Structure of the nerve myelin membrane: Proof of the low-resolution profile. *J Mol Biol* **56**: 35–52.

Blaurock AE, Caspar DLD. (1971) The structure of the myelin sheath: Electron density distribution. Fourier plot. *Neurosci Res Program Bull* **9**: 535–539.

Blaurock AE, Caspar DLD, Schmitt FO. (1971) X-ray diffraction of living nerve. Small-angle diffraction. *Neurosci Res Program Bull* **9**: 512–515.

Boulin CJ, Kempf R, Gabriel A, Koch MHJ. (1988) Data acquisition system for linear and area X-ray detectors using delay-line readout. *Nucl Instrum Methods* **A269**: 312–320.

Caspar DLD, Kirschner DA. (1971) Myelin membrane structure at 10 Å resolution. *Nat New Biol* **231**: 46–52.

Eisenberg H. (1981) Forward scattering of light, X-rays and neutrons. *Q Rev Biophys* **14**: 141–172.

Elder M, Hitchcock P, Mason R, Shipley GG. (1977) A refinement analysis of the crystallography of the phospholipid,1,2-dilauroyl-DL-phosphatidylethanolamine, and some remarks on lipid–lipid and lipid–protein interactions. *Proc R Soc London* **A354**: 157–170.

Finean JB, Burge RE. (1963) The determination of the Fourier transform of the myelin layer from a study of swelling phenomena. *J Mol Biol* **7**: 672–682.

Franks NP, Lieb WR. (1979) The structure of lipid bilayers and the effects of general anaesthetics. An X-ray and neutron diffraction study. *J Mol Biol* **133**: 469–500.

Gabriel A. (1977) Position sensitive X-ray detectors. *Rev Sci Instrum* **48**: 1303–1305.

Geren BB. (1954) The formation from the Schwann cell surface of myelin in the peripheral nerves of chick embryos. *Exp Cell Res* **7**: 558–562.

Inouye H, Ganser AL, Kirschner DA. (1985) Shiverer and normal peripheral myelin compared: Basic protein localization, membrane interactions, and lipid composition. *J Neurochem* **45**: 1911–1922.

Inouye H, Karthigasan J, Kirschner DA. (1989) Membrane structure in isolated and intact myelins. *Biophys J* **56**: 129–137.

Inouye H, Kirschner DA. (1988) Membrane interactions in nerve myelin: II. Determination of surface charge from biochemical data. *Biophys J* **53**: 247–260.

Inouye H, Tsuruta H, Sedzik J, Uyemura K, Kirschner DA. (1999) Tetrameric assembly of full-sequence protein zero myelin glycoprotein by synchrotron X-ray scattering. *Biophys J* **76**: 423–437.

Kirschner DA. (1974) Comparative X-ray and neutron diffraction from nerve myelin membranes. In Yip S, Chen S-H (eds). *Spectroscopy in Biology and Chemistry*, pp. 203–233. Academic Press, New York.

Kirschner DA, Blaurock AE. (1992) Organization, phylogenetic variations and dynamic transitions of myelin structure. In Martenson RE (ed). *Myelin: Biology and Chemistry*, pp. 3–78. CRC Press, Boca Raton.

Kirschner DA, Caspar DLD. (1977) Diffraction studies of molecular organization in myelin. In Morell P (ed). *Myelin*, pp. 51–89. Plenum Press, New York.

Kirschner DA, Hollingshead CJ. (1980) Processing for electron microscopy alters membrane structure and packing in myelin. *J Ultrastruc Res* **73**: 211–232.

Kirschner DA, Caspar DLD, Schoenborn BP, Nunes AC. (1975) Neutron diffraction studies of nerve myelin. *Brookhaven Symp Biol* III68–III76.

Kosaras B, Kirschner DA. (1990) Radial component of CNS myelin: Junctional subunit structure and supramolecular assembly. *J Neurocytol* **19**: 187–199.

Lazzarini RA (ed). (2004) *Myelin Biology and Disorders*. Elsevier/Academic Press, Amsterdam.

Luo X, Cerullo J, Dawli T, Priest C, Haddadin Z, Kim A, Inouye H, Suffoletto BP, Avila RL, Lees JP, Sharma D, Xie B, Costello CE, Kirschner DA. (2008) Peripheral myelin of *Xenopus laevis*: Role of electrostatic and hydrophobic interactions in membrane compaction. *J Struct Biol* **162**: 170–183.

Luo X, Inouye H, Gross AA, Hidalgo MM, Sharma D, Lee D, Avila RL, Salmona M, Kirschner DA. (2007b) Cytoplasmic domain of zebrafish myelin protein zero: Adhesive role depends on beta-conformation. *Biophys J* **93**: 3515–3528.

Luo X, Sharma D, Inouye H, Lee D, Avila RL, Salmona M, Kirschner DA. (2007a) Cytoplasmic domain of human myelin protein zero likely folded as β-structure in compact myelin. *Biophys J* **92**: 1585–1597.

Luzzati V, Mateu L. (1990) Order-disorder phenomena in myelinated nerve sheaths. I. A physical model and its parametrization: Exact and approximate determination of the parameters. *J Mol Biol* **215**: 373–384.

Mateu L, Luzzati V, Vargas R, Vonasek E, Borgo M. (1990) Order-disorder phenomena in myelinated nerve sheaths II. The structure of myelin in native and swollen rat sciatic nerves and in the course of myelinogenesis. *J Mol Biol* **215**: 385–402.

McIntosh TJ, Worthington CR. (1974) Direct determination of the lamellar structure of peripheral nerve myelin at low resolution (17 Å). *Biophys J* **14**: 363–386.

Moody MF. (1963) X-ray diffraction pattern of nerve myelin: A method for determining the phases. *Science* **142**: 1173–1174.

O'Brien JS, Sampson EL. (1965) Lipid composition of the normal human brain: Gray matter, white matter and myelin. *J Lipid Res* **6**: 537–544.

Pascher I, Sundell S. (1982) The crystal structure of cholesteryl dihydrogen phosphate. *Chem Phys Lipids* **31**: 129–143.

Robertson JD. (1958) Structural alterations in nerve fibers produced by hypotonic and hypertonic solutions. *J Biophys Biochem Cytol* **4**: 349–364.

Schmitt FO, Bear RS, Palmer KJ. (1941) X-ray diffraction studies on the structure of the nerve myelin sheath. *J Cell Comp Physiol* **18**: 31–42.

Scott SC, Bruckdorfer KR, Worcester DL. (1980) The symmetrical distribution of cholesterol across the myelin membrane bilayer determined by deuterium labeling *in vivo* and neutron diffraction. *Biochem Soc Trans* **8**: 717.

Uzman BG, Hedley-Whyte ET. (1968) Myelin: Dynamic or stable? *J Gen Physiol* **51**: 8S–17S.

Verwey EJW, Overbeek JTG. (1948) *Theory of the Stability of Lyophobic Colloids*. Elsevier, New York.

Wrabetz L, D'Antonio M, Pennuto M, Dati G, Tinelli E, Fratta P, Previtali S, Imperiale D, Zielasek J, Toyka K, Avila RL, Kirschner DA, Messing A, Feltri ML, Quattrini A. (2006) Different intracellular pathomechanisms produce diverse *MPZ*-neuropathies in transgenic mice. *J Neurosci* **26**: 2358–2368.

Chapter 6

Two-Dimensional Crystallization of Biological Macromolecules

*Hans Hebert**

Biological macromolecules can be arranged periodically in single layers as two-dimensional (2D) crystals. This enables crystallographic structure analysis using transmission electron microscopy. Periodic repeat of a large number of unit cells contributes to significant information in images or diffraction patterns from unstained specimens. Structural details at a resolution of a few Ångströms in all directions can be obtained. Such three-dimensional maps are used for building atomic models. The techniques have been used to determine structure and function relationships for membrane proteins. The dense packing obtained in 2D crystals can also be used for constructing devices at the molecular level.

Keywords: 2D crystallization; membrane proteins; crystals.

Introduction

In 1975 Richard Henderson and Nigel Unwin showed that three-dimensional (3D) structural information in the subnanometer region could be obtained from unstained protein molecules arranged periodically in a plane as a two-dimensional (2D) crystal (Henderson, Unwin, 1975). Their results spurred a great deal of effort in many laboratories to produce 2D crystals for structural studies. Since then several membrane proteins have been characterized, several at high resolution, with the use of this strategy.

*Karolinska Institutet, Department of Biosciences and Nutrition and Royal Institute of Technology, School of Technology and Health, Novum, S-141 57 Huddinge, Sweden. E-mail: hans.hebert@ki.se.

In this chapter some basic principles for obtaining 2D crystals of both membrane-bound and soluble proteins will be described together with their analysis and properties.

Why Two-Dimensional Crystals?

In order to observe scattering of X-ray frequency range photons giving rise to diffraction spots with sufficiently high signal-to-noise ratio at a resolution which allows building of atomic models of proteins, it is required that the molecules are arranged into well-ordered crystals in three-dimensions (3D). For some specimens this may be difficult to achieve. Factors that may hamper 3D crystallization include heterogeneity, hydrophobicity and tendency of aggregation at high concentration. Even if the protein can form 3D crystals resulting in a native data set, it may be difficult to produce heavy atom data derivatives or selenomethionine incorporation if that is needed for phasing or to study structure-function relationships related to ligand binding which may require an environment resembling the natural state.

Since electrons interact more strongly with matter than X-rays do, electron diffraction patterns and images can be observed in a transmission electron microscope (TEM) from protein molecules periodically arranged in a single layer. This arrangement is called a two-dimensional (2D) crystal. Thus, 2D crystals combined with electron crystallography presents an interesting alternative methodological approach for systems where 3D crystallization is likely to fail. Unfortunately, it is difficult to predict the outcome of crystallization trials and consequently there is no strict rule-of-thumb for selecting one or the other of the two techniques. Since structure solution by X-ray diffraction has become very efficient and well established it is not unusual to try to circumvent initial problems by modifications of the protein such as introducing mutations or performing truncations. However, with improvements in electron crystallography both with regard to data collection and data processing it may be equally interesting to set up extensive 2D crystallization experiments and to solve the structure using the TEM.

Electron Crystallography: Initial Characterization

Whether or not 3D crystallization under certain conditions results in crystals can easily be checked initially with light microscopy but the properties of crystals can only be determined by X-ray diffraction. Monitoring 2D crystallization requires a TEM. Normally, this is done after negative staining of the specimen. This preparation method is very quick and simple and produces a long-lasting specimen which is stable under the high vacuum conditions in the TEM column and which also tolerates a high electron dose (for a review, see Harris and Horne (1994)). The contrast between the staining material and the protein is high but the resolution from crystalline material is limited to about 15 Å. Consequently, this technique does not give a definite answer to the question whether or not the crystals give rise to high resolution information, but from the size of the coherent crystalline areas and the sharpness of the diffraction spots in calculated power spectra it is nevertheless possible to obtain a qualitative indication regarding the size and order of the crystals.

Even if the negative stain preparation procedure is fast, checking of the crystallinity from a large number of specimens is time consuming. Every EM grid has to be scanned by the operator. Images from interesting specimen areas are collected on a CCD camera attached to the TEM followed by calculation of the corresponding power spectra online and storage of representative information. Efforts have been launched in order to speed up the whole procedure. Multi-grid specimen holders have been constructed and automatic image acquisition systems have been tried (Lefman *et al.*, 2007). However, it is difficult to make the complete procedure fully automatic, since in particular the final evaluation step is not simple to transfer from human judgement to computer software.

To summarize this part, it is extremely advantageous to have a TEM facility in the vicinity of the lab where the 2D crystallization experiments are going to be performed. It is sufficient to have a simple microscope, and specimen preparation and instrument operation does not require a very experienced electron microscopist. On the other hand, it is recommended to have some crystallographic knowledge since already at this stage it may be possible to draw important conclusions with regard to unit cell size, symmetry and possible stacking.

How to Make 2D Crystals of Proteins

For the purpose of discussing 2D crystallization of proteins it is appropriate to subdivide these biological macromolecules into two categories: soluble proteins and membrane proteins. The reason for this is that two completely different strategies are dominating for handling the two subtypes. However, the subdivision is not static. Although all soluble proteins can be dissolved in a water solution, some of them, like pore-forming bacterial toxins, may be able to change conformation for accommodation in a lipophilic environment. Integral membrane proteins of type I and type II with only one membrane spanning segment may be transferred into a soluble protein by cleavage of the membrane anchor. Since 2D crystallization has been most frequently applied to membrane proteins, those will be treated first and with most detail in the following.

Membrane proteins

2D crystallization of membrane proteins can proceed in two different ways depending on the specimen. Most frequently the starting material is a purified membrane protein dissolved in a detergent-containing buffer. The other category is proteins which are embedded in a natural membrane and suspended in a buffer. The dominating purified specimens will be treated first.

The basic idea behind forming 2D crystals from membrane proteins in detergent solutions is to add phospholipids solubilized in detergent and reduce the detergent concentration of the mixture (for a recent review, see Schmidt-Krey (2007)). Under proper conditions phospholipid bilayers will form with inserted protein molecules which arrange themselves into a crystalline sheet (Fig. 1). The reduction of the detergent concentration is normally done either by dialysis or adsorption onto biobeads. Dialysis can be performed with different types of containers such as dialysis tubes,

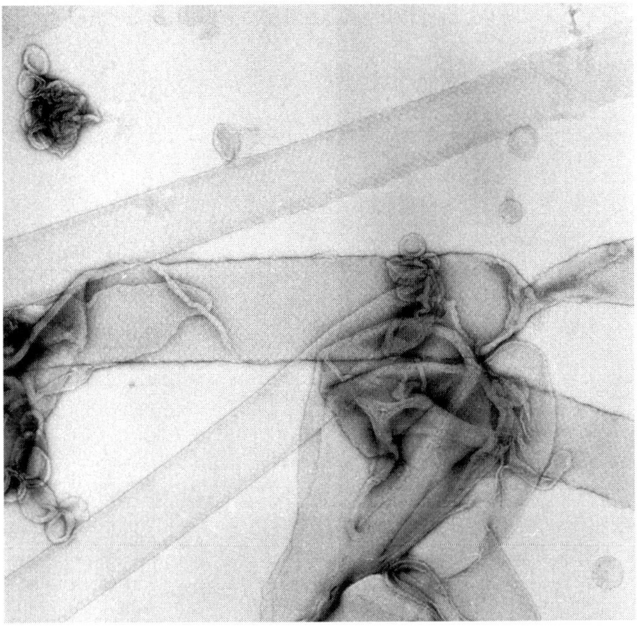

Fig. 1. 2D crystals of the sugar transporter melibiose permease from *E. coli*. The crystals grow as tubes several micrometers in length.

buttons enclosed by a dialysis membrane, dialysis slides and bent capillaries. In order to control conditions special dialysis machines have been constructed (Jap *et al.*, 1992).

A large number of parameters can be varied in order to find conditions at which a particular purified membrane protein forms 2D crystals. Among those listed in Table 1 a few can be pointed out as being of special importance. Even if the detergent will be depleted or completely removed, its properties can influence initial stages in formation of the proteolipid complexes. The critical micellar concentration (CMC) of the detergent will influence dialysis time when this method is used for detergent removal. Low CMC detergents like Triton X-100 will require several days of dialysis. The choice of phospholipids normally follows the principle that it should mimic the natural environment of the protein. Chain length and degree of saturation are important properties. The ratio between

Table 1. Factors that may Influence 2D Crystallization

The protein itself	Size of hydrophobic/hydrophilic parts, purity, concentration, glycosylation
Type of detergent	Ionic, non-ionic, zwitterionic, length of hydrophobic tail, head group size, CMC
Type of phospholipid	Fatty acid chain length, degree of saturation, phase transition temperature
Lipid-to-protein ratio	Range 1–500 (molar ratio)
Temperature	40–4°C and slow annealing
Salt concentration	0–0.5 M
pH	

lipid and protein in method descriptions called LPR and given as either w/v or molar ratios may have a pronounced impact on crystal formation, size and order. The molar ratio may vary from below ten to several hundred. Large proteins with many transmembrane domains will require a high LPR. The temperature is a variable that will influence crystallization time. In order to speed up low CMC detergent removal, it may be advantageous to dialyze at room temperature if the protein tolerates that. In fact, many membrane proteins seem to be more stable once they are inserted into a lipid environment than in detergent solutions. In some cases a low temperature may be important to slow down the process and decrease the number of nucleation sites for the 2D crystal growth.

As mentioned in the previous section, definite evaluation of crystallization trials can only be made using a TEM. However, initial hints can be obtained simply by looking at the crystallization solution. Often an increased turbidity is a sign of formation of proteolipid complexes and membranes with inserted protein. Indeed, many successful 2D crystallization conditions result not only in useful areas but also in precipitate. Often the crystalline areas grow as extensions from heavy aggregates.

In some cases a membrane protein may be so tightly packed in its native membrane that it packs into a 2D crystal. A well-known example is bacteriorhodopsin from the purple membrane of *Halobacterium halobium*. Using such arrays and electron crystallography, Unwin and Henderson (1975) were able to obtain the first 3D structural information of a membrane protein. More frequently, a membrane may well be dominated by one species of an integral membrane protein but with additional proteins that are more loosely attached. Those can be removed by centrifugation following detergent treatment and conditions can be applied which stimulate contacts leading to periodic packing. An example is the purified Na^+, K^+-pump, which forms 2D arrays in its native membrane after incubation with vanadate (Hebert *et al.*, 2001).

Soluble proteins

The basic principle behind making 2D crystals of soluble proteins is to concentrate them to the fluid-air interface like an oil film spreading on a water surface of a compartment (Uzgiris, Kornberg, 1983). The protein should have an affinity for the surface. A phospholipid monolayer may act as the scaffold for the protein. In some cases, like for bacterial toxins, the protein may have an intrinsic affinity for the lipid. Insertion of the protein can be measured in a Langmuir trough as an increase in lateral pressure. For some proteins lipid interaction is dependent on concentration of ions, e.g. annexins have increased lipid affinity in the presence of Ca^{2+} (Brisson *et al.*, 1991). Attaching a special chemical group to the lipid head groups could be used to induce specific interaction with proteins having affinity for that particular chemical group. A general approach along this line for proteins containing a histidine tag has been to synthesize Ni-chelated phospholipids (Kubalek *et al.*, 1994).

The 2D crystallization trials can be performed in small containers holding a few milliliters of solution. After spreading the phospholipid at the buffer-air interface, the protein is injected into the subphase. The container should be kept in a humid environment.

Properties of 2-D crystals

Symmetry

Two-dimensional crystals are objects which are periodic in a plane with an extension of one or a few molecular layers along the normal to this plane. In publications dealing with such crystals, the *xy*-plane refers to the layer whereas the *z*-direction is the perpendicular direction, where *x*, *y* and *z* are the directions of the "lab" coordinate system. The *z*-direction is parallel to the direction of the electron beam. Like for 3D crystals the smallest repeating unit of the 2D counterpart is the unit cell defined by the *a*, *b* and *c* unit cell vectors (Fig. 2). In other words, *a*, *b* and *c* define the crystal coordinate system. For untilted specimens, the *z*- and *c*-directions are parallel so that the *xy*- and *ab*-planes coincide. The length of *c* corresponds to the thickness of the 2D crystal.

(A)

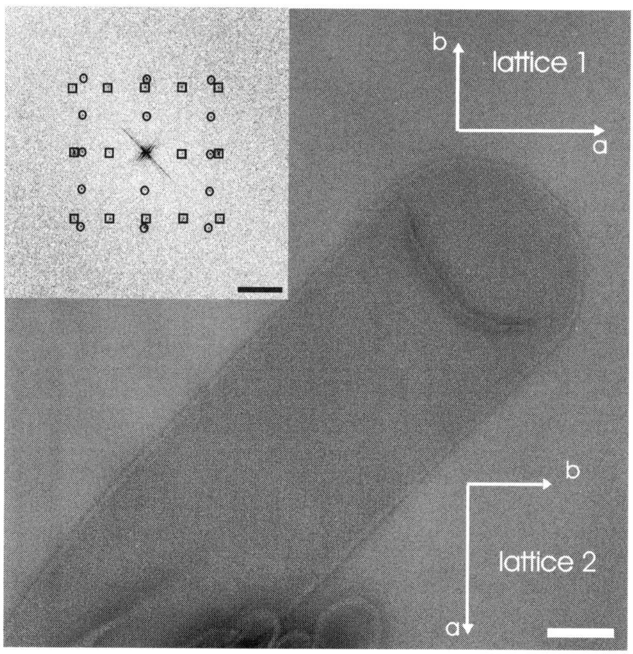

Fig. 2. Properties of a melibiose permease 2D crystal. (A) The collapsed tube shows two lattices with a relative in-plane rotational difference close to 90° (calculated diffraction pattern in the inset). The depicted in-plane unit cell vectors *a* and *b* show the directions relative to the crystal. Scale bars: 1/75 $Å^{-1}$ in the inset and 100 nm. (B) The projection map shows two unit cells and the packing of four protein molecules in one unit cell. These are in the $p222_1$ two-sided plane group related by a two-fold rotation perpendicular to the plane of the membrane, a two-fold rotation parallel to the short *b*-axis and a two-fold screw axis along the *a*-axis direction, respectively.

(B)

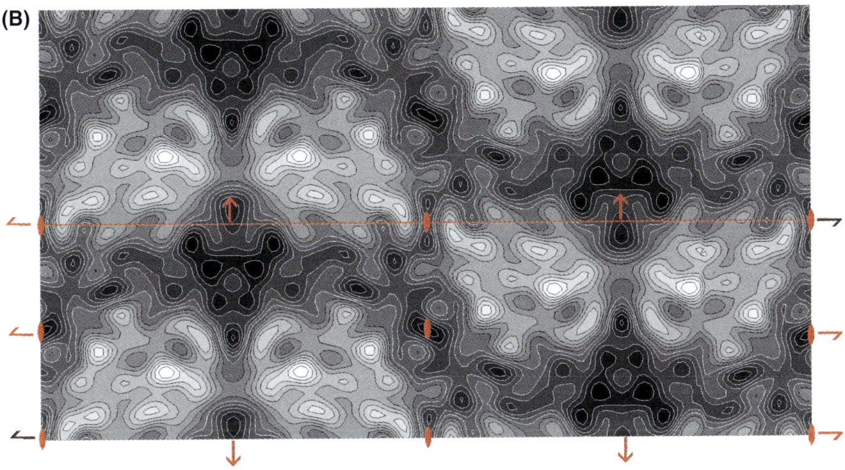

Fig. 2. (*Continued*)

Objects in space forming 3D crystals or in a pure plane without any extension in the third dimension can be arranged with 230 space-group and 17 projection-group symmetries, respectively. Biological macromolecules in a 2D crystal on the other hand will form one of 17 two-sided plane groups (Holser, 1958). Unlike pure 2D objects they may have symmetry operations in the plane of the membrane in such a way that one molecule may be facing upwards while its neighbor is in the opposite direction. In fact, this is very common for reconstituted membrane proteins. Obviously, it does not form a viable system for a membrane protein having vectorial transport as its main function but for 2D crystallization it may be an advantage since the layer will be symmetrical and less prone to bending. Flatness is often a problem for membrane proteins crystallized in their native membrane.

Size

The size of the coherent crystalline area will directly influence the signal-to-noise ratio in diffraction data either as calculated from images or as acquired directly in electron diffraction patterns. In fact, diffraction spots can only be detected directly in TEM diffraction mode if the crystals have a rather isotropic extension of about 1 μm (Fig. 3). For images it may be sufficient to have areas which include a few thousand unit cells in micrographs recorded at a magnification of about 50 000 times.

Order

The ultimate resolution that can be obtained from 2D crystals will depend not only on the size of the crystals but also on the order. Deviation from perfect crystallinity will show up as blurred diffraction spots. In images, problems of continuous limited in-plane disorder can be reduced by unbending (see below). An approach to any kind of disorder is to treat one or a few unit-cells as separate entities and apply single-particle processing (Koeck *et al.*, 2007).

Flatness

The requirement to work with large crystalline areas puts a critical demand on the flatness of the layer, which to some extent may depend on the support film. Deviation from perfect planarity will cause most severe distortions for data collected at high tilt angles. Successful attempts have been reported for improving flatness and other mechanical and conducting properties of the specimen support (Gyobu *et al.*, 2004; Rhinow, Kühlbrandt, 2007).

Stacking, multilayers

Ideally, the 2D crystals should consist of a single layer of the protein. For membrane proteins the arrangement may often be a double layer. This is for instance the case when the crystals grow as large tubes which flatten on the support film (Figs. 1 and 2). The two layers may be in perfect register with a relative in-plane rotation. In such cases the two layers may be separated during processing since two sets of diffraction spots appear. However, two layers may also be crystallographically arranged relative to

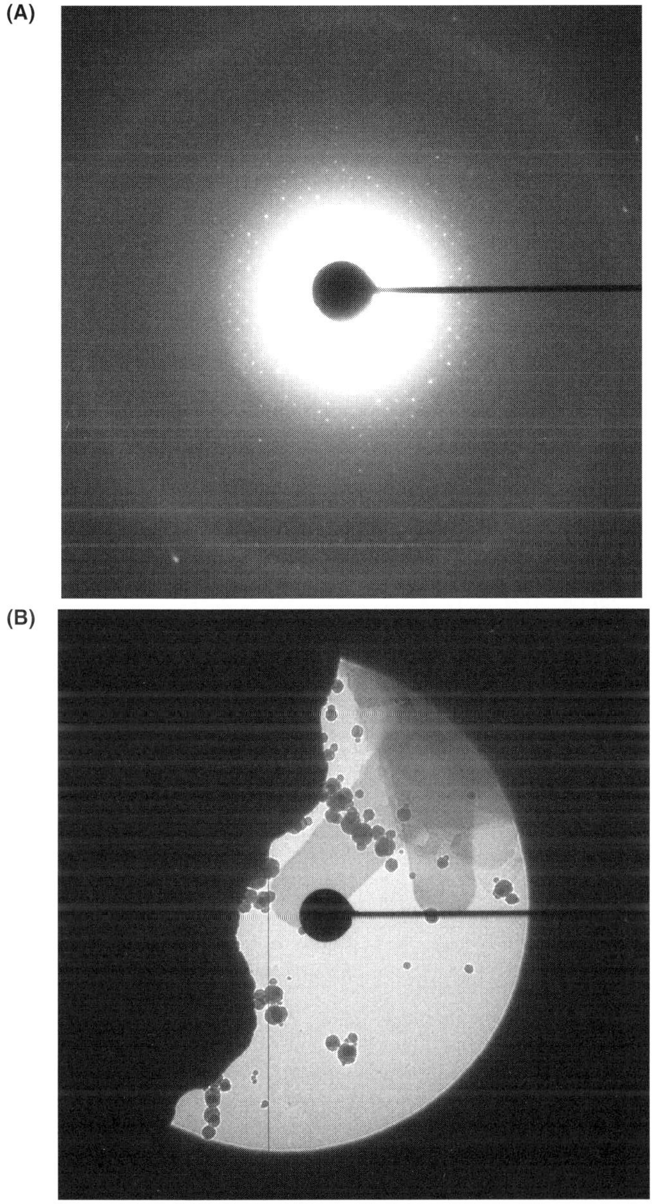

Fig. 3. Electron diffraction pattern from a 2D crystal. The pattern in (A) was recorded from the crystal in (B), which is partly hidden by the beam stop used to protect the CCD camera from the primary beam. The black dots in the image originate from ice contamination.

each other. This can be considered as a 3D crystal just consisting of two layers. However, such thin arrangements still fall into the category of 2D crystals of interest for electron crystallography, where they can be handled, rather than 3D crystals for X-ray crystallography since they do not give rise to detectable X-ray diffraction spots. If there is a tendency for the protein to form many layers, the electron crystallographic approach runs into a problem since the diffraction intensity along the *z*-dimension will be modulated by the thickness of the crystal. A tendency for crystalline areas to stack during 2D crystallization can create an additional severe problem. It will show up in diffraction patterns as an overlay of multiple lattices that may be difficult to separate.

Electron Crystallography: Data Collection and Processing

Specimen preparation and temperature

In order to collect high resolution data from 2D crystals, it is necessary to work at low specimen temperature using cryo-electron microscopy (cryoEM) and avoid contrast enhancing additives such as negative staining. The specimen support is normally a copper or molybdenum grid covered with a carbon layer which, in contrast to most other cryoEM applications, is continuous in order to ensure flatness. A drop of typically 2–5 μl of the 2D crystal suspension is applied to the grid. Stabilization by freezing can proceed in two different ways. The grid can be plunged frozen in liquid ethane as in single particle cryoEM or embedded in a low molecular weight, non-volatile medium like glucose, tannin or trehalose followed by freezing, which in many cases can be done in liquid nitrogen. Most high resolution studies of 2D crystals have been made using the embedding technique, which also allows application of a second carbon layer to form a sandwich preparation (Gyobu *et al.*, 2004). The grid is then mounted in a conventional side entry liquid nitrogen cryo-holder and transferred into the TEM or alternatively placed in a special cartridge if the analysis is going to be performed with a cryo-TEM which allows specimen cooling using liquid helium. In fact, most studies which have resulted in atomic models have been performed with liquid helium cryo-TEMs (Fujiyoshi *et al.*, 1991). It has been demonstrated that electron diffraction spots fade away slower when the specimen temperature is decreased towards a few Kelvin, indicating reduced beam damage. Equally important for improved information limits may be the stability of liquid helium specimen stages.

Electron diffraction

It is a major advantage if the amplitude information of the Fourier components can be obtained from electron diffraction data since the intensities will not be modulated by the influence of the objective lens like amplitudes calculated from images. A recording using a CCD camera is preferable. As compared to photographic film these have larger linear dynamic range and initial evaluation of quality can be made online.

Images

If no approximate phase information is available such as from the structure of a very similar protein or from a different construct/complex of the same protein, experimental phases have to be calculated from TEM images of the 2D crystals. If images are recorded on film, a medium which combines a large detection area with high resolution, the crystalline areas have to be digitized using a high resolution scanner. Important concepts to consider for the data collection are adjusting the dose in order to avoid radiation damage without loss of too much contrast, nominal magnification to adapt to the grain size of the film, the spot size of the scanner, and the expected resolution.

Data processing

The basic programs for processing both electron diffraction data and images are from the MRC package (Crowther *et al.*, 1996). This includes indexing of diffraction patterns, extracting electron diffraction amplitudes, unbending lattices in images, correcting for the contrast transfer function, merging data in 2D and 3D, lattice line adaption and finally ending up with a complete set of Fourier components consisting of the Miller indices, amplitude values, standard deviations of the amplitudes, phases and figures-of-merit of the phases. Alternative program systems have been developed which may or may not use the MRC programs in the background at the computational level (Gipson *et al.*, 2007; Philippsen *et al.*, 2007). The complete 3D dataset can subsequently be transferred to a conventional X-ray crystallography package like CCP4 for further processing, map calculation, model building and refinement.

2D or 3D Crystallization?

For the purpose of structural studies, crystallization in general is just a means of improving signal-to-noise ratio in recorded diffraction/image data. Ideally, the crystalline arrangement should as closely as possible resemble the natural *in situ* environment and not induce irrelevant structural changes of the studied macromolecule. Furthermore, the system should preserve activity and allow induction of conformational changes that may be directly coupled to the function.

For soluble proteins, 2D crystallization generates an artificial system except in those special cases when proteins have a natural tendency to interact with membranes. Moreover, even if production of large well-ordered 2D crystals on lipid monolayers has been demonstrated, there is no example yet of a structural study that has resulted in an atomic model.

For membrane proteins, however, 2D crystals resemble the natural environment and there are even examples of naturally occurring crystalline sheets of very densely packed proteins. Many membrane proteins expose active sites to either side of the lipid bilayer and these areas are accessible also in 2D crystals, implying that the effects of ligand binding can be studied. Furthermore, conformational states can be frozen and studied separately (Unwin, 2003). Membrane bound enzymes may also have hydrophobic substrates that are interacting with an active site in the lipophilic region also accessible in 2D crystals. Furthermore, two or more different membrane proteins may form complexes which are necessary for natural activity and these may form co-crystals that can be analyzed to determine protein-protein interactions.

In order to understand more about one specific or a group of membrane proteins it can be worthwhile because of preserved functional competence as described above, to perform structural studies using 2D crystals and cryoEM. In many cases X-ray and electron crystallography can provide complementary information. Some of the initial bottlenecks of a structural study such as expression, purification, stability, and so on are common to the two techniques. Thus, working together gaining experience of the properties of a protein will be an advantage.

Summary

For structural biology applications, 2D crystallization together with cryoEM present an interesting alternative in particular in cases where a system resembling the natural state is needed. However, inducing 2D crystals may not only be of interest in structural biology. Biotechnology applications can be envisaged that profit from an extreme concentration of proteins which are capable of performing functions such as transport of signals, ions or small molecules, receptors functioning as specific detectors, etc. (Birge *et al.*, 1999; Hillebrecht, 2004). Even non-biological applications have been developed. For example, 2D crystallization of ferritin has recently been used for fabrication of floating nanodot gate memories (Miura *et al.*, 2007), having advantages such as low power consumption, high operation speed and with possibilities to scale down the dimensions of memory cells.

Acknowledgments

This work is partially supported by grants from the Swedish Research Council, EU Network of Excellence "3DEM", and VINNOVA (Swedish Governmental Agency for Innovation Systems) program for Multidisciplinary-Bio as bilateral cooperation with Japan.

References

Birge RR, Gillespie NB, Izaguirre EW, Kusnetzow A, Lawrence AF, Singh D, Song QW, Schmidt E, Stuart JA, Seetharaman S, Wise KJ. (1999) Protein-based associative processors and volumetric memories. *J Phys Chem B* **103**: 10746–10766.

Brisson A, Mosser G, Huber R. (1991) Structure of soluble and membrane-bound human annexin V. *J Mol Biol* **220**: 199–203.

Crowther RA, Henderson R, Smith JM. (1996) MRC image processing programs. *J Struct Biol* **116**: 9–16.

Fujiyoshi Y, Mizusaki T, Morikawa K, Yamagishi H, Aoki Y, Kihara H, Harada Y. (1991) Development of a superfluid helium stage for high-resolution electron microscopy. *Ultramicroscopy* **38**: 241–251.

Gipson B, Zeng X, Zhang ZY, Stahlberg H. (2007) 2dx — user-friendly image processing for 2D crystals. *J Struct Biol* **157**: 64–72.

Gyobu N, Tani K, Hiroaki Y, Kamegawa A, Mitsuoka K, Fujiyoshi Y. (2004) Improved specimen preparation for cryo-electron microscopy using a symmetric carbon sandwich technique. *J Struct Biol* **146**: 325–333.

Harris JR, Horne RW. (1994) Negative staining: A brief assessment of current technical benefits, limitations and future possibilities. *Micron* **25**: 5–13.

Hebert H, Purhonen P, Vorum H, Thomsen K, Maunsbach AB. (2001) Three-dimensional structure of renal Na,K-ATPase from cryo-electron microscopy of two-dimensional crystals. *J Mol Biol* **314**: 479–494.

Henderson R, Unwin PN. (1975) Three-dimensional model of purple membrane obtained by electron microscopy. *Nature* **257**: 28–32.

Hillebrecht JR, Wise KJ, Koscielecki JF, Birge RR. (2004) Directed evolution of bacteriorhodopsin for device applications. *Methods Enzymol* **388**: 333–347.

Holser WT. (1958) Point groups and plane groups in a two-sided plane and their subgroups. *Z Kristallogr* **110**: 266–281.

Jap BK, Zulauf M, Scheybani T, Hefti A, Baumeister W, Aebi U, Engel A. (1992) 2D crystallization: From art to science. *Ultramicroscopy* **46**: 45–84.

Koeck PJB, Purhonen P, Alvang R, Grundberg B, Hebert H. (2007) Single particle refinement in electron crystallography: A pilot study. *J Struct Biol* **160**: 344–352.

Kubalek EW, Le Grice SF, Brown PO. (1994) Two-dimensional crystallization of histidine-tagged, HIV-1 reverse transcriptase promoted by a novel nickel-chelating lipid. *J Struct Biol* **113**: 117–123.

Miura A, Uraoka Y, Fuyuki T, Kumagai S, Yoshii S, Matsukawa N, Yamashita Y. (2007) Bionanodot monolayer array fabrication for nonvolatile memory application. *Surface Science* **601**: L81–L85.

Lefman J, Morrison R, Subramaniam S. (2007) Automated 100-position specimen loader and image acquisition system for transmission electron microscopy. *J Struct Biol* **158**: 318–326.

Philippsen A, Schenk AD, Signorell GA, Mariani V, Berneche S, Engel A. (2007) Collaborative EM image processing with the IPLT image processing library and toolbox. *J Struct Biol* **157**: 28–37.

Rhinow D, Kühlbrandt W. (2008) Electron cryo-microscopy of biological specimens on conductive titanium-silicon metal glass films. *Ultramicroscopy* **108**: 698–705.

Schmidt-Krey I. (2007) Electron crystallography of membrane proteins: Two-dimensional crystallization and screening by electron microscopy. *Methods* **41**: 417–426.

Uzgiris EE, Kornberg RD. (1983) Two-dimensional crystallization technique for imaging macromolecules, with application to antigen- antibody-complement complexes. *Nature* **301**: 125–129.

Unwin N. (2003) Structure and action of the nicotinic acetylcholine receptor explored by electron microscopy. *FEBS Lett* **555**: 91–95.

Chapter 7

Crystallization of Proteins: Principles and Methods

*Lata Govada**

Crystallization of biological macromolecules is governed by a variety of variables. Since protein crystals are stabilized by rather weak interactions, they are extremely sensitive to subtle variations in solution conditions; consequently, protein crystal growth presents particularly challenging problems. In macromolecular crystallography, the success of a project ultimately resides in the characteristics of the crystals utilized for the study. The growth of suitable protein crystals is therefore the most crucial step in the structure determination of a protein. Physico chemical parameters such as protein and precipitant concentration, pH, solubility, temperature, ionic strength, pressure and viscosity can play a crucial role in determining the optimal conditions for perfect crystal formation. Biochemical properties such as particle purity and conformational homogeneity and the presence and nature of impurities and additives can also influence crystallization. These parameters affect macromolecular solubility and are the major factors behind the complexity of crystallization and therefore make predicting suitable conditions laborious. A phase diagram is a useful tool to design experiments for crystal growth. Applying a phase diagram can provide some guidelines on the various possible conditions that can influence crystallization processes. This can be constructed by setting up many crystallization trials, varying at least two conditions and plotting their outcomes, after a certain period of time, on a two- or many-dimensional parameter grids.

Keywords: Supersaturation; phase diagram; solubility and supersolubility curve; batch and diffusion.

*Department of BioMolecular Medicine, Division of Surgery, Oncology, Reproductive Biology and Anaesthetics (SORA), Faculty of Medicine, Sir Alexander Fleming Building, Imperial College London, Exhibition Road, South Kensington, London SW7 2AZ, UK. E-mail: l.govada@imperial.ac.uk.

Introduction

Protein crystallization has gained a new strategic and commercial relevance in the post-genomics era because of its pivotal role in structural genomics (Chayen, 2005). Crystallization is a phase transition phenomenon and occurs under non-equilibrium conditions.

The principle of inducing protein crystallization depends on the basic strategy of bringing a system into a state of limited supersaturation, which is the ratio of the protein concentration over its solubility value (Ducruix, Giege, 1999). Knowledge of protein solubility in the presence of a crystallizing reagent enables one to effectively use the crystallizing reagent in a way that will promote crystallization (Anderson *et al.*, 2006).

Principles of Crystallization

Biocrystallization follows the same rules of crystallization as inorganic or organic small molecules, but is multiparametric, making it a more complex process. The fact that proteins are extremely sensitive to external conditions accentuates this complexity. The three classical steps for crystal formation include nucleation, growth and cessation of growth. Nucleation is a prerequisite and the first step to crystal growth.

The crystallization phase diagram

The multiparametric nature of crystallization and the limited knowledge of the mechanisms of crystal growth of proteins (McPherson, 1995) have restricted most investigations to empirical work. Studies on model macromolecules show that phase diagrams can be useful for this purpose but they have only infrequently been applied to find high-quality crystals of proteins for structure determination (Ataka, Tanaka, 1986; Feher, Kam, 1985; Saridakis *et al.*, 1994).

During crystallization, the parameters characterizing the solution may change with time, and the system will follow a particular path which can be illustrated by a phase diagram. The phase diagram is a map which represents the state of a material (e.g. solid and liquid) as a function of the ambient conditions (e.g. temperature and concentration) (Asherie, 2004). Phase diagrams form the basis for the design of crystal growth. Knowledge about them is central to understanding the principles of protein crystallization (Haas, Drenth, 1999).

Phase diagrams can be constructed by setting up a number of crystallization trials where a minimum of two parameters are varied, and plotting the outcome after a set period of time (Sauter *et al.*, 1999a). Figure 1 represents a 2D solubility phase diagram using two of the most commonly varied parameters, the concentrations of protein and the crystallizing agent.

The two curves of the phase diagram are:

The solubility curve

This divides the phase diagram into two major concentration zones.

(a) The undersaturated zone, where crystallization does not occur as the solution is thermodynamically stable.
(b) The supersaturated zone, which corresponds to the regions above the solubility curve. Crystals can only form in supersaturated solutions.

The supersolubility curve

This is less well defined than the solubility curve but can be found experimentally to a reasonable approximation much more easily. It can be further subdivided into three zones:

Precipitation zone

Precipitation occurs at very high supersaturation. Excess protein molecules immediately separate from solution to form amorphous aggregates.

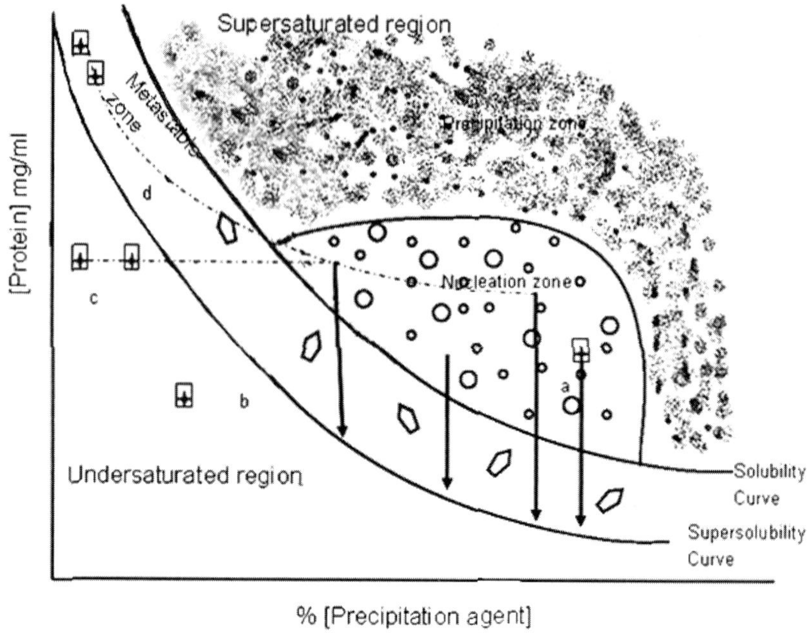

Fig. 1. A classical phase diagram; the extent of supersaturation is mapped with regards to the system parameters. Knowledge of phase diagrams is a prerequisite to the design of crystallization experiments. The figure represents the phase diagram showing the different routes of reaching supersaturation by different methods. The four major crystallization methods are represented, showing their different routes to the nucleation and metastable zones. The starting conditions are represented by: (a) batch, (b) vapor diffusion, (c) dialysis, and (d) free interface diffusion (FID). Two alternative starting points are shown for FID and dialysis because the undersaturated protein solution can contain either protein alone or protein mixed with a low concentration of the precipitating agents. The solubility is defined as the concentration of protein in the solute that is in equilibrium with crystals. The supersolubility curve is defined as the line separating conditions where spontaneous nucleation (or phase separation, precipitation) occurs from conditions where the crystallization solution remains clear if left undisturbed (Chayen, 2005).

Nucleation zone

This is the zone of the phase diagram where crystals appear. Nucleation is the process by which molecules or non-crystalline aggregates which are free in solution come together in such a way as to produce thermodynamically stable aggregates with repeating lattices. Nucleation requires a lower supersaturation than precipitation.

Metastable zone

A supersaturated solution in this zone does not nucleate at all, as the critical supersaturation is not yet reached. Crystals may grow in the metastable region if nuclei are somehow present, but nuclei will not form (McPherson, 1999).

Construction of Supersolubility Curve

The supersolubility curve is a distinct line of the supersaturated zone separating conditions where spontaneous nucleation (such as phase separation, precipitation) occurs from those where the crystallization solution remains clear in the liquid phase for an indefinite length of time if it is left undisturbed. Construction of the supersolubility curve requires setting up many crystallization trials by altering at least two conditions. One of the variables is typically the protein concentration. The outcomes can then be plotted after three to four weeks, on a two- or many-dimensional parameter grid. These experiments may be called "static," by contrast with the "dynamic" experiments involving seeding or some process of nucleation inducement (Saridakis, Chayen, 2003). The static experiments will show where on the grid (i.e. for which conditions) the crystallization solution remains clear and where it spontaneously nucleates or precipitates. When the supersolubility curve has been determined, the metastable zone can be accessed for the growth of nuclei into crystals. Knowledge of the supersolubility curve means that informed decisions can be made as to which conditions are metastable.

Crystallization Methodologies

During crystallization, the parameters characterizing the solution may change with time, and the system will follow a particular path on the phase diagram (Fig. 1). Although all methods aim at bringing the macromolecular solution to a supersaturated state, the means by which this is achieved varies in each of these techniques.

The main techniques for protein crystallization include:

Batch crystallization

This is the oldest method used for crystallizing proteins. The sample is directly mixed with the precipitant creating a homogeneous crystallization medium requiring no equilibration with a reservoir as the supersaturation is immediately reached (Fig. 2).

Over the years this method has been modified and is now carried out using microlitre amounts of the protein in either glass capillaries or in micro droplets under oil (Fig. 3).

Figure 4 shows protein crystals grown under oil.

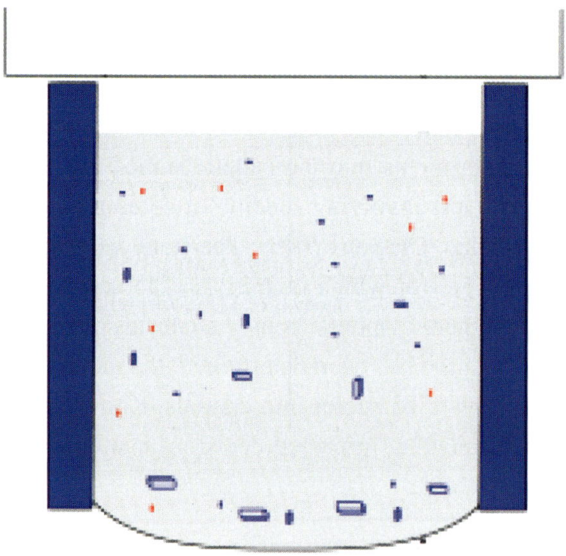

Fig. 2. Macromolecular crystals grown using the oldest method of protein crystallization, the batch method using millilitre volumes, where the samples are directly mixed with each other.

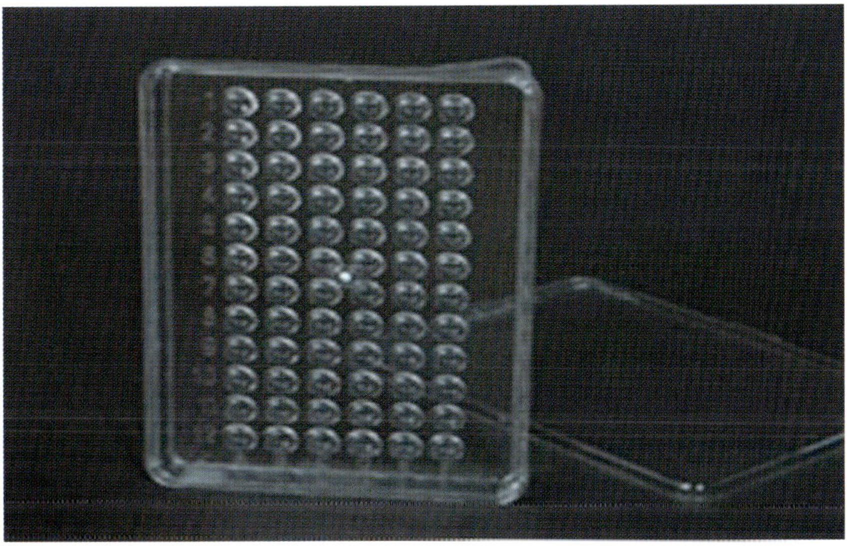

Fig. 3. 72-well HLA plate also known as a Terizaki plate [Nunc cat. #470378 or Molecular Dimensions Ltd. MD3-13ROWORLD or from Hampton Research Ltd. (HR3-120, 122, 086 and 087)] used in the microbatch under oil method.

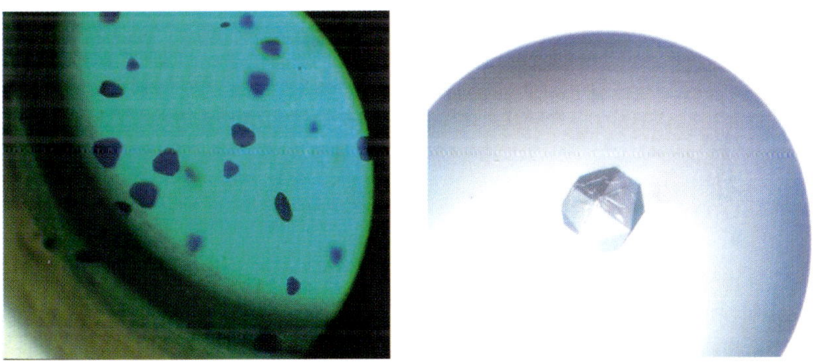

(a) (b)

Fig. 4. Crystals obtained using the microbatch method: (a) 0.5 × 0.4 × 0.3 mm lysozyme crystals with a diffraction limit of up to 1.2 Å (4% NaCl and 0.1 mM CH_3COONa pH 4.5). (b) 0.05 × 0.05 × 0.03 mm blue coloured crystals of a 16-mer complex of α-crustacyanin diffracting up to 5 Å resolution (work under progress).

Modified microbatch

This is similar to the method mentioned above, the only difference being the use of either silicone fluid, a polymeric compound composed of repeating dimethylsiloxane units in place of the paraffin oil, or alternatively a 1:1 ratio of silicon oil and paraffin oil.

Diffusion methods

In diffusion methods the samples are equilibrated against the crystallization solutions. During this equilibration process the sample becomes more concentrated, increasing the relative supersaturation of the sample in the drop.

Vapor diffusion

Vapor diffusion (Fig. 5) is by far the most widely employed crystallization method. The principle of evaporation and diffusion of the volatile species (water or organic solvent) is utilized by this technique. The initial reagent concentration in the droplet is less than that in the reservoir. Vapor exchange occurs between the precipitant and the drop until equilibration is reached. During this equilibration process the sample becomes more concentrated, increasing the relative supersaturation of the sample in the drop. Vapor diffusion trials can be carried out using any one of the following three techniques:

Hanging drop

A small droplet of the sample mixed with the precipitant is pipetted out onto a siliconized glass cover slide and inverted over a reservoir containing the precipitant.

Sitting drop

A small droplet of the sample is mixed with the precipitant and placed on a platform to allow the drop to undergo equilibration against a reservoir solution.

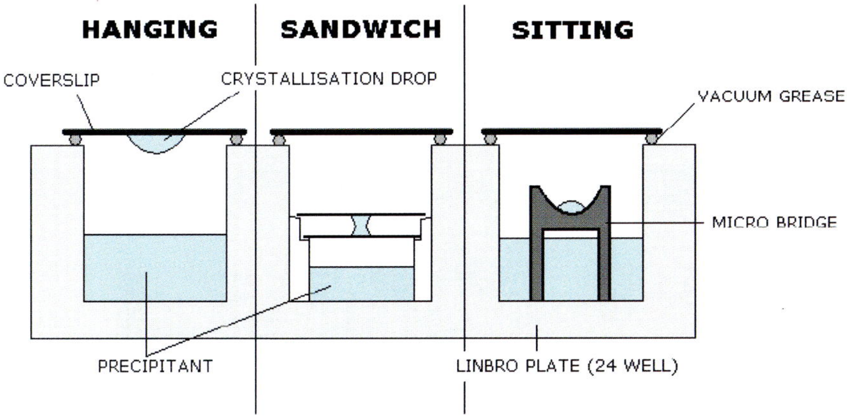

Fig. 5. Schematic representation of the three techniques of vapor diffusion: the hanging drop, where the crystallizing drop is hung over a reservoir of crystallizing solution; the sandwich drop, where the crystallizing drop is sandwiched between two coverslips over a reservoir of crystallizing agent; the sitting drop, where the crystallizing drop is placed on a bridge in the crystallizing solution.

Sandwich drop

The sample solution is mixed with the precipitant and placed in the middle of two cover slides with different dimensions, such that the drop is sandwiched between the glass cover slides.

Figure 6 illustrates a 24-well plate used for vapor diffusion trials. The volume of reservoir in usually 0.5–1 ml. The rims of the wells are greased with vacuum grease. The protein and the crystallizing agent are normally mixed in a 1:1 ratio on a cover slip. The volumes of the drop can vary between 2 and 10 μl. The cover slip is then inverted and sealed over the well.

Liquid-liquid interface diffusion

Liquid-liquid interface diffusion also referred to as free interface diffusion (FID), crystallization is another technique used for growing protein crystals. This technique is commonly performed in small capillaries, the geometry of which reduces undesirable convective effects. It consists of a

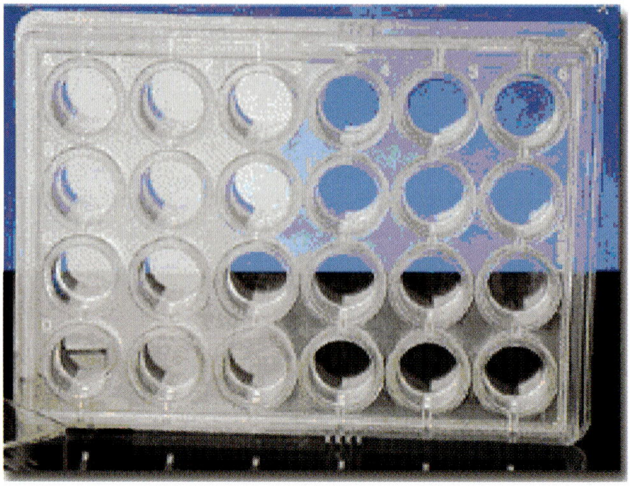

Fig. 6. A 24-well Linbro plate used in vapor diffusion from Molecular Dimensions Ltd. (MD3-11ROWORLD) or Hampton Research Ltd. (HR3-110 or HR3-112).

protein solution and precipitating solution separated by a liquid interface. As the two solutions diffuse into each other, the varying supersaturation conditions lead to crystal growth at the interface between the two liquids in the capillary. Free interface diffusion is one of the methods favored by NASA for their microgravity crystallization research.

Granada crystallization box® (GCB)

This is a simple device designed by Prof. Garcia-Ruiz and his co-workers (Garcia-Ruiz *et al.*, 2002), used to crystallize protein and other biological macromolecules by the counter-diffusion method (Fig. 7).

The bottom of the body of the box is filled with a layer of buffered agarose gel at 0.5% w/v concentration. Capillaries of different inner diameters ranging from 0.2 to 1.5 mm are used as the protein chamber. Once the gel is set, the capillaries are filled with the protein and inserted through the hole of the guide and punctuated to a given depth into the gelled buffer layer. Finally, the crystallizing agent is poured on top of the buffer gel.

Fig. 7. The GCB available from Hampton Research (HR3-194) and Triana Science & Technology (GCB-A-20) consists of three parts made of polystyrene: the box body consisting of a narrow box open on one side, the capillary holder that fits into the body, and the cover lid that closes the box. The box body has been designed to be small enough for experiment stowage and transport. It can hold up to six different crystallization experiments. The box is very narrow and transparent, allowing microscopic observation of experiments.

Dialysis

Dialysis is among the oldest, most useful and most effective techniques for dynamically altering the degree of supersaturation of a protein solution and in accomplishing crystal growth. It is a variation of a typical vapor diffusion method and utilizes the principle of slow diffusion by selective permeability. This method allows for an easy variation of the different parameters which influence the macromolecular crystallization process (Ducruix, Giege, 1999). The macromolecule solution to be crystallized is equilibrated against a large volume of precipitant using a semi-permeable membrane, allowing free passage of small molecules and ions but preventing circulation of macromolecules by the small pore size of the membrane. Equilibration against the solvent facilitates supersaturation, eventually resulting in crystallization.

Bulk dialysis

This procedure is carried out in a large dialyzing bag. Although this method is very convenient for successive re-crystallization, the major drawback is that large volumes of the sample are required (Ducruix, Giege, 1999).

Microdialysis

The microdialysis method is an adaptation working on the same principle of bulk dialysis but was designed to use very small sample volumes thus overcoming the drawback of the earlier method (Fig. 8). Microdialysis with Zeppezauer cells are capillary tubes closed with dialysis tubes or polyacrylamide gel plugs. Microdialysis is carried out using commercially available dialysis buttons (Fig. 9). The buttons come in varying sizes and are provided with a chamber in which the sample to be crystallized is placed. A dialysis membrane with an appropriate molecular weight cut off is placed over the top of the button and held by an O-ring. The button

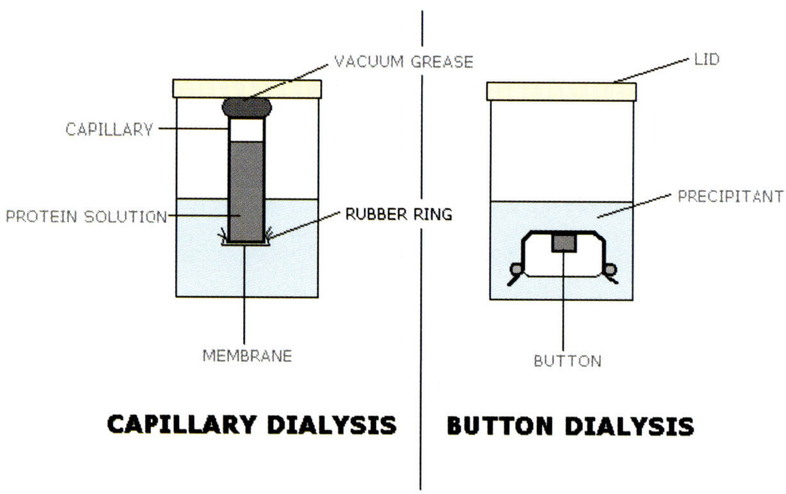

Fig. 8. Dialysis cell for protein crystallization. In the microdialysis method, the sample is separated from the "precipitant" by a semi-permeable membrane which allows small reagent molecules to pass but prevents biological macromolecules from crossing.

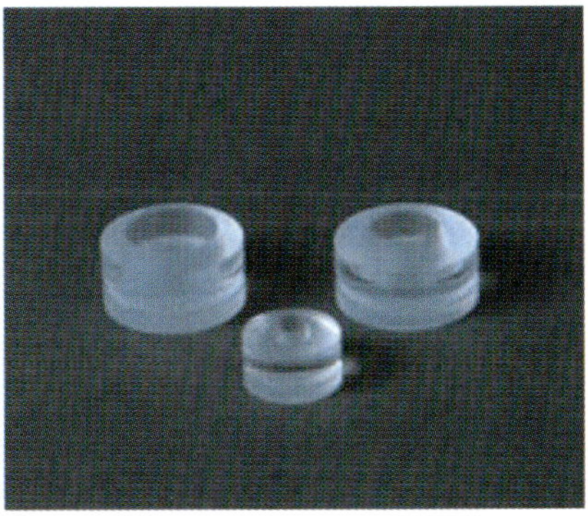

(a)

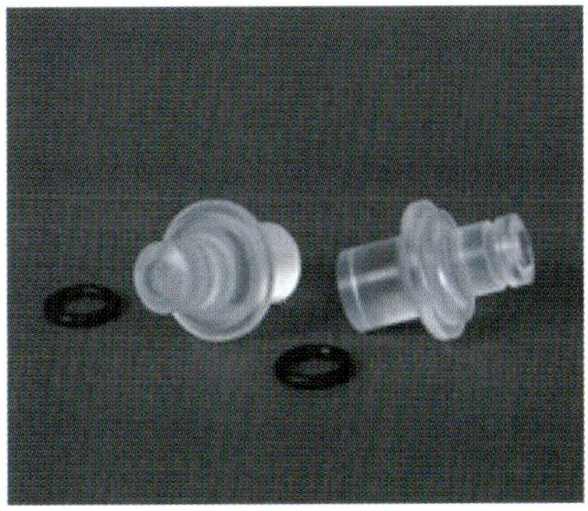

(b)

Fig. 9. Dialysis buttons are either machined from transparent perspex or injection moulded from polystyrene, and are the size of a small button. The sample is placed in this chamber so as to create a slight dome of liquid at the top of the button. A dialysis membrane is placed over the top of the button/sample and is held in place with an O-ring. The O-ring is held in place by a groove in the dialysis button. (a) Microdialysis buttons from Hampton research (HR3-336). (b) Microdialysis rods from Hampton research (HR3-348).

along with the membrane is then immersed into a well containing the precipitant to allow for equilibration.

A modified microdialysis button called the Microdialysis Rod (Fig. 9b), has been designed and manufactured for use in protein crystallization. This new apparatus allows easier handling and storage, and provides versatility to users (Lee, Cudney, 2004).

Conclusions

Before embarking on crystallization trials, systematic studies on the macromolecule is an important factor to be considered for success.

A phase diagram can form the basis of the design of crystallization experiments. It will enable the experimenter to manipulate the phase diagram and actively control the crystallization environment in order to lead to crystal growth in the direction that will produce the desired results.

Any of the methods discussed above can be used to grow macromolecular crystals. Keeping an open mind is essential as no single method is superior to another.

Acknowledgments

The author would like to acknowledge Prof. Naomi Chayen for a critical reading of the manuscript and valuable discussions, and Dr. Sahir Khurshid for providing some of the figures. The author would like to thank Barnali and Bhaskar for their invaluable help in the illustration of the phase diagram. The UK Engineering and Physical Sciences Research Council (EPSRC) and the European Commission OptiCryst Project (SHG-CT-623-2006-037793) are acknowledged for financial support.

References

Anderson MJ, Hansen CL, Quake SR. (2006) Phase knowledge enables rational screens for protein crystallization. *Proc Nat Acad Sci USA* **103**: 16746–16751.

Asherie N. (2004) Protein crystallization and phase diagrams. *Methods* **34**: 266–272.

Ataka M, Tanaka S. (1986) The growth of large single-crystals of lysozyme. *Biopolymers* **25**: 337–350.

Chayen NE. (2005) Protein Crystallization — The current challenge. *Euro Pharma Rev* **3**: 106–111.

Ducruix A, Giege R. (1999) *Crystallization of Nucleic Acids and Proteins: A Practical Approach*, 2nd Edn. Oxford University Press, Oxford.

Feher G, Kam Z. (1985) Nucleation and growth of protein crystals: General principles and assays. *Methods Enzymol* **114**: 77–112.

Garcia-Ruiz JM, Gonzalez-Ramirez LA, Gavira JA, Otalora F. (2002) Granada Crystallization Box: A new device for protein crystallization by counter-diffusion techniques. *Acta Crystallogr D* **58**: 1638–1642.

Haas C, Drenth J. (1999) Understanding protein crystallization on the basis of the phase diagram. *J Cryst Growth* **196**: 388–394.

Lee SSJ, Cudney R. (2004) A modified microdialysis button for use in protein crystallization. *J Applied Cryst* **37**: 504–505.

McPherson A. (1995) Increasing the size of microcrystals by fine sampling of pH limits. *J Applied Cryst* **28**: 362–365.

McPherson A. (1999) *Crystallization of Biological Macromolecules*. Cold Spring Harbor Laboratory Press, Cold Spring Harbor, NY.

Saridakis E, Chayen NE. (2003) Systematic improvement of protein crystals by determining the supersolubility curves of phase diagrams. *Biophys J* **84**: 1218–1222.

Saridakis EEG, Stewart PDS, Lloyd LF, Blow DM. (1994) Phase-diagram and dilution experiments in the crystallization of carboxypeptidase-G(2). *Acta Crystallogr D* **50**: 293–297.

Sauter C, Lorber B, Kern D, Cavarelli J, Moras D, Giege R. (1999a) Crystallogenesis studies on yeast aspartyl-tRNA synthetase: Use of phase diagram to improve crystal quality. *Acta Crystallogr D* **55**: 149–156.

Sauter C, Ng JD, Lorber B, Keith G, Brion P, Hosseini MW, Lehn JM, Giege R. (1999b) Additives for the crystallization of proteins and nucleic acids. *J Cryst Growth* **196**: 365–376.

Chapter 8

The Role of Oil in Protein Crystallization

*Naomi E. Chayen**

The crystal growth of proteins is a complicated process which is dependent on numerous factors. This chapter highlights the unique contribution of oil as a major parameter in protein crystallization. Oils are used in screening to find initial crystallization conditions as well as for optimizing crystallization by control of nucleation and growth. A variety of techniques using oils is described.

Keywords: Crystallogenesis; nucleation; protein crystallization; solubility; vapour diffusion; microbatch; oil.

Introduction

Searching for crystallization conditions for a new protein or any macromolecule has been compared with looking for a "needle in a haystack." Once a condition which looks promising for crystallization is obtained, conditions can generally be optimized by making variations to the parameters (precipitant, pH, temperature, etc.) involved. However, there is generally no indication that one is close to crystallization conditions until the first crystals are obtained. Consequently, crystallization breaks down naturally into two phases: screening, where one tries to obtain crystals of any description, and optimization, where one tries to improve the size and diffraction-quality of the crystals (Chayen *et al.*, 1996).

*Department of BioMolecular Medicine, Division of Surgery, Oncology, Reproductive Biology and Anaesthetics, Faculty of Medicine, Imperial College London, Sir Alexander Fleming Building, London SW7 2AZ, UK. E-mail: n.chayen@imperial.ac.uk.

When trying to crystallize a protein, the initial aim of the crystallizer is to screen rapidly numerous conditions with a minimum of labor, and using as little material as possible.

To ease the laborious and time-consuming task of crystallization, an automated technique known as microbatch crystallization was developed at Imperial College in 1990, which fulfils the following requirements:

(a) dispensing of very small samples,
(b) high level of accuracy,
(c) rapid screening to determine solubility properties,
(d) homing in on conditions that produce crystallization,
(e) automatic execution in a quick and simple way, and
(f) flexible changes of operation mode for different types of survey.

To reduce the consumption of material, very small samples must be dispensed. This leads to two major problems: (a) evaporation and (b) inaccuracy of dispensing.

The Microbatch Techniques

The problem of evaporation is solved by dispensing and incubating the crystallization samples under oil, while high accuracy is achieved by using a micro-dispenser (IMPAX, sold by Douglas Instruments, UK) comprising a bank of Hamilton syringes driven by stepper motors under computer control to set up the samples for crystallization. The components of the trial (protein, precipitating agents, etc.) are mixed in their final concentrations, and a specially designed multi-bore microtip allows very small volumes (0.5 μl or less) to be dispensed ready mixed and with good precision. The crystallization samples are typically dispensed into paraffin-filled microtitre plates. The oil used is paraffin liquid light (density $\rho \sim 0.83$ g cm^{-3}), a purified mixture of liquid saturated hydrocarbons obtained from petroleum (Molecular Dimensions UK, Cat. no. MD2-03). Paraffin was chosen after trying a variety of oils, many of which were not suitable due to their interaction with the crystallization trials (e.g. causing precipitation of proteins). The IMPAX software facilitates rapid design and execution of experiments. (Chayen *et al.*, 1990, 1992, 1994). Delivery from a fine tip under oil cleanly removes the delivered volume as the tip is drawn out from the oil, and eliminates "carry-over."

The mechanism of crystallization under oil

Figure 1 illustrates the mechanism of dispensing a crystallization trial under oil. A crystallization drop (which contains the macromolecule to be crystallized and the crystallizing agents) is dispensed either manually or automatically, into a container, under the surface of a thick layer of oil. As the dispensing tip is withdrawn from the oil, the drop detaches from it and the tip is wiped clean by the oil, thereby preventing any carry-over from one trial to another. The aqueous drop sinks to the bottom of the vessel since it

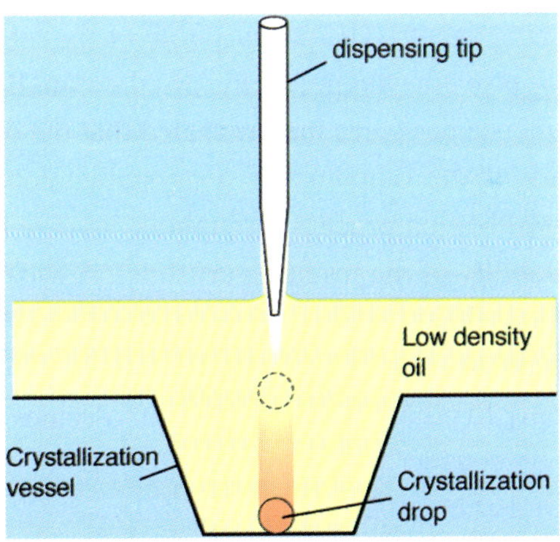

Fig. 1. The mechanism of dispensing a crystallization trial under oil. A crystallization drop (which contains the macromolecule to be crystallized and the crystallizing agents) is dispensed into a container, under the surface of a layer of oil. As the pipette tip is withdrawn from the oil, the aqueous drop detaches from it and sinks to the bottom of the vessel.

is heavier than the oil and settles at the bottom of the container. Setting up batch trials is simpler and speedier than other methods especially when trials are dispensed automatically (Chayen *et al.*, 1990, 1992, 1994; Luft *et al.*, 2001; DeLucas *et al.*, 2003).

The IMPAX equipment is usually operated in two main modes: one mode is used to screen a wide range of crystallization conditions, while the other generates a matrix survey, to optimize the precise conditions for obtaining the best crystals.

The IMPAX apparatus was the first robot to dispense microbatch experiments. Today, there are several devices for doing that (Luft *et al.*, 2001; DeLucas *et al.*, 2003; Stock *et al.*, 2005). A considerable number of proteins and other macromolecules have been successfully crystallized using the microbatch technique (e.g. Kleywegt *et al.*, 1994; Meyer *et al.*, 2004; Malkowski *et al.*, 2007).

The utilization of oil for protein crystallization was originally initiated as described above in order to enable the dispensing and incubation of very small crystallization samples using the microbatch method in which crystals are grown in nanolitre volume drops of a mixture of a protein and crystallizing agents. The primary role of the oil was to act as an inert sealant to prevent evaporation of the small-volume trials. Experimental evidence has since then revealed that the oil itself can play an important part in the outcome of a crystallization experiment by affecting the crystallization process throughout its stages (of nucleation, growth and the stability of the resulting crystals). A wide scope of experiments which exploit the presence of oil to aid protein crystal growth are presented. This article focuses on protein crystals, although the methods described also apply to other biological macromolecules. Figure 2 shows a view of a 2 μl drop under oil containing a crystal of an alcohol dehydrogenase (Korkhin *et al.*, 1996).

The microbatch method is essentially a batch experiment in which the macromolecule and the crystallizing agents are mixed at their final concentrations at the start of the experiment; thus, supersaturation is achieved upon mixing. Consequently, the volume and the composition of a trial remain constant. This is in contrast to all other crystallization

Fig. 2. Photograph of a crystal of an alcohol dehydrogenase under oil. Courtesy of Y. Korkhin. Scale: 1.5cm = 0.5 mm.

methods in which the conditions gradually change until equilibrium is reached. The stability of the batch is an important benefit for conducting diagnostic studies on the process of crystal growth since the history of the sample can be followed reliably. However, this benefit may become a handicap in the case of screening for crystallization conditions since it is conceivable that the gradual change of conditions (en route to equilibrium), which occurs by the other methods, may be the crucial factor for the formation of crystals (Chayen, 1996, 1997a, 1997b; D'Arcy *et al.*, 1996).

A major element in making the microbatch experiment a batch, as opposed to a diffusion system, is the sealing of the samples by the paraffin oil. Paraffin oil has proved to be a good sealant, allowing only a negligible amount of water evaporation through it during the average time required for a crystallization experiment.

Experience has shown that although oil and water are thought to be immiscible, water can evaporate at different rates from different oils. Paraffin oil allows for little or no diffusion of water through it, while a drop incubated under silicone fluid (a polymer of repeating dimethylsiloxane units) can dry up within 24 hours. Paraffin and silicone oils are miscible, and it was shown by D'Arcy *et al.*, (1996) that by mixing different ratios of these two oils the evaporation from a trial drop can be regulated so that the ingredients in a trial become more concentrated with time until the eventual drying of the drop. This modification of the microbatch method provides a means of simultaneously retaining the benefits of a microbatch experiment combined with the inherent advantage of the self-screening process of a diffusion trial. The authors reported that using a combination of paraffin and silicone oils to cover microbatch trials for screening experiments resulted in the appearance of crystals within a shorter space of time compared with trials which were situated under paraffin oil. These results were confirmed by Baldock *et al.* (1996). A similar effect can be accomplished by using silicone oil alone (D'Arcy *et al.*, 2004) or by varying the thickness of the oil layer covering the trials (Chayen, Saridakis, 2002). Obviously, when using the combinations of oils and/or a thin layer of oil, frequent monitoring of the trials is imperative. Once crystals are observed, more oil must be applied to prevent the drops from drying up.

It is interesting to note that not only can the type and the quantity of the oil dictate the outcome of a crystallization experiment, but also the time of incubation. The effect of time as an additional factor was revealed when crystallizing β-crustacyanin, a protein of the lipocalin family, which could only be crystallized by the microbatch method under paraffin oil, yet it took four months to produce diffraction quality crystals (Cianci *et al.*, 2002). No crystals were produced if trials were set under a mixture of paraffin and silicone.

It transpired that in spite of being covered by paraffin oil, some evaporation was taking place due to the lengthy time of incubation. In a typical microbatch experiment, crystallization takes place within a week or two, and since water and paraffin oil are essentially immiscible, evaporation during this time is negligible. However, given ample time, slow evaporation can occur (as there is no absolute immiscibility), which can proceed until the drop dries out. It is apparent that the β-crustacyanin underwent a very gradual concentration until it reached the certain point suitable for its nucleation and subsequent growth of crystals (Chayen, 1998).

The application of oil to improve vapour diffusion experiments

The microbatch method and variations of it have established a new concept in protein crystallization. However, vapor diffusion, which is still the most popular method of crystallization, does have advantages that are not fulfilled by microbatch. One such advantage is the possibility to affect the equilibration rate of the trials (without the risk of the trials drying out) and thus approach supersaturation more slowly by varying the distance between the reservoir and the crystallization drop (Luft *et al.*, 1996). However, this cannot be achieved in the popular Linbro and CrysChem (Hampton Research, USA) plates since a change in the drop-to-reservoir distance is not sufficient to affect the equilibration rate in such plates (Mikol *et al.*, 1990). A means to slow down the equilibration rate and thus approach supersaturation more slowly was devised (Chayen, 1997a) by the introduction of an oil barrier over the reservoir of conventional vapor diffusion trials (hanging or sitting drops) in Linbro and Cryschem plates (Fig. 3a). It was demonstrated that the type of oil and the thickness of the oil layer situated above the reservoir dictated the speed of crystallization. In trials containing an oil barrier, crystals required over a week to grow to full size, yet their number was reduced and their size and diffraction quality were far superior (Fig. 3c) to crystals which grew overnight in control trials which had no barrier (Fig. 3b) (Chayen, 1997a, 1997b).

The contribution of oil to the control of heterogeneous nucleation

Nucleation is a prerequisite and the first stage of any crystallization. The ability to control this first stage would be a big step forward in designing crystallization experiments, and hence studies concerning nucleation are of high priority in the field of crystal growth.

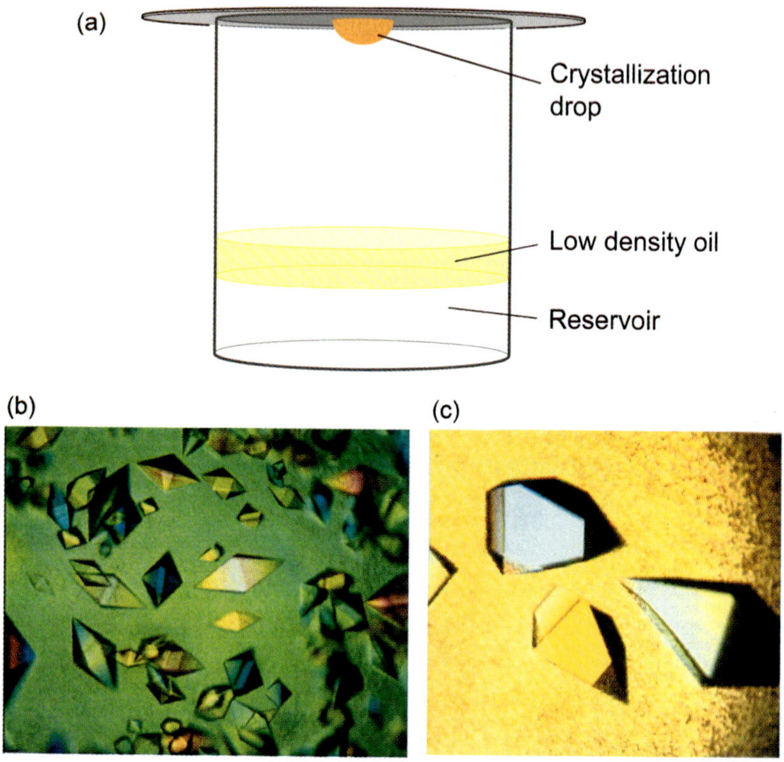

Fig. 3. Application of oil to improve vapour diffusion experiments: (a) set-up of a hanging drop experiment containing an oil barrier, (b) thaumatin crystals grown overnight in a standard hanging drop, and (c) thaumatin crystals grown over eight days under the same conditions as in (b), the only difference being the presence of 500 microlitres of oil (equal volumes of paraffin and silicone) over the reservoir. Scale: 1.5 cm = 0.25 mm.

Cleanliness of trials

Control of nucleation requires extremely clean solutions. In microbatch where the drops are maintained under oil the samples are never exposed to air and are therefore protected from airborne contamination. The combination of filtration methods, which allow the removal of particles as small as 100 nm from small volumes with dispensing and incubating of trials under oil, provides an ideal environment for controlled heterogeneous nucleation experiments (Blow *et al.*, 1994; Chayen, 1996; Chayen *et al.*, 1993).

Filtration of a crystallization trial through a 0.1 micrometer filter can prevent nucleation under conditions previously considered standard (Chayen *et al.*, 1993; Hirschler *et al.*, 1995). Providing the trial remains under oil, nucleants can be inserted in a controlled manner since one is ensured that the oil prevents any other contaminant entering the trial, other than that which the experimenter wishes to add. Experiments have been performed in which the nucleation, and consequently the number and size of several protein crystals, was determined at will by the addition of different quantities of a nucleant to filtered trials containing these proteins (Blow *et al.*, 1994; Chayen *et al.*, 1993, 2006). The cleanliness of such trials produces highly reproducible results.

Effect of surface contact

It has been reported that heterogeneous nucleation which is often detrimental to the production of diffraction quality crystals can be induced by the contact of a crystallization trial with the walls of its supporting vessel (Blow *et al.*, 1994; Yonath *et al.*, 1982). The nucleation properties of such solid surfaces can be manifested even after filtration. A series of experiments shown in Fig. 4 demonstrates how the application of oil can determine the contact area between the trial and its supporting vessel, thus enabling the experimenter to monitor the nucleation and reduce or increase its level at will. The figure illustrates three situations: (a) represents a drop which has been dispensed onto the floor of a vial and then covered by a layer of oil; the drop spreads out and flattens over the floor of the container. (b) shows a drop dispensed into oil as performed in the normal microbatch procedure (shown in Fig. 1); the drop forms a round shape with just a small part of it touching the floor. (c) represents a situation of "containerless crystallization" in which a crystallization drop is suspended between two oils of different densities as described by Chayen (1996) and by Lorber (1996). The oils are not miscible and the drop floats at the interface, thereby not touching the container walls.

The number of crystals produced by procedures (b) and (c) is significantly reduced and their size is much larger compared with those

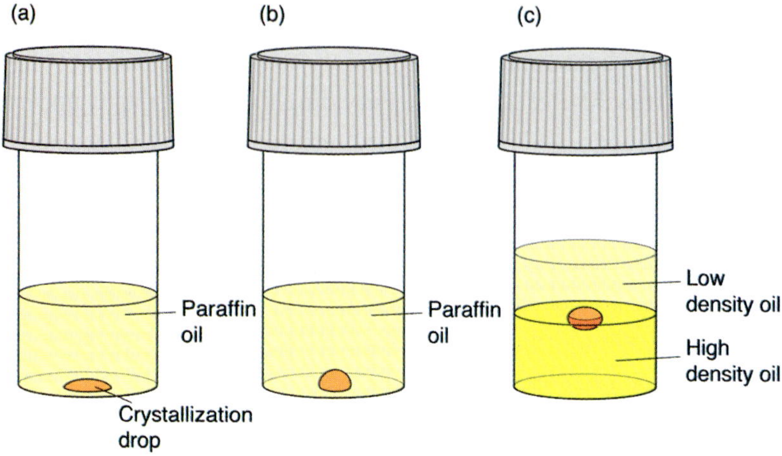

Fig. 4. Determination of the contact area between a crystallization trial and its supporting vessel by application of oil. (a) Large contact area: a crystallization drop dispensed onto the floor of a vial and then covered by a layer of oil; the drop spreads out and flattens over the floor of the container. (b) Small contact area: the result of dispensing by the normal microbatch procedure (shown in Fig. 1); a rounded drop with only a small part of it touching the floor. (c) "Containerless crystallization": a crystallization drop is suspended between two oils of different densities; the drop floats at the interface thereby not touching the container walls.

grown by procedure (a), where the drop has the largest contact area with its vessel (Blow *et al.*, 1994; Chayen, 1996). Nucleation is not totally eliminated by procedure (c) and surprisingly, the difference between the number of crystals grown by the normal microbatch procedure (b) and the containerless situation (c) was not marked. The number of crystals in the containerless procedure was merely 10% less than the number obtained by normal microbatch dispensing. This indicates that the interface between the two oils also acts as a surface but with somewhat reduced nucleation properties compared with that of a solid material. A simpler way to conduct such "containerless crystallization" is by replacing the bottom layer with a layer of Vaseline above which the low density oil is dispensed (Chayen, Saridakis, 2002). A kit for such experiments is available from Molecular Dimension (Cat. no. MD1-12ROWORLD).

Protection of crystallization samples and crystals by the oil

It is generally believed that external disturbances such as vibration can cause excess nucleation and lead to the formation of smaller crystals or to crystal imperfections. Trials under oil are preserved from physical shock since the nuclei and the forming crystals are buoyed and cushioned by the viscous oil, making trials resistant to vibration and allowing unmounted crystals to be easily transportable.

The presence of the oil can offer the additional benefit of protecting crystals which have formed in the oil from dissolution. Using vapor diffusion one often encounters problems concerning changes in drop volume, particularly when precipitants such as polyethylene glycol and volatile solvents are used. The absorbance of a volatile precipitating agent (Yonath *et al.*, 1982) or a slight change in temperature can cause enlargement of the drops, thus diluting the protein and causing dissolution of crystals; this can occur during the short space of time when crystals are being observed under the microscope. Provided the crystals are incubated under a sufficiently thick layer of paraffin oil, the volume of the drops remains constant and no dissolution occurs (Conti *et al.*, 1996) unless the solubility of the protein is temperature dependent [e.g. Zagari *et al.*, (1994)].

Limitation of Crystallizing Under Oil

Application of organic molecules as precipitants and/or additives

Experimental data has indicated that most macromolecules which were tested could be crystallized under oil [e.g. Chayen (1998), and references therein]. The oils described above do not interfere with the common precipitants such as salts, polyethylene glycol (PEG), Jeffamine and 2-methyl-2,4-pentanediol (MPD). Moreover, samples containing detergents have also been crystallized under oil (Snijder 2003; Stock *et al.*, 1999). However, not every case is suitable for crystallization under oil. A limitation of this method is that it cannot be applied at all in cases where volatile organic molecules (which are soluble in the oils, e.g. dioxane, phenol, thymol) are required in the crystallization medium, nor with organic precipitants or additives which interact with the oil.

Harvesting and mounting of crystals

Harvesting crystals from under oil is somewhat more difficult than harvesting from a coverslip or a Cryschem plate; hence, a detailed protocol for harvesting and mounting crystals from the oil has been reported by Shaw, Stewart (1995). A common problem is the sticking of crystals to their supporting surface. The standard procedure is to use microtools to gently release the crystals. Growing crystals in suspended drops between two oils (as described above and in Fig. 4c) would solve the problem of sticking but may cause other difficulties. An aid to harvesting crystals which seems to have been overlooked (at least in harvesting from microbatch), was reported back in 1984 by Ray, Puvathingal (1984), who found that a layer (2 mm thick) of high-vacuum silicone grease (which has a gel-like texture) provided an excellent support on which to grow protein crystals and facilitated subsequent harvesting. The authors also used the silicone grease to orient seed crystals of phosphoglucomutase.

Crystallization of membrane proteins under oil

Crystallization of membrane proteins under oil has not been widely attempted due to doubts about the suitability of an oil-based method for crystallizing lipophilic compounds. Surprisingly, a considerable number of proteins have been successfully crystallized under oil (e.g. Snijder, 2003; Stock *et al.*, 1999), some which could not be produced by other crystallization methods. It is possible that the oil is essential in driving the process, by slowly absorbing the detergent from the aqueous drop, thereby encouraging the protein to gradually come out of solution and crystallize (Hankamer, personal communication).

Summary

The different facets of the utilization of oil demonstrate that an individual oil and/or combinations of different oils can influence the outcome of crystallization experiments. The oil can play a part in the control of nucleation, affect the rate of equilibration and consequently determine the size of the resulting crystals. Whether used for microbatch, vapor diffusion or for control of nucleation, the presence of oil is a parameter which can contribute to the accuracy, cleanliness and consequently to the reproducibility of experiments. Furthermore, the oil has a role in the protection of the trial during the course of its duration and to the increase in the stability of the resulting crystals.

Acknowledgment

This chapter was updated based on an article published as Chayen NE. (1997) The role of oil in macromolecular crystallization, *Structure* **5**(10): 1269–1274, Copyright Elsevier 1997. The UK Engineering and Physical Sciences Research Council (EPSRC) and the European Commission OptiCryst Project LSHG-CT-623 2006-037793 are acknowledged for financial support.

References

Baldock P, Mills V, Stewart PS. (1996) A comparison of microbatch and vapour diffusion for initial screening of crystallization conditions. *J Cryst Growth* **168**: 170–174.

Blow DM, Chayen NE, Lloyd LF, Saridakis E. (1994) Control of nucleation of protein crystals. *Protein Sci* **3**: 1638–1643.

Chayen NE. (1998) Comparative studies of protein crystallization by vapour-diffusion and microbatch techniques. *Acta Crystallogr D Biol Crystallogr* **54**: 8–15.

Chayen NE. (1996) A novel technique for containerless protein crystallization. *Protein Eng* **9**: 927–929.

Chayen NE. (1997) A novel technique to control the rate of vapour diffusion, giving larger protein crystals. *J Appl Cryst* **30**: 198–202.

Chayen NE. (1997) The role of oil in macromolecular crystallization. *Structure* **5**: 1269–1274.

Chayen NE, Boggon TJ, Cassetta A, Deacon A, Gleichmann T, Habash J, Harrop SJ, Helliwell JR, Nieh YP, Peterson MR, Raftery J, Snell EH, Hadener A, Niemann AC, Siddons DP, Stojanoff V, Thompson AW, Ursby T, Wulff M. (1996) Trends and challenges in experimental macromolecular crystallography. *Q Rev Biophys* **29**: 227–278.

Chayen NE, Radcliffe JW, Blow DM. (1993) Control of nucleation in the crystallization of lysozyme. *Protein Sci* **2**: 113–118.

Chayen NE, Saridakis E. (2002) Protein crystallization for genomics: Towards high-throughput optimization techniques. *Acta Cryst D Bioll Crystallogr* **58**: 921–927.

Chayen NE, Saridakis E, Sear RP. (2006) Experiment and theory for heterogeneous nucleation of protein crystals in a porous medium. *Proc Nat Acad Sci USA* **103**: 597–601.

Chayen NE, Stewart PDS, Baldock P. (1994) New developments of the IMPAX small-volume automated crystallization system. *Acta Crystallogr D Biol Crystallogr* **50**: 456–458.

Chayen NE, Stewart PDS, Blow DM. (1992) Microbatch crystallization under oil — A new technique allowing many small-volume crystallization trials. *J Cryst Growth* **122**: 176–180.

Chayen NE, Stewart PDS, Maeder DL, Blow DM. (1990) An automated system for microbatch protein crystallization and screening. *J Appl Cryst* **23**: 297–302.

Cianci M, Rizkallah PJ, Olczak A, Raftery J, Chayen NE, Zagalsky PF, Helliwell JR. (2002) The molecular basis of the coloration mechanism in lobster shell: Beta-crustacyanin at 3.2 Å resolution. *Proc Nat Acad Sci USA* **99**: 9795–9800.

Conti E, Lloyd LF, Akins J, Franks NP, Brick P. (1996) Crystallization and preliminary diffraction studies of firefly luciferase from *Photinus pyralis*. *Acta Crystallogr D Biol Crystallogr* **52**: 876–878.

D'Arcy A, Sweeney AM, Haber A. (2004) Practical aspects of using the microbatch method in screening conditions for protein crystallization. *Methods* **34**: 323–328.

D'Arcy A, Elmore C, Stihle M, Johnston JE. (1996) A novel approach to crystallizing proteins under oil. *J Cryst Growth* **168**: 175–180.

DeLucas LJ, Bray TL, Nagy L, McCombs D, Chernov N, Hamrick D, Cosenza L, Belgovskiy A, Stoops B, Chait A. (2003) Efficient protein crystallization. *J Struct Biol* **142**: 188–206.

Hirschler J, Charon MH, Fontecilla-Camps JC. (1995) The effects of filtration on protein nucleation in different growth media. *Protein Sci* **4**: 2573–2547.

Kleywegt GJ, Bergfors T, Senn H, Le Motte P, Gsell B, Shudo K, Jones TA. (1994) Crystal structures of cellular retinoic acid binding proteins I and II in complex with all-trans-retinoic acid and a synthetic retinoid. *Structure* **2**: 1241–1258.

Korkhin Y, Frolow F, Bogin O, Peretz M, Kalb (Gilboa) A J, Burstein Y. (1996) Crystalline alcohol dehydrogenases from the mesophilic bacterium *Clostridium beijerinckii* and the thermophilic bacterium *Thermoanaerobium brockii*: Preparation, characterization and molecular symmetry. *Acta Cryst D* **52**: 882–886.

Lorber B, Jenner G, Giege R. (1996) Effect of high hydrostatic pressure on nucleation and growth of protein crystals. *J Cryst Growth* **158**: 103–117.

Luft JR, Albright DT, Baird JK, DeTitta GT. (1996) The rate of water equilibration in vapor-diffusion crystallizations: Dependence on the distance from the droplet to the reservoir. *Acta Crystallogr D Biol Crystallogr* **52**: 1098–1106.

Luft JR, Wolfley J, Jurisica I, Glasgow J, Fortier S, DeTitta GT. (2001) Macromolecular crystallization in a high throughput laboratory — The search phase. *J Cryst Growth* **232**: 591–595.

Malkowski MG, Quartley E, Friedman AE, Babulski J, Kon Y, Wolfley J, Said M, Luft JR, Phizicky EM, DeTitta GT, Grayhack EJ. (2007) Blocking S-adenosylmethionine synthesis in yeast allows selenomethionine incorporation and multiwavelength anomalous dispersion phasing. *Proc Nat Acad Sci USA* **104**: 6678–6683.

Meyer P, Prodromou C, Liao C, Hu B, Roe SM, Vaughan CK, Vlasic I, Panaretou B, Piper PW, Pearl LH. (2004) Structural basis for recruitment of the ATPase activator Aha1 to the Hsp90 chaperone machinery. *EMBO J* **23**: 1402–1410.

Mikol V, Rodeau JL, Giege R. (1990) Experimental determination of water equilibration rates in the hanging drop method of protein crystallization. *Anal Biochem* **186**: 332–339.

Ray WJ, Puvathingal JM. (1984) The use of silicones in protein crystallizations involving single or multiple growth cycles. *J Applied Cryst* **17**: 370–371.

Shaw SPD, Conti E. (1995) Harvesting crystals from microbatch for cryocrystallography. Application Note 3, Douglas Instruments, www.douglas.com.

Snijder A, Barrends T, Dijkstra D. (2003) Crystallization of phospholipase A in two biological oligomerization states. In Iwata S. (ed.) *Methods and Results in Crystallization of Membrane Proteins*, pp. 265–278. International University Line, La Jolla, California.

Stock D, Leslie AG, Walker JE. (1999) Molecular architecture of the rotary motor in ATP synthase. *Science* **286**: 1700–1705.

Stock D, Perisic O, Lowe J. (2005) Robotic nanolitre protein crystallization at the MRC Laboratory of Molecular Biology. *Prog Biophys Mol Biol* **88**: 311–327.

Yonath A, Mussig J, Wittmann HG. (1982) Parameters for crystal growth of ribosomal subunits. *J Cell Biochem* **19**: 145–155.

Zagari A, Sica F, Scarano G, Vitagliano L, Bocchini V. (1994) Crystallization of a hyperthermophilic archaeal elongation factor 1 alpha. *J Mol Biol* **242**: 175–177.

Chapter 9

Introduction to Crystallization of Fine Chemicals and Pharmaceuticals

*Åke C. Rasmuson**

This chapter presents an introduction to crystallization of organic fine chemicals and pharmaceutical compounds, written for newcomers to the field. The coverage includes the fundamental concepts of solubility, supersaturation, nucleation, growth, and polymorphism. We will also discuss the control of crystal size distribution, crystal shape and purity, and specifically address cooling crystallization and reaction crystallization processes.

Keywords: Crystallization; solubility; supersaturation; nucleation; growth; polymorphism; size; shape; purity; control.

Introduction

In a crystallization process, the substance is separated in solid crystalline form from a solution or a melt. Crystallization is a unit operation of great industrial importance. Production of pure sugar and table salt are old and well-known crystallization processes, and large volumes of inorganic salts such as chlorides, sulphates and chlorates are produced via crystallization. In addition, crystallization is a frequently used unit operation in the pharmaceutical industry and in the production of organic fine and specialty chemicals.

Crystallizations are complicated processes, since they involve discrete solid particles, and the physical course of events is particularly sensitive

*Department of Chemical Engineering and Technology, Royal Institute of Technology, SE-10044 Stockholm, Sweden. E-mail: rasmuson@ket.kth.se.

to process variables such as concentration, temperature, agitation, and impurities, as well as being strongly dependent on the properties of the substance itself. It is therefore not reliable to directly extrapolate the results for one substance to another, from one set of conditions to another, or from laboratory scale to full scale. However, understanding the mechanisms and phenomena that decide various product properties make it possible to interpret the results of one process under certain conditions in a more qualified way, and thereby facilitate work on other processes, or on the same process under different conditions. Crystallization of fine chemicals and pharmaceuticals tends to be more complicated since a third of all organic molecules may occur in different crystalline structures: polymorphs and an additional third form solvates, i.e. the solvent appears in the crystalline lattice. In addition, often such molecules are large and flexible, and impurity molecules can be quite similar to the crystallizing molecule. Usually, equipment is not specialized for crystallization of a particular compound, and the agitation is not specialized for the particular crystallization process.

A crystallization process is evaluated in terms of production capacity of crystals, purity of the product, and various solid state properties of the solid material, such as crystal size distribution, crystal shape, crystal structure and agglomeration. The solid state properties are important for downstream processing, e.g. filtration, washing and drying, as well as for the end-use properties like tableting, and dissolution rate.

Fundamentals

The fundamental requirement for a substance to be crystallized from solution is that the concentration in the solution must be higher than the solubility of the substance under the conditions in question, i.e. the solution has to be supersaturated with respect to the compound to be crystallized. In a supersaturated solution, the compound may form nuclei, extremely small solid particles having a crystalline structure, and crystals may grow. During the process the crystals may also agglomerate into larger particles.

Solubility and Supersaturation

The *solubility* of a substance in solution is the concentration that is reached in the solution after a long period of contact with the substance in the solid form. The solubility does not depend on how well you stir the solution, nor on the amount of solid material present: these things only affect the time it takes to reach the final equilibrium concentration. The solubility does, on the other hand, depend on temperature and composition of the solution. In the literature, solubility data are usually published for pure solvents and substances. In industry, however, that kind of purity is rarely achieved, and it is therefore advisable to verify data in the literature with experiments on the actual solutions. The solubility of small crystals, <10 μm, depends on the particle size. The smaller the crystal, the higher the solubility. This results in large crystals in solution growing at the expense of small ones. The phenomenon is called "Ostwald ripening" and complicates the production of small crystals of uniform size, which is sometimes of importance in the pharmaceutical industry.

The *supersaturation* denotes the concentration, c [kg/kg inert], of a substance in excess of its solubility, c^* [kg/kg inert]. The supersaturation can be expressed as a concentration difference Δc [kg/kg inert], as the supersaturation ratio s, or as relative supersaturation σ:

$$\Delta c = c - c^*, \qquad s = \frac{c}{c^*}, \qquad \sigma = \frac{\Delta c}{c^*}. \tag{1}$$

The base for concentration is chosen as the mass of non-crystallizing compounds (often mainly the solvent) and is called inert. All expressions for supersaturation above are approximations only, the true driving force being the difference in chemical potential. Normally, the solubility increases with increasing temperature, as is illustrated in Fig. 1. For all states below the solubility curve the solution is undersaturated, and crystallization is not possible. All states above the curve signify a supersaturated solution, for which crystallization is possible (but not necessarily probable). Suppose a solution is initially at the temperature and concentration of point A in the figure. Making the solution supersaturated corresponds to moving to a point above

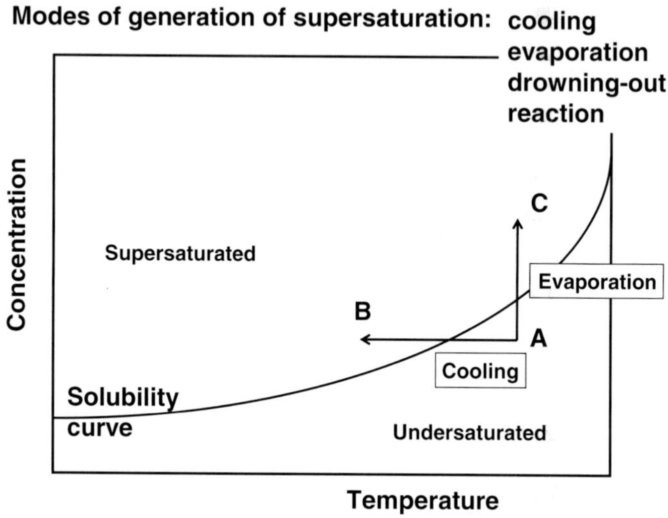

Fig. 1. Schematic description of change in the solution conditions at different modes of generation of supersaturation. The diagram shows solute concentration in the solution, e.g. in units of [kg solute/kg solvent], versus solution temperature in degrees centigrade or Kelvin. Depending on the compound and the process, the temperature range covered in a cooling crystallization is typically 10–40°C. Supersaturation and undersaturation can be described, e.g. by $\Delta c = c - c^*$ [kg/kg solvent], where c is the concentration and c^* is the solubility. The arrows describe how the solution conditions change at cooling and evaporation, respectively.

the solubility curve, either by moving the state of the solution or by moving the corresponding solubility. In principle, there are four ways of accomplishing this. If the solution is cooled, the state of the solution is moved from point A horizontally towards point B. After passing the solid solubility line, the solution is supersaturated. On the other hand, if the solvent is evaporated, the concentration is increased, and the state of the system is moved vertically upwards from point A towards point C.

An alternative is to add another soluble substance or solvent, which lowers the solubility of the substance to be crystallized, i.e. in principle moving the solubility curve so that point A ends up being above it. Finally, in a reaction crystallization, the substance is formed by a chemical reaction in concentrations exceeding the solubility.

Nucleation

Nucleation denotes the formation of new crystals, and is divided into primary and secondary nucleation. A moderately supersaturated solution can be termed metastable, which refers to the fact that, in the absence of crystals in solution, nucleation is either non-existent or very slow. At sufficient supersaturation, the solution becomes unstable, and the nucleation becomes very fast — *primary nucleation*. Primary nucleation is the result of the molecules merging into crystal nuclei, small entities having crystalline structure. For weakly soluble substances, a high relative supersaturation is often required for primary nucleation. If the solubility is high, however, primary nucleation takes place at low relative supersaturation.

The *metastable zone* denotes the supersaturation interval within which primary nucleation is insignificant or improbable, and is the operative region for many crystallization processes. In the metastable zone, crystal growth and secondary nucleation take place (see below). The width of the metastable zone is affected by a number of factors, the most important of which are: temperature, solution purity, thermal history of the solution, the presence of additives (Mullin, Jancic, 1979), and hydrodynamics. Nyvlt *et al.* (1985) provided values of the metastable zone width for several inorganic substances. Primary nucleation can be catalyzed by foreign particles in the solution (e.g. dust particles), called heterogeneous primary nucleation, which leads to a narrowing of the metastable zone.

Primary nucleation is often described by an empirical relationship:

$$B_{\mathrm{p}} = k_{\mathrm{p}} \Delta c^{np}, \tag{2}$$

where B_{p} is the primary nucleation rate [no/kg inert, s], k_{p} [no (kg inert/kg)^{np}/kg inert, s] is the primary nucleation rate constant, and Δc [kg/kg inert] denotes the supersaturation driving force. The exponent np — is generally larger than five, i.e. Eq. (2) describes a very nonlinear relationship, as is illustrated in Fig. 2 (right). This is what we see as a zone with little or no nucleation, followed by the onset of a very fast nucleation.

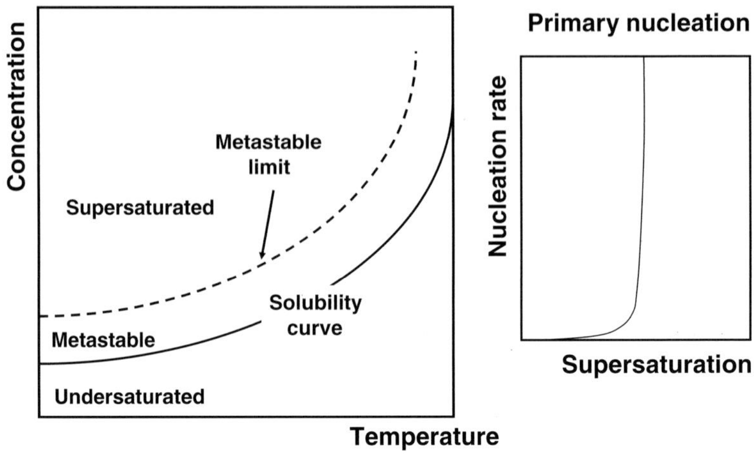

Fig. 2. Schematic description of the conditions of nucleation (left) and the behavior of primary nucleation (right). The diagram on the left shows solute concentration in the solution, e.g. in units of [kg solute/kg solvent], versus solution temperature in degrees centigrade or Kelvin. The solid line is the solubility curve, and the dashed line represents the metastable limit. When the solution conditions reach the metastable limit, i.e. when the level of supersaturation reaches the metastable limit, usually an enormous number of new crystals are created very rapidly — primary nucleation [no/kg solvent, s]. The diagram on the right shows that the rate of primary nucleation is strongly dependent on the supersaturation.

The term *secondary nucleation* denotes nucleation by mechanisms that require the presence of crystals in the solution. Secondary nucleation depends on the amount of crystals in the solution, and is often considerable even within the metastable zone. One of the most important mechanisms of secondary nucleation is attributed to collisions in the system. Crystals collide with one another and with the equipment (walls and impeller), causing the formation of new nuclei. In cases where there is visible damage to the mother crystal, this is called fragmentation, otherwise the term "contact nucleation" can be used. Furthermore, more or less ordered molecular clusters can be present at the crystal-solution interface. Collisions, but also shear forces in the liquid, may dislodge these structures, which may then turn into nuclei. Different substances show varying degrees of proclivity towards secondary nucleation. Secondary nucleation increases with increasing stirring rate (Larson, 1984). An increase in the stirring rate leads to more collisions, and higher collision energy and shear forces. Even the material of the impeller is important (Shah *et al.*, 1973). Secondary nucleation is generally modeled by a relationship of the form:

$$B_s = k_s \, W_T \, N_s^p \, \Delta c^{ns} \qquad (3)$$

where B_s [no/kg inert, s] is the secondary nucleation rate, k_s is the secondary nucleation rate constant [no (kg inert/kg)$^{ns+1}$(s)^p /kg inert, s], W_T [kg/kg inert] is the magma density, i.e. the mass of crystals per unit mass of solvent (or sometimes per unit mass of solution), N_s is the agitation rate [1/s], and Δc [kg/kg inert] denotes the supersaturation driving force. The exponent ns is in general — in the range 2–4. p is a dimensionless exponent often about 3–4, i.e. the rate of secondary nucleation increases strongly with increasing stirring rate.

In batch processes, primary nucleation can be considerable, particularly at the beginning. It is either generated deliberately, or unwanted. Later, as the crystal mass accumulates, the importance of secondary nucleation increases. In many processes with substances of normal solubility, the relative supersaturation is typically a few percent, but rarely more than ten percent, whereas in reaction crystallization, very high supersaturation often arises locally at the feed point, and primary nucleation dominates. Particularly for sparingly soluble substances, the relative supersaturation can reach many orders of magnitude.

Crystal Growth

At the molecular level, crystals grow because molecules or possibly molecular clusters are incorporated into the crystal lattice. This incorporation takes place primarily at kinks and "steps" on the surface, causing crystal growth to progress through incomplete layers growing to completion, and new growth layers being initiated, which in turn grow to completion, and so on. The result is that the crystal surface gradually grows in a direction perpendicular to the surface.

The growth rate of a crystal as a whole is normally discussed in terms of the linear growth rate G [m/s], i.e. the rate of change in a characteristic linear dimension of the crystal, L [m], with time t [s]:

$$G = \frac{dL}{dt}. \tag{4}$$

This characteristic linear dimension can be an actual dimension of the crystal, or for example the diameter of a sphere with the same volume as the crystal. The growth rate will depend on the choice of the characteristic linear dimension. Growth rates are often of the magnitude 10^{-7}–10^{-9} m/s, and the values for some substances are available in the literature (Mullin, 1979, 1993).

Crystal growth is governed by two sources of resistance in series. Molecules have to be transported from the bulk solution to the crystal surface — *boundary layer transport*. At the surface, the molecules are inserted into the crystal lattice — *surface integration*. Normally a certain amount of energy — the heat of crystallization — is released, which has to be transferred out into the bulk solution. In general, though, the amount of energy is moderate, and the heat transfer is much faster than the mass transfer. Crystal growth in solution is sometimes governed by the resistance to mass transfer through the boundary layer. This step can be considered as a normal diffusive/convective mass transfer from the solution to the surface of the solid particle. In the literature, a large number of studies of mass transfer from solution to solid particles are available. Different relationships are given for different particle sizes, hydrodynamic conditions and material properties. In the case of a stirred tank, the equation of Levins and Glastonbury (1972) is recommended. Note that the mass transfer coefficient increases with decreasing particle size and increasing mixing intensity. The resistance to surface integration depends on the properties of the crystal surface and is normally assumed to be independent of the hydrodynamics. At increasing molecular size and flexibility, the surface integration resistance is expected to increase. Impurities can easily block growth sites, however, and sometimes concentrations in the parts-per-million range can dramatically reduce the growth rate. The rate of surface integration differs for the different faces of the crystal.

The total growth rate, including both steps, is usually described by a simple expression:

$$G = k_g \Delta c^g \tag{5}$$

where G [m/s] is the linear growth rate, k_g is the linear growth rate constant [m(kg inert/kg)^g/s], and g is a dimensionless exponent. g frequently has a value between 1 and 2 (Mullin 1979, p. 249). g is equal to unity when the diffusion through the boundary layer dominates, and often higher than unity if the surface integration dominates. For weakly soluble substances, the surface integration is often slow and the growth rate is relatively independent of hydrodynamic conditions (Mullin, 1979, p. 249). For highly soluble substances the reverse is often true.

Usually, there are a large number of small crystals (<50 μm) in a crystallizer, which grow relatively slowly. This is probably a case of *dispersion* in the growth rate, which signifies that crystals of equal size in the same environment can grow at different rates. The growth rate of a particular crystal can vary with time, e.g. as a result of collisions or because of intermittent blocking of growth sites (adsorption of impurities). Another explanation is that crystals are "born" with a certain growth activity, depending on minute structural variations, different for each crystal (Zumstein, Rousseau, 1987). If different crystals grow at different rates, those that grow slowly will become small and those that grow fast will become relatively large (Girolami, Rousseau, 1985).

The Product

Productivity

The production in batch crystallization is essentially determined at the planning of the process. The amount of crystals produced, Prod [kg] is

$$\text{Prod} = V\,(c_0 - c_f) \tag{6}$$

where V denotes the mass of solvent in the suspension [kg inert], c_0 is the initial concentration [kg/kg inert] and c_f is the final concentration [kg/kg inert]. For example, for a cooling crystallization, c_f is normally close to the solubility at the final temperature, which means that the production is determined by the initial concentration and by the choice of final temperature. There are often limitations on how far it is desirable or possible to cool. One such limitation could be that it is difficult to maintain purity when cooling to high yield and low temperature. During a batch cooling or evaporation crystallization, the concentration of impurities relative to the concentration of the substance increases, thereby gradually increasing the tendency towards crystal impurity incorporation. In addition, the viscosity of the solution may be higher nearer the end of the process since viscosity generally increases with decreasing temperature. This also leads to an increased tendency towards product impurity.

Solid phase

A substance may crystallize in more than one solid phase, which may either differ with respect to the chemical composition, e.g. *hydrates* and *solvates*, or have the same composition but different crystal structure: *polymorphism* (such as in graphite and diamond). Polymorphism is common among organic substances. Since the crystal structure itself differs, different polymorphs vary in terms of physical properties such as melting point, density and solubility. At any given temperature and pressure, only one form is thermodynamically stable, except at the transition temperature where two polymorphs may be stable at the same time (Fig. 3). Other forms are unstable and have the potential to transform into the stable form. It is also possible to show that in all solvents at a certain temperature, the stable form always has a lower solubility than the unstable ones. A system where each of two forms are stable in their respective temperature intervals is called enantiotropic, and transition in stability occurs at a specific temperature. A system where one form is stable throughout the entire temperature interval is termed monotropic.

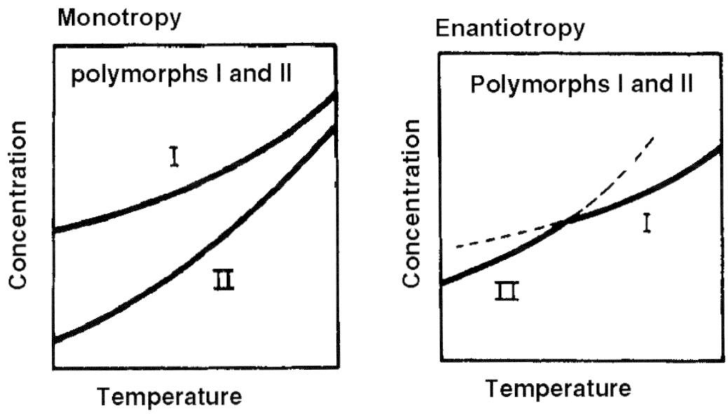

Fig. 3. Polymorphism: the same compound may appear as different solid materials having different crystal structures (a classic example is carbon in the form of graphite and diamond). The figure shows two main types of polymorphism in terms of solubility curves for two different polymorphs of the same compound. The stable polymorph always has the lowest solubility. In a monotropic system, one structure is the stable structure over the entire temperature range and hence always has the lower solubility. In enantiotropic systems, there is a change in relative stability at a certain temperature below the melting point, and hence there is a change in relative solubility.

When crystallization is undertaken in polymorphic systems, it is by no means uncommon for a metastable form to appear initially. This has even been called the *Ostwald Rule of Stages*. Accordingly, the form that crystallizes first is not only a question of thermodynamics but also depends on the rate of nucleation and growth of the respective polymorphs. The rate of nucleation and growth can be slower for the stable form than for the metastable one. A crystallized metastable phase can transform more or less quickly into the thermodynamically stable form. The phase transformation will only occur with certainty if the stable phase is present, either as a result of nucleation or by seeding. When dealing with problems of polymorphism, it is important to consider both thermodynamic and kinetic aspects.

Particle size

Size is not always the most important product property, but there are few processes where size does not matter. Particle size can be important for

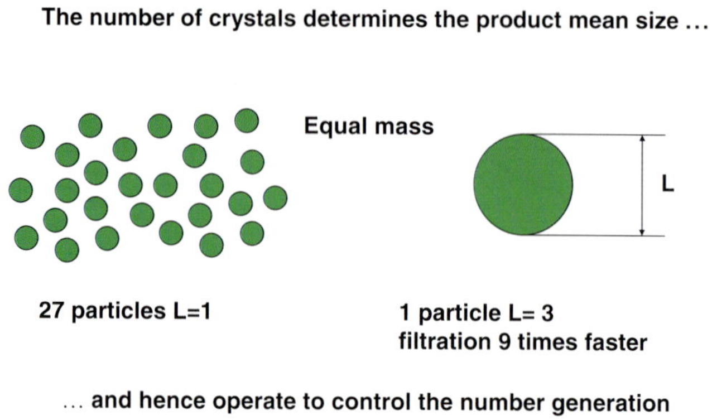

Fig. 4. Control of product crystal mean size. The product crystal mean size from a crystallization process is determined by the number of crystals that shares the total mass of crystalline material. The number of crystals is mainly controlled by the nucleation in the process. The figure compares how many particles that are required to carry the same mass if one particle with a relative diameter of three is replaced by particles with a relative diameter of unity. Filtration rate is proportional to size raised to power two and accordingly filtration is much faster when the particles are larger.

the end-use, and is definitely of importance for the processing downstream of the crystallization, e.g. filtration, washing and drying. The possibility of separating the particles from the filtrate and washing liquid is of importance for the purity. The size of the crystals from a crystallization process is controlled by the number of particles sharing the total crystal mass produced (Fig. 4).

The total number of crystals in turn depends on the number added in the form of seeds and the number formed during the process as a result of nucleation and crystal breakage. The formation of new crystals is governed by kinetic factors. In principle then, it is possible to obtain any substance at any particle size. In practice, it is of course easier or cheaper to obtain large particles for certain substances.

Crystals are not spherical, and the measured particle size will depend on the method used for determination: different techniques measure different dimensions of the particle. In addition, particles are usually not of a uniform size, but are found in a more or less wide distribution of sizes. The crystal size distribution (CSD) can be given as a cumulative distribution based on the total number, mass or surface area of the particles, as a density distribution, or as a histogram. The average crystal size can be represented by the mean size, the median size or the peak value. The mean size of crystals can refer to the number, surface area or volume (mass) distribution of the particles. The median size is the size where the cumulative distribution passes 50% of the total amount. The spread in the distribution is given as the coefficient of variation (CV), defined as the ratio of standard deviation to mean size.

Crystal shape

Together with particle size, particle shape, or *habit*, is of major importance to the filtering and washing of the crystals, and thus also to the purity of the product. Filtering of thin, flat particles is very time-consuming in comparison to filtering of more or less spherical particles. The habit of crystals can also be important depending on the use of the product. Crystal habit depends on two main factors. The inner, crystalline structure determines the faces the crystal can present; this is covered in the field of crystallography. Growth conditions, however, determine the relative size of each face, and hence also the overall shape.

Based on crystalline structure, crystalline materials are divided into 32 symmetry classes. These classes are generally divided into seven different systems, designated regular (cubic), tetragonal, orthorhombic, monoclinic, triclinic, trigonal and hexagonal. In order to be able to discuss how the crystal surfaces are related to one another, a system of coordinates is introduced with the axes x, y, z. These axes are generally not perpendicular to one another, but are aligned with the edges of the *unit cell*, a geometrically defined block encapsulating a unit of the crystal structure by which the entire crystal structure can be constructed through repetition. The seven crystal systems differ with regard to the angles between the axes and the relative length of the linear dimensions of the unit cell. Using the crystallographic axes as reference, the various crystal faces are named according to the so-called Miller indices.

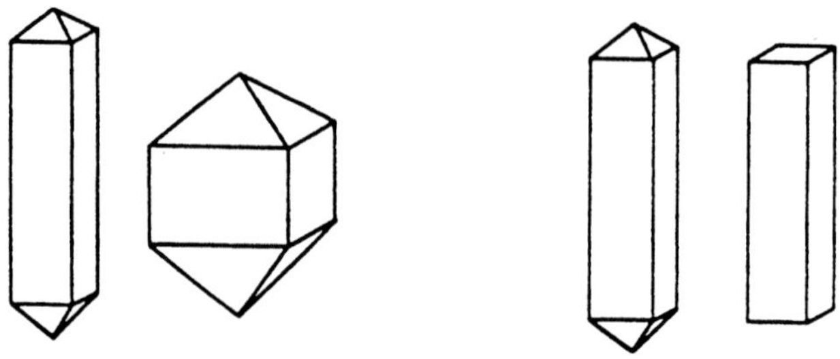

Fig. 5. Crystal habit and morphology. Habit is the same as shape and refers to the overall appearance of the particle. Morphology refers to the faces that are exposed.

The individual particles in a sample of crystalline material normally differ with regard to both size and shape. After careful analysis, however, it is generally found to be the case that all particles present the same faces (even though for certain particles, some surfaces may be so small that they are invisible). Figure 5 shows the difference between the concepts of morphology and shape (habit).

As shown in the figure, the same morphology means that the crystal structure is identical and accordingly the same faces in principle are expressed, even though their relative sizes may differ. The same shape or habit means that the overall shape of the crystals appears to be similar.

The three particles in Fig. 6 have equivalent morphology but different shape. The three particles all have the same surfaces present, but they vary in relative size. The arrows aim to illustrate the growth vectors of the faces. A long arrow means fast growing. The particle on the left is dominated by the two large, horizontal surfaces at the top and bottom, which results in a flat, tablet-like shape. The particle on the right is dominated by the elongated vertical surfaces, giving the particle a rod-like, almost needle-like, shape. The horizontal surfaces of the particle on the left are relatively large because the growth of these surfaces (perpendicular to the surface) is slow, as is illustrated by a short growth vector. The particle on the right becomes elongated because the same surfaces grow quickly in relation to the lateral surfaces. In this manner, the shape of a crystal is

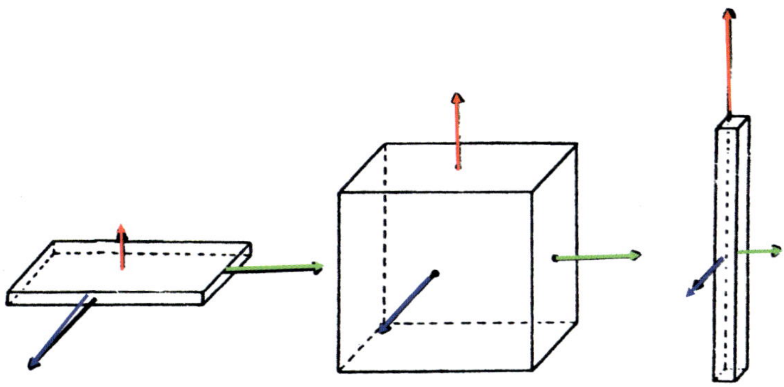

Fig. 6. Development of crystal shape. The shape is governed by the relative growth rate of different faces and is dominated by those faces having the lowest growth rate. An arrow describes the gradual movement direction of the surface when it is growing. Different arrow colors refer to the growth rates of different faces and the arrow length is proportional to the rate of growth.

largely related to the relative growth rates of the individual surfaces. Note that the growth rate of a surface relates to the linear translation of the phase boundary perpendicular to the plane of the surface. A surface that accumulates new material relatively quickly may gradually decrease its size and disappear into an edge or a point, whereas slow-growing surfaces become large and will dominate the outer shape of the crystal.

Supersaturation, temperature, impurities, pH, choice of solvent, and so on: all influence the growth rate of the individual surfaces. Hence, the relation between the growth rates of the various surfaces can be affected, and thus the shape of the final crystals may differ depending on conditions. The sensitivity of the particle shape to various variables is such that laboratory experiments can even be misleading with regard to what can be expected on an industrial scale. Frequently, impurities in very low concentrations can affect the particle shape significantly. Conversely, this means that additives can be used to control particle shape.

There is a growing interest in using additives to control the result of crystallization processes. Additives can be used to affect the shape, i.e. the relative growth rates of different surfaces, but also to influence nucleation and overall growth rate. Additives are divided into *tailor-made* and *multifunctional* types (van Rosmalen *et al.*, 1989). By identifying the molecular structure of the crystal, it is possible, at least in principle, to design additives having a specific function. The effects of tailor-made additives depend on two properties: structural similarity and specific difference. The dosage of additive that is required is quite large: about 5 wt-% if a change in shape is desired, and 10 wt-% for growth inhibition. Tailor-made additives are suitable for organic substances, which have complicated structures. Often by-products from previous reaction steps can act as tailor-made additives. Ionic compounds often have relatively simple and unspecific structures, and for such compounds, multifunctional additives, such as phosphonic acids, polycarboxylic acids polysulphonic acids and polymers with different acid groups, are used. This type of additive is generally active at concentrations in the range 10–50 ppm. Some metal ions are known for causing strong effects in crystallization processes.

Product purity

For many substances, a high purity is desired. Impurities can occur in the product in at least four different ways (de Jong, 1984): (i) process solution can adhere to the outer surfaces of the crystal, (ii) impurity molecules can be incorporated into the crystal lattice, (iii) solution can be encapsulated in cavities inside the crystal (inclusions), and (iv) impurities can be adsorbed into lattice channels and cavities (Fig. 7). In addition, solution can be encapsulated in agglomerate formations (occlusions). Depending on the way in which impurities are present in the product, different kinds of remedies are called for.

Incorporation of impurity molecules in the crystal lattice is caused by similarities between substance and impurity, and is the most difficult kind to remedy. Possibly, a reduction in growth rate can be beneficial. In a crystallizer, the occurrences of inclusions in crystals are frequently the result of too high a growth rate, or alternating growth and dissolution. One example is repeated growth and dissolution in a heat exchanger, e.g. in an oversized evaporation crystallizer. Inclusions are often the result when

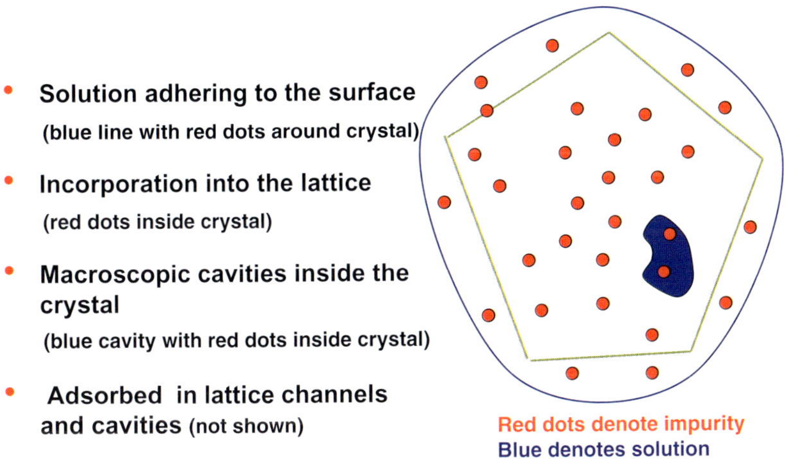

Fig. 7. Modes of impurity incorporation into a crystalline material. After separation of crystals from the solution, there is solution with impurities that remains on the surface of the crystal. Because of poor molecular separation, impurities can be integrated into the lattice. Under poor crystal growth conditions, cavities can be formed trapping impure mother solution inside the crystal. Sometimes the crystal structure contains channels and cavities in which impurities can be trapped.

damaged edges and corners are regenerated. A change in the design of the stirrer can reduce the mechanical damage the crystals are subjected to (de Jong 1984). Inclusion formation can to some extent be prevented by the presence of ionic impurities, change of solvent, increased viscosity or the usage of ultrasound.

The possibility of easily washing away the process solution adhering to the surfaces of the crystals is strongly linked to the filterability of the product. If it is hard to filter the crystals, it will be hard to wash them. As mentioned earlier, the filterability mainly depends on particle size and shape. Larger, more spherical crystals improve the possibility of washing off process solution adhering to the surface.

For some processes, agglomeration is desirable since the product particles become larger and thus easier to filter. However, the agglomerates may contain process solution, which leads to an impure product, and occlusions may be difficult to wash away. Occlusion problems must primarily be dealt with by trying to reduce the degree of agglomeration, or changing the nature of the agglomerates, making them easier to wash. Unfortunately, our knowledge concerning agglomeration is incomplete and it is not possible to give general guidelines for reducing or increasing the agglomeration in a process. Some studies have been reported in the literature, however (Åslund, Rasmuson, 1992). Agglomerates are formed as a result of (i) particles colliding, (ii) particles held together in aggregates, and (iii) the aggregates being cemented by continued crystal growth. Agglomeration generally increases with increasing supersaturation. Increased supersaturation often leads to more particles per unit volume, and increases the growth rate of crystalline "bridges." Normally, aggregation of particles is increased as the particle concentration goes up. Agglomeration depends on crystal size, and is particularly important for particles from 1 μm up to about 30 μm. In general, agglomeration benefits from reduced agitation, but in some cases agglomeration increases with increasing stirring rate, and in others there is no dependence at all. The size of agglomerates generally decreases with increased agitation. Some other important factors for agglomeration are temperature, pH, and additives/impurities. In crystallization of pharmaceuticals and fine chemicals, often the solvent can have a profound influence on the tendency for agglomeration and the properties of the agglomerates (Ålander, Rasmuson, 2007).

The Process

Supersaturation can be generated by four different methods: cooling, evaporation, addition of a substance or solvent that lowers the solubility, and chemical reaction — alone or in combination. Cooling and evaporation crystallizations are the most frequently occurring types.

In a *cooling crystallization*, the fact that the solubility normally decreases with reduced temperature is utilized. Cooling can be achieved by an indirect heat exchanger, by vacuum cooling, or through the use of direct contact cooling. In vacuum cooling, the pressure is reduced so that some solvent evaporation occurs. The heat of evaporation is taken from the solution, which is thereby cooled. Direct contact cooling has the advantage of avoiding heat exchanger surfaces, where incrustation can occur. The technique is not often used, however (Bohlin, Rasmuson, 1986). If the temperature dependence of the solubility is weak, cooling is an inefficient way of generating supersaturation, and the yield will be low. In that case, one usually turns to *evaporation crystallization* (Nyvlt, 1982). The process is similar to a normal evaporation process, but the purpose is to obtain the substance in the form of crystals. Evaporation crystallizations are normally performed under vacuum.

In another type of crystallization, a second solvent or substance is added, which causes the substance to be supersaturated. "Salting-out" is the term for when, for example, potassium chloride is made to crystallize by adding sodium chloride. The chloride concentration is increased until the solubility product is exceeded. "Drowning-out" denotes a process where a solvent is added, such as water to an alcohol, so that the solubility is lowered. In a *reaction crystallization*, the substance to be crystallized is generated *in situ* in concentrations exceeding the solubility. An example is an organic substance precipitated through the addition of hydrochloric acid. In such processes, the local supersaturation at the feed point is often very high, which makes mixing conditions particularly important.

A crystallization process can be continuous, semi-continuous or batch-wise operated (Nyvlt, 1982). Often, a batch process is simpler, easier to scale up, and less demanding in terms of maintenance and training of staff. Any remaining incrustations vanish when a new batch is started, and relatively big crystals can be obtained. When particularly narrow size distributions are desired, batch processes have the greatest potential (Larson, 1978). A batch crystallization is more flexible than a continuous one, but also more sensitive to changes in the process. Even quite small changes in solubility, supersaturation, and rate of cooling and evaporation, etc., can cause large changes in the product size. This can lead to significant variations from one batch to the next. Furthermore, batch processes are more labor intensive. In batch processes, the equipment can be thoroughly cleaned between batches, thereby avoiding contamination or seeding of unwanted polymorphs. In the pharmaceutical industry, crystallizations are generally carried out using the same equipment as is used for other unit operations and processes, i.e. an agitated tank in a multipurpose, multi-product plant. For the production of fine chemicals, there is usually some margin for adapting the batch equipment to a particular crystallization.

The product from a crystallizer is generally filtered, washed and dried. A significant degradation of the product size may occur during these steps (Jancic, Grootscholten, 1984) and it is necessary to take this into consideration when planning and designing crystallization.

The role of agitation

The purpose of stirring in a crystallizer is, among other things, to suspend the crystals from the bottom in order to promote crystal growth. Crystals piled up in stagnant zones in the equipment tend to agglomerate. In addition, stirring affects the distribution of supersaturation in the tank, perhaps mainly through the distribution of supersaturation-consuming crystals. Stirring is also important for the transport of heat from cooling surfaces: gradients can be reduced and hence the risk of incrustation. Mixing conditions affect the mass transfer between the fluid and the particles, and exert a strong influence on nucleation. Normally, in cooling or evaporation crystallization, agitation should be limited to the level where all crystals are just suspended from the bottom. As the stirring is increased, secondary nucleation increases faster than crystal growth, resulting in a lowering of the product mean size. Depending on the stirrer type, the fractions of the energy supplied as pumping and as shear and turbulence, respectively, differ (Oldshue, 1983). Suspension of particles is best achieved with an axial flow stirrer pumping downwards. The mixing energy is used more efficiently if so-called baffles are used. Some good reviews of particle suspending have been done by Nienow (1968), Mersmann *et al.* (1975) and Chapman *et al.* (1983). The equation of Zwietering (1958) is still very useful for calculating the necessary stirring for complete off-bottom suspension.

For reaction crystallizations, the situation is somewhat different. In the first place, the two reactant solutions need to be brought into contact with each other for the chemical reaction to occur. Secondly, the crystals are often relatively small, and suspension is not necessarily a major problem. Finally, the result is governed less by secondary nucleation than by primary nucleation. It is common in these discussions to distinguish the mixing action into *macro-* and *micro-mixing*, and sometimes also into *meso-mixing*. Macro-mixing comprises large-scale mixing, whereby the reactant streams are split into smaller liquid fractions, which are distributed in the vessel. Meso-mixing denotes mixing at the same approximate level as the dimensions of the feed pipe, and is of particular relevance when the process involves feeding of reactants. Micro-mixing comprises the small-scale turbulence, shearing and straining of fluid layers and the molecular diffusion that eventually brings the reactants together (Baldyga, Bourne, 1986). The greater the proportion of the supplied energy that is used to create turbulence, the better the micro-mixing becomes.

Control of the product crystal mean size

As shown earlier, the product size is governed by the number of crystals that shares the total crystal mass. The crystals in a size distribution are created by nucleation and possibly by fragmentation, or are added in the form of seeds, and grow in size in the crystal growth process. The nucleation rate is primarily decided by two factors. Both primary and secondary nucleation depend strongly on the supersaturation. Secondary nucleation also depends strongly on the hydrodynamic conditions. Both the nucleation rate and the crystal growth rate increase with the supersaturation, but the nucleation rate generally increases faster. Supersaturation is generated through cooling, for example, and is consumed through crystal growth by an increasing number of particles. Whether the total mass is divided between a small number of large particles or a large number of small particles is decided by the interplay, or competition, between nucleation and crystal growth.

The *supersaturation balance* is a mass balance with a saturated solution as reference, and it describes how the driving force for crystallization varies with the process conditions. The supersaturation in a batch process varies with time as a result of variations in consumption, and in generation. At a given supersaturation, the consumption rate increases with time since the total crystal surface area increases with time. The total crystal surface area increases partly because each crystal increases in size due to growth and partly because new crystals are generated (nucleation). If the rate of consumption is low in comparison to the rate of generation, the supersaturation increases, resulting in increasing growth and nucleation rates. This in turn leads to an increase in total crystal surface area, which together with the increased linear growth rate causes the consumption of supersaturation to increase. A mass balance for the substance in solution gives

$$\frac{d\,\Delta c}{dt} = -\frac{dc^*}{dt} + W_R - W_G, \tag{7}$$

where Δc [kg/kg inert] is the supersaturation driving force, t is time [s], and c^* is the solubility [kg/kg inert]. W_R [kg/kg inert, s] represents the change in concentration of the crystallizing compound due to processes like evaporation of the solvent or chemical reaction, and W_G [kg/kg inert, s] denotes the decrease in concentration because of crystal growth, as can be described by

$$W_G = 3\frac{k_v}{k_a}\rho_c A_T G, \tag{8}$$

where G is the linear crystal growth rate [m/s], A_T is the total surface area of crystals in the suspension [m^2/kg inert], and ρ_c is the density of the crystals [kg/m^3]. k_v — and k_a — are the volume and area shape factors, respectively, of the crystals. Normally, the consumption of supersaturation caused by the actual nucleation can be neglected. For a normal batch cooling crystallization, W_R is zero, and

$$\frac{dc^*}{dt} = \frac{dc^*}{dT}\frac{dT}{dt}, \tag{9}$$

where T is the temperature [K]. In *batch cooling crystallization* practice, the cooling is often achieved by pumping coolant water of constant temperature through the cooling jacket: *natural cooling*. The driving force for heat transfer, i.e. the temperature difference, is largest at the beginning, and declines towards the end of the crystallization. This leads to the temperature falling quickly at the beginning, and slower towards the end. The normal thing therefore is for a massive, uncontrolled primary nucleation to occur early, resulting in a small product size. Large variations will be observed from one batch to another. If large crystals are desired, a careful generation of supersaturation is called for, in particular at the beginning of the process. As the crystal surface area consuming the supersaturation increases, it is possible to generate supersaturation faster. This procedure is called *controlled cooling* (Mullin, Nyvlt, 1971) and the main principle is that careful cooling should be employed initially, followed by a gradual increase in cooling rate as the total crystal surface area increases. In Fig. 8, a few different controlled cooling profiles are shown, together with two examples of natural cooling profiles.

In the case of natural cooling, supersaturation peaks very high and marked, leading to a powerful primary nucleation early on in the process.

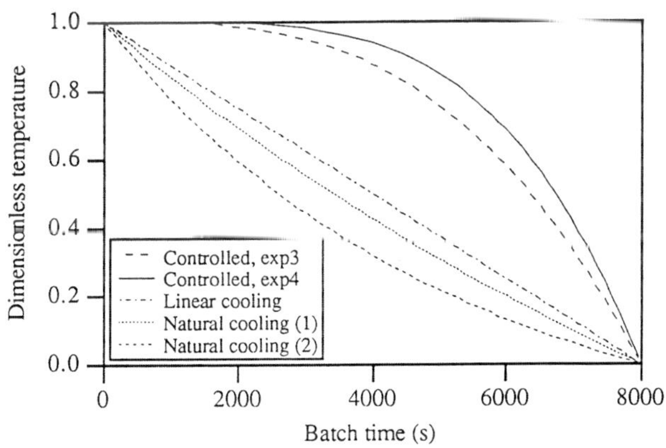

Fig. 8. Different cooling profiles in batch cooling crystallization. The diagram shows the temperature of the solution as $(T_0-T)/(T_0-T_f)$ versus time when the batch is cooled from the initial temperature, T_0, to the final temperature, T_f. T is the current temperature. In controlled cooling, different exponents, α, can be used in the equation describing the cooling profile: $(T_0-T)/(T_0-T_f) = (t/t_f)^\alpha$, where t_f is the total cooling time. Natural cooling is described in the text and is given in the figure for two different final temperatures at equal cooling water temperature.

After the initial nucleation peak, the solid mass increases rapidly since the supersaturation is still high. The supersaturation is quickly consumed, and the primary nucleation subsides. Due to the increasing magma density, secondary nucleation starts, but is stabilized at a fairly low level. With controlled cooling, the cooling rate is better adjusted to the surface area that is available for consumption of the supersaturation. For a fairly long time, a crystal surface is gradually developed that is able to counteract a fast increase in supersaturation. The peak in supersaturation occurs late, and is much lower. Due to higher supersaturation late in the process in comparison with natural cooling, the secondary nucleation is substantially higher towards the end. With seeding, the early peak in supersaturation is further reduced, and thereby also the primary nucleation. The seed crystals provide area at the beginning of the process and contribute to keeping the supersaturation level under control. With seeding, the size distribution may become bimodal, i.e. it will consist of two distinct peaks corresponding to the seeded crystals and the crystals formed through nucleation. The mean weight size and the CV depend on the amount of seeds added.

For controlled cooling, a long period of time is used for cooling just a few degrees early in the process. It is therefore crucial that the controlled cooling starts rather precisely when the solution is saturated. When seeds are used, it is important that these are added when the solution is within the metastable zone. If the solution is undersaturated, the seeds will simply dissolve. In practice, it is often hard to know exactly when the solution is saturated, as the solubility can vary from one batch to another, and the precision of temperature measurement is not always sufficiently good in terms of supersaturation. Simulations of batch crystallizations show that the process is very sensitive to initial supersaturation, the shape of the cooling curve, and seeding (Bohlin, Rasmuson, 1992).

Seeding is used to control the product size in batch crystallizations, and to improve the reproducibility. There are three different principles for choosing the number and size of the seed crystals. One strategy is to add just the number of crystals that are desired in the end product, of a size that will lead to crystals of the required size after growth. This method demands that the process is operated so that nucleation is avoided. A second strategy is to add seed crystals which, through secondary nucleation, will generate the desired number of crystals. In this case, the seed crystals have to be large enough to spawn nuclei efficiently (Mullin, 1993). The number of seeds to be added depends on nucleation kinetics and on the way the process is run. Primary nucleation is undesirable. The third strategy is to render the primary nucleation more reproducible, by so-called induction seeding (Moore, 1994). Usually, in practice, it is recommended to add a proportion of seeds corresponding to a few percent by mass of the production.

Analogously to cooling crystallization, "controlled conditions" can be achieved by controlling the rate of evaporation (Larson, Garside, 1973) or by controlling the feed rate in, for example, *drowning-out crystallizations* (Tavare, Chivate, 1980). The supersaturation should be generated very slowly to start with, and increase along with the available crystal surface area. In drowning-out crystallization, there is a possibility that local feed point conditions affect the product. It is also important to bear in mind that in such processes the chemical composition of the solution changes gradually during the process.

A reaction crystallization process is often significantly more complicated. During a reaction crystallization, very high supersaturation levels can frequently arise in the region where the reactants are mixed. In reaction crystallization, either one reactant solution is added to a stirred solution of the other reactant, or both solutions are added simultaneously. The effect of process variables such as the type of stirrer, stirring rate, location of the feed point, reactant concentrations, and the rate of addition, have been studied for the former process type by Åslund and Rasmuson (1992). Benzoic acid was crystallized by the addition of hydrochloric acid to a solution of sodium benzoate. The results show that the average size of the crystals produced increases with increasing agitation rate, passes through a maximum, and then decreases. The produced crystals are larger if the feed point is located close to the impeller instead of at the surface. At any given level of supplied agitation energy, different impellers give the same product size. It has been found that it is possible to show all the results of the effects of hydrodynamic conditions on one single diagram, where the product size is plotted against the feed point mixing energy dissipation rate. In addition, the results show that low rates of reactant addition, and more importantly a low reactant concentration, result in a larger product size. The fact that the product size increases if the reactant concentrations are reduced can be explained by the fact that lower reactant concentrations lead to lower supersaturation at the feed point, which in turn leads to a lower nucleation rate. Fewer particles will be sharing the crystallized mass. The local mixing intensity is at the maximum close to the impeller. At the surface (where addition frequently takes place in industrial processes) the mixing is often poor. However, too much mixing will result in fragmentation and breakage of the crystals.

Crystallizers and crystallization processes cannot be designed simply theoretically, any more than they can be designed only on the basis of experiment. At present, the knowledge required to design, construct and run a full-scale crystallizer must be obtained from well-designed laboratory experiments on the system in question, an interpretation of the results based on fundamental crystallization principles, relevant half-scale experiments, and sometimes even from supplementary full-scale experiments.

Conclusions

In this chapter, we have presented the fundamentals of crystallization as it applies to processing of fine organic chemicals and low molecular pharmaceuticals. In addition some aspects on processing have been discussed. Crystallization of larger molecules and proteins are governed by the same fundamental mechanisms. Hence, the presentation here may contribute to a mechanistically sound analysis and development of the crystallization of proteins.

References

Baldyga J, Bourne JR. (1986) Principles of micromixing. In NP Cheremisinoff (ed). *Encyclopaedia of Fluid Mechanics*, 1.

Bohlin M, Rasmuson ÅC. (1986) Direct Contact Cooling Crystallization; 1. C. Report B4 to IKF, Department of chemical engineering, KTH.

Bohlin M, Rasmuson ÅC. (1992) Application of controlled cooling and seeding in batch crystallization. *Can J Chem Eng* **70**: 120–126.

Chapman CM, Nienow AW, Cooke M, Middleton JC. (1983) Particle-gas-liquid mixing in stirred vessels. Part I: Particle-liquid mixing. *Chem Eng Res Design* **61**: 71–81.

Girolarni MW, Rousseau RW. (1985) Size-dependent crystal growth — A manifestation of growth rate dispersion in the potassium alum-water system. *Am I Chem Eng J* **31**: 1821–1828.

Jancic SJ, Grootscholten PAM. (1984) Effect of external classification on product size distribution in large crystallizers. *Industrial Crystallization*. Delft University Press.

Jong de EJ. (1984) The one or the other. In SJ Jancic & EJ de Jong (eds). *Industrial Crystallization* 84, Elsevier.

Larson MA. (1978) Guidelines for selecting a crystallizer. *Chem Eng* Feb **13**: 90–102.

Larson MA. (1984) Advances in the characterization of crystal nucleation. *Am I Chem Eng Symp* **Ser. 80**(240): 39–44.

Larson MA, Garside J. (1973) Crystallizer design techniques using the population balance. *The Chemical Engineer* **274**: 318–327.

Levins DM. Glastonbury JR. (1972). Particle-liquid hydrodynamics and mass transfer in a stirred vessel. 2. Mass transfer. *Trans I Chem* **E50**(2): 132–146.

Mersmann A, Einenkel WD, Käppel M. (1975) Design and scale-up of agitators. *Chem Ing Techn* **47**: 953–963.

Moore WP. (1994) Optimize batch crystallization. *Chem Eng Prog* **90**: 73–79.

Mullin JW. (1979) *Crystallization*. In *Kirk-Othmer Encyclopedia of Chemical Technology*, 3rd ed., p. 249. John Wiley & Sons.

Mullin JW. (1993) *Crystallization*. Butterworths, London.

Mullin JW, Jancic SJ. (1979) Interpretation of metastable zone widths. *Trans I Chem Eng* **57**: 188–193.

Mullin JW, Nyvlt J. (1971) Programmed cooling of batch crystallizers. *Chem Eng Sci* **26**: 369–377.

Nienow AW. (1968) Suspension of solid particles in turbine agitated baffled vessels. *Chem Eng Sci* **23**: 1453–1459.

Nyvlt J. (1982) *Industrial Crystallization, The State of the Art.* Verlag Chemie.

Nyvlt J, Söhnel O, Matuchova M, Broui M. (1985) *The Kinetics of Industrial Crystallization.* Chemical Engineering Monographs, Vol 19. Elsevier, Amsterdam.

Oldshue JY. (1983) *Fluid Mixing Technology.* McGraw-Hill.

Rosmalen van GM, Witkamp WJ, Vreugd de CH. (1989) In J Nyvlt & S Zacek (eds). *Industrial Crystallization* 87, p. 15. Elsevier.

Shah BC, McCabe WL, Rousseau RW. (1973) Polyethylene vs. stainless steel impellers for crystallization processes. *Am I Chem Eng J* **19**: 194.

Tavare NS, Chivate MR. (1980) CSD analysis from a batch dilution crystallizer. *J Chem Eng Japan* 13: 371-379.

Zumstein RC, Rousseau RW. (1987) Growth rate dispersion by initial growth rate distributions and growth rate fluctuations. *Am I Chem Eng J* **33**: 121–129.

Zwietering ThN. (1958) Suspending of solid particles in liquid by agitators *Chem Eng Sci* **8**: 244–253.

Ålander E, Rasmuson ÅC. (2007) Agglomeration and adhesion free energy of paracetamol crystals in organic solvents. *Am I Chem Eng J* **53**: 2590–2605.

Åslund B, Rasmuson ÅC. (1992) Semibatch reaction crystallization of benzoic acid. *Am I Chem Eng J* **38**: 328–342.

Chapter 10

Myelin Basic Protein, A Saucy Molecule With High Responsiveness to the Environment or Just an Unusual Membrane Protein?

*Paolo Riccio**

Myelin basic protein (MBP) is a major component of the myelin sheath in the central nervous system. Since its purification in 1969, MBP has been studied very extensively for its role in myelin compaction and its possible involvement in multiple sclerosis. All attempts to crystallize this protein have failed and its tertiary structure remains undefined. Circular dichroism, NMR and EPR spectrometry, SAXS, SANS and other techniques have clarified that MBP might have a compact C-shaped form with a core element of β-sheet structure when associated with lipids. Nevertheless, classic MBP is still considered a water-soluble and natively unfolded peripheral myelin protein, which is very flexible and prone to many post-translational modifications.

In this chapter, we review the properties of MBP and postulate that MBP is a very special membrane protein that can exist in several forms and in different myelin domains, and not an intrinsically unstructured protein as it appears to be in water. Relevant for the multiple MBP forms and the different locations and functions are the post-translational modifications of MBP, rather than its intrinsic malleability and high responsiveness to environmental changes.

Keywords: Myelin; myelin basic protein; lipids; lipid rafts; protein structure; crystallization; intrinsically unstructured proteins; central nervous system.

*Dipartimento di Biologia D.B.A.F., University of Basilicata, Via Ateneo Lucano 10, 85100 Potenza, Italy, and Istituto Nazionale Biostrutture e Biosistemi, Consorzio Interuniversitario, Viale Medaglie d'Oro 305, 00136 Roma, Italy. Email: paolo.riccio@unibas.it.

Introduction

The physiological environment of MBP is the myelin membrane. About 95% of MBP is associated with lipids and requires a detergent for its extraction. The membrane-bound MBP can be found within or outside lipid rafts, depending on methylation, phosphorylation, deimidation and other chemical modifications of its residues. Part of MBP cannot be extracted even with SDS and urea. In all lipid-bound MBP preparations, the protein appears to be structured. Thus, MBP structure could be best analyzed after reincorporation of the protein in liposomes made of myelin lipids. In conclusion, rather than being an intrinsic unstructured protein, as suggested in recent years, MBP is a membrane protein with an easily modifiable structure.

The classic acid-extracted, water-soluble MBP is really a mixture of different isomers, and this might explain why MBP has never been crystallized. New crystallization attempts could be made using as starting material the protein purified from specific membrane microdomains, since in this case MBP may have the same post-translational modifications and be more homogeneous. Then, the knowledge of MBP structure will lead to a better understanding of myelin structure and of its degradation in demyelinating diseases.

Myelin Basic Protein: What Is It and What Does It Do?

The myelin sheath of the central nervous system (CNS) is the lipid-rich, multi-lamellar membrane process extending from the oligodendrocytes to cover all around segments of the nerve axon and to facilitate rapid and efficient nerve conduction (Kirschner, Blaurock, 1992; Moscarello, 1996).

Myelin basic protein (MBP) is the second major protein in CNS myelin and perhaps the most studied among its components (Smith, 1992; Harauz *et al.*, 2004). Human MBP is present in different isoforms — of 17, 18.5, 20.2, 21.5 kDa — deriving from the alternate splicing of the mRNA transcript (Givogri *et al.*, 2000). The MBP gene, which is made of seven exons, belongs to the larger Golli gene (Givogri *et al.*, 2001).

In particular, the 18.5 kDa MBP, the major human isoform, is an interesting protein to study for different reasons:

(i) It is believed to have the key role of biological "glue" in the formation, compaction and maintenance of the multi-lamellar structure of the myelin sheath (Privat *et al.*, 1979; Riccio *et al.*, 1986; Readhead *et al.*, 1987; Riccio *et al.*, 2000).
(ii) It is a candidate autoantigen in the context of multiple sclerosis (MS) research, since it can induce the Experimental Allergic Encephalomyelitis (EAE), an animal model of MS (Alvord *et al.*, 1984; Massacesi *et al.*, 1993; Gold *et al.*, 2006).
(iii) Its possible relationship with the IUPs, the "intrinsically unstructured proteins" or "natively unfolded proteins" (Harauz *et al.*, 2004) and the difficulty to achieve its crystallization (Sedzik, Kirschner, 1992).

The IUPs

IUPs (intrinsically unstructured proteins) are proteins lacking a precise folded structure (Wright, Dyson, 1999; Tompa, 2002). Under physiological conditions (the native and functional state), IUPs are present with a highly flexible, random coil structure but are prone to adopt a folded conformation upon binding to different ligands and consequently switch to distinct, unrelated functions (Tompa, 2005). Structural disorder is then the basis for a protein to fulfil different tasks. In general, IUPs are mainly hydrophilic, charged molecules, and do not possess a hydrophobic core. Their amino acid composition is different from the average amino acid distribution of globular proteins reported in the Protein Data Bank (Dawson *et al.*, 2003).

The assignment of MBP to the IUPs was made on the basis of its high net charge, low hydrophobicity and the claimed absence of structure in the native form (Harauz *et al.*, 2004). However, with respect to amino acid usage, MBP differs from IUPs in the content of A, R, E, G, H, F and V.

MBP Ligands and Functions

On the other hand, when looking to its capability to bind a number of different ligands (Campagnoni, Skoff, 2001), MBP appears to be a good candidate as an IUP. MBP can bind many divalent cations, in particular zinc, a physiological ligand present in myelin (Earl *et al.*, 1991; Cavatorta *et al.*, 1994; Riccio *et al.*, 1995; Tsang *et al.*, 1997). MBP can also bind small molecules such as the heme group (Vacher *et al.*, 1984; Morris *et al.*, 1987); azo compounds (Liebes *et al.*, 1976); serotonin and hallucinogens (Alivisatos *et al.*, 1971; Carnegie *et al.*, 1972) and GTP (Chan *et al.*, 1988).

Binding of MBP to detergent and lipids has been studied in depth, although MBP was considered to be an extrinsic, lipid-free and water-soluble protein (Riccio *et al.*, 1984; Smith 1992; Riccio, Quagliariello, 1993; Beniac *et al.*, 1997; Hu *et al.*, 2004; Harauz *et al.*, 2004). MBP can also bind to other myelin proteins: myelin proteolipid protein [(PLP) (Golds, Braun, 1978; Boggs, Wang, 2004)], and myelin 2′,3′-cyclic nucleotide 3′-phosphodiesterase [(CNP) (Dyer, Benjamins, 1989; Richter-Landsberg, 2001)]. MBP can bind to polyanionic proteins: calmodulin [(CaM) (Chan *et al.*, 1994; Polverini *et al.*, 2004)]; actin (Barylko, Dobrowolski, 1984; Boggs *et al.*, 2005); tubulin (Modesti, Barra, 1986; Gendreau *et al.*, 2003) and clathrin (Boggs, 2006). MBP can also bind to alfa$_2$-macroglobulin (Gunnarson *et al.*, 1998, 2003) and heat shock protein 70 (Aquino *et al.*, 1998; Cwiklinska *et al.*, 2003). Adherence of MBP to T cells has been reported (Bobba *et al.*, 1991).

According to their ability to interact with different ions, molecules and cells, all MBP isoforms seem to fulfil different functions in myelin (Boggs, 2006), and seem to be involved also in apparently unrelated activities outside the CNS (Harauz *et al.*, 2004). In fact, besides the main role in compacting myelin, MBP may be involved in various different functions, such as modulation of signal transduction pathways (Dyer, 1997; Campagnoni *et al.*, 2003), and induction of insulin and glucagon release from the pancreas (Kolehmainen, Sormunen, 1998).

Post-Translational Modifications of MBP

Up to the present time, very little is known about the native three-dimensional MBP conformation. This can be ascribed also to the difficulty to crystallize MBP, even after an exhaustive number of experiments [(Sedzik, Kirschner, 1992); BIOMED-2 (1996–1999)]. As we will discuss later, this is certainly due to the high heterogeneity of MBP isoforms as many charge and mass isomers derive from numerous post-translational modifications [N-terminal acylation (Ala 1), phosphorylation (Ser 7, 12, 19, 56, 71, 102, 115, 136, 151, 161, 163, 165; Thr 17, 20, 95, 98, 149), deimination (Arg 5, 9, 25, 31, 33, 43, 49, 54, 65, 79, 97, 113, 122, 130, 159, 162, 169, 170), deamidation (Gln 8, 81, 103, 121, 147), methylation (Arg 107, 159) oxidation (Met 21), insertion of GalNac (Thr 95, 98) and ADP ribosylation (Arg 9, 54)] with reference to the human MBP sequence (Harauz *et al.*, 2004; DeBruin, Harauz, 2007).

The main result of post-translational modifications of the 18.5 kDa MBP (pI = 10.6) is the occurrence of charge isomers termed C1-C2-C3-C4-C5-C6-C8 and the consequent reduction of net positive charge from +19 to +13 from C1 to C8. C1 and C2 are the least modified and most positively charged isomers. The deiminated (citrullinated) component C8 occurs in greater amounts in patients with MS (Moscarello *et al.*, 1994). Besides post-translational modifications, association with different types of ligands (including lipids) and type of location are a further cause of great diversification of MBP.

Post-translational modifications may serve to direct MBP to different environments: for example, the 20.2 and 21.5 kDa isoforms of MBP move to the nucleus if they become phosphorylated (Pedraza *et al.*, 1997), whereas the distribution of the 18.5 kDa MBP in the membrane is dependent on the type of modification (phosphorylation, methylation or deimination) (DeBruin *et al.*, 2005; Carlone *et al.*, 2005; DeBruin, Harauz, 2007).

Is MBP a Natively Unfolded Protein?

The absence of structure in the acid-extracted molecule

To decide whether MBP is a natively unfolded protein, it is necessary to establish what its native environment is. MBP was purified for the first time in 1969 after extraction from defatted brain powder with HCl at pH below 3 (Oshiro, Eylar, 1970). The final procedure was based on treatment with urea and use of ion-exchange chromatography (Deibler *et al.*, 1972, 1984). Purified MBP — now referred to as classic MBP — was shown to be an unfolded, water-soluble, and lipid-free protein (LF-MBP). In the first report on the properties of the acid-extracted MBP, the protein was described as a very asymmetric and extended ellipsoidal molecule with an axial ratio near 10:1 (Eylar, Thompson, 1969). MBP appeared to be very resistant to denaturation, with no significant role for hydrophobic and hydrogen bonding in maintaining its structure. Other studies on the structure of LF-MBP in aqueous solution were in agreement with a flexible coil conformation of the protein (Krigbaumand, Hsu, 1975; Gow, Smith, 1989; Smith, 1992). On these grounds, the protein extracted with HCl might be assigned to the family of IUPs.

The question may arise now whether this image of MBP may be a reasonable representation of a membrane-associated, but possibly extrinsic, protein or, in general, of a protein in a functional form. In other words, one is left with the question whether the unfolded state and the absence of any detectable biological activity in water correspond to the native physiological environment of MBP, or whether this state is a consequence of the drastic conditions of the extraction procedure from the myelin membrane. In fact, there is no doubt that MBP is a protein associated to myelin and should be considered, as it is, a membrane protein.

MBP as a membrane protein: The discovery of lipid-bound MBP

Several studies have shown that LF-MBP can interact with detergents, lipids, and other molecules, thereby assuming a more ordered structure (Anthony, Moscarello, 1971; Burns *et al.*, 1981; Smith, 1982, 1992; Beniac *et al.*, 1997; Haas *et al.*, 1998; Facci *et al.*, 2000; Polverini *et al.*, 2003; Hu *et al.*, 2004; Cristofolini *et al.*, 2005; Rispoli *et al.*, 2007; Haas *et al.*, 2007). Binding of MBP to lipids is in accord with the view that lipid-protein interactions are critical for the stability of the myelin sheath (Boggs *et al.*,

1982; Smith, 1992; Staugaitis *et al.*, 1996; Riccio *et al.*, 2000), and that both components work synergistically to provide the adhesion and overall structure (Hu *et al.*, 2004).

In the 1980s, we introduced the idea that, if MBP is a membrane protein, it should be extracted with detergents and not with strong acids. Furthermore, we suggested that MBP should be extracted directly from myelin and not from whole brain.

Following the treatment of myelin with mild detergents, MBP was extracted together with its natural lipid environment. Purified MBP was found to be associated to almost all myelin lipids and was called lipid-bound MBP (LB-MBP) (Riccio *et al.*, 1984). MBP extracted with the detergent CHAPS was found to be associated mainly with phosphoglycerides (Riccio *et al.*, 1994). After the discovery of myelin rafts, it was clear that the CHAPS-extracted LB-MBP was MBP present in the non-raft region of myelin.

LB-MBP was found to differ in various structural and immunological aspects from the LF-MBP (Bobba *et al.*, 1991; Lolli *et al.*, 1993; Massacesi *et al.*, 1993; Liuzzi *et al.*, 1996; Vergelli *et al.*, 1997; Mazzanti *et al.*, 1998). Circular dichroism measurements showed that LB-MBP has a much higher proportion of ordered secondary structure than LF-MBP, which was a substantially random coil protein (Polverini *et al.*, 1999). Small angle X-ray scattering (SAXS) studies on LB-MBP in solution confirmed the CD studies (Haas *et al.*, 2004). According to the results obtained with CD and SAXS measurements, the structural model of Beniac *et al.* (1997) and Ridsdale *et al.* (1997) was modified. The Beniac–Ridsdale model of the human MBP in the presence of lipids (entry 1qcl in the Protein Data Bank), built up using both experimental (electron microscopy) and computational techniques, was a C-shaped structure. The main differences between our model and the 1qcl model are represented by the replacement of two coil segments (residues 61–66 and 131–136), which lie at the two ends of the C-shaped model with α-helical structures, although the characteristic C-shape is maintained.

With LB-MBP and lipids, self-organization of stable, myelin-like membranes could be induced under conditions in which lipids alone remained poorly organized (Riccio *et al.*, 1986, 2000). A comparison between LF-MBP and LB-MBP has been discussed (Riccio, Quagliariello, 1993).

Partitioning of MBP in the Myelin Membrane

According to the new model of biological membranes proposed by Simons (Simons, Ikonen, 1997), it was shown that myelin is made of lipid platforms consisting of glycosphingolipids and cholesterol microdomains called *lipid rafts*, moving in a more fluid *non-raft* region consisting prevalently of glycerophospholipids (Taylor *et al.*, 2002). Lipid rafts can be involved in a series of biological process, such as signal transduction pathways, apoptosis, cell adhesion, and protein sorting (Brown, London, 1998; Simons, Toomre, 2000).

Raft and non-raft microdomains are usually recognized by their different solubility in detergents. Non-raft microdomains are soluble in mild detergents as CHAPS or Triton-X100, whereas lipid rafts are not. Accordingly, purified myelin proteins extracted with CHAPS have been found to be associated mainly with glycerophospholipids such as PS and PE. LB-MBP was the first example of a myelin protein belonging to the non-raft microdomains.

Further studies have shown that MBP is distributed in all different microdomains of myelin (DeBruin *et al.*, 2005; DeBruin, Harauz, 2007). Accordingly with this finding, we have extracted different MBPs from myelin under different conditions. First of all, MBP was extracted with salts (0.5 M NaCl) (*salt-extracted MBP*). The second treatment was based on the use of CHAPS to extract MBP from the salt-treated myelin residue (*non-raft LB-MBP*). Thereafter, the CHAPS residue was treated with SDS to extract MBP from the raft domains (*soluble raft LB-MBP*). Part of MBP was not extracted at all and remained in the so-called *hyper-raft* domain. As assessed by high performance thin layer chromatography (HPTLC), all MBPs extracted with detergents or present in the detergent-insoluble *hyper-raft* domain, were found to be associated to lipids, but the lipid pattern was different. Non-raft MBP (the classical LB-MBP) was extracted and purified with phosphoglycerides, the typical lipids of the non-raft domain, whereas MBP belonging to the raft and the hyper-raft domains was mainly associated to glycosphingolipids and cholesterol.

What changes MBP distribution of MBP in the myelin membrane are some post-translational modifications. For example, raft MBP is phosphorylated and non-raft MBP is methylated (DeBruin, Harauz, 2007).

Selective extraction of MBP demonstrates that MBP has different partitioning and, as a consequence, different roles in the myelin membrane. Classic acid-extracted MBP is actually a mixture of MBP with high heterogeneity, and this might be the reason why the protein has been never crystallized.

Incorporation of myelin basic protein in liposomes made of myelin lipids

All our studies on the MBP extracted from the membrane within its native lipid environment and purified with bound lipids indicate that the protein is structured. All detergent-extracted, lipid-bound MBPs can unfold following a denaturing treatment. Moreover, it can be affirmed that the MBPs distributed in different membrane domains have been subjected to different chemical modifications and have different structures and functions.

On these grounds, MBP can no longer be seen as an IUP or as a natively unfolded protein, but rather as a flexible, malleable and multifunctional protein that can be adapted to different environments following precise chemical modifications.

Time is now ripe for the study of MBP structure but restricting the investigation to MBP extracted and purified from the different myelin domains. However, since detergents have without doubt an influence on protein structure, we suggest that MBP structure should be studied after removal of the detergent used for its extraction and incorporation in the original lipid mixture.

Therefore, we have tried to reconstitute the MBP in its original environment, reincorporating the purified protein into liposomes made by the original natural myelin lipids, after removing the detergent (Riccio *et al.*, 2002). Myelin lipids were recovered by treatment of brain powder with organic solvents according to the procedure of Deibler *et al.*, (1972), or by removal of lipids from LB-MBP using a $CHCl_3/CH_3OH$ 2:1 mixture.

Reconstitution of MBP into liposomes was achieved following slightly different procedures, all based on the replacement of detergent (removed) by myelin lipids (added). Classic, acid-extracted LF-MBP and non-raft LB-MBP were studied. After the reconstitution procedure, SDS-PAGE showed that both the LB-MBP and the LF-MBP remain intact. Densitometric analysis of the lipids, extracted from both MBP forms and fractionated by HPTLC, revealed that the reconstituted LB-MBP was enriched in lipids with respect to the same protein in solution. In contrast, LF-MBP was found to bind only a very low percentage of lipids.

Using electron microscopy and spectroscopic techniques, we have investigated the structure of both lipid-free and lipid-bound MBP, after reconstitution in liposomes made of myelin lipids, in order to verify the differences in their interaction with liposomes and to evaluate the effects on their structure of the reconstitution in a lipid microenvironment very similar to the native one. The aim of this study was to understand not only the kind of protein-lipid interaction, but also how much this interaction affects the correct functional folding of the protein.

Electron microscopy showed that LF-MBP was adherent to the liposome surface without penetrating the membrane. The CD measurements carried out to investigate the secondary structure of LF-MBP (Table 1), did not show relevant structural changes on LF-MBP in the presence of myelin liposomes. In contrast, comparing the electron microscopy results obtained on the LB-MBP in solution and in the presence of myelin liposomes, the protein seemed to penetrate into liposomes. In addition, the CD spectra showed a higher amount of ordered secondary structure than in the LB-MBP with the detergent, indicating a transition to a higher content of β structure (Table 1). Interaction of LF-MBP or LB-MBP in liposomes made of dimyristoyl-lα-phosphatidic acid (DMPA) increased the α-helix content in both proteins (not shown). A high percentage of random coil was observed only in the lipid-free MBP.

Table 1. Summary of Circular Dichroism (CD) Experiments Carried Out as shown in Polverini *et al.*, (1999)[a]

	% α-helix	% β-sheet (antiparallel)	% β-turn	% Random Coil	% Other Structures
LF-MBP dialyzed versus 20 mM Tris, pH = 8.0	10.0	9.0	15.4	56.8	8.8
LF-MBP in myelin liposomes	14.1	12.7	24.3	43.9	4.9
LF-MBP in 20 mM HEPES pH = 7.0	5.1	6.0	35.9	52.7	0.3
LF-MBP + CHAPS	8.6	7.2	32.4	51.7	0.1
LB-MBP dialyzed versus 0.5% CHAPS/20 mM Tris + additives, pH = 8.0	28.1	9.6	29.4	23.3	7.5
LB-MBP in myelin liposomes	17.6	22.6	27.7	29.7	2.3
LB-MBP + CHAPS	23.8	20.3	27.9	28.0	0.0

[a] The minimum lipid content of the non-raft LB-MBP after dialysis was 1.5 mg phospholipids/mg protein. The relative amounts of secondary structure in the proteins were evaluated with the convex constraint analysis (CCA) method, developed by Perczel *et al.* (1991, 1992).

Concluding Remarks

Knowledge of the structure of myelin proteins and in particular of MBP is very important to understand the complex myelin structure, but also to understand myelin degradation in MS and to devise effective strategies for therapeutic management. However, we do not know the native three-dimensional MBP conformation since the protein has never been crystallized. This failure can be ascribed to the fact that the starting material was not homogeneous but rather a mixture of MBPs (classic, acid-extracted and water-soluble MBPs), never being able to re-attain their inherent structure, even when reconstituted with lipids.

It is now clear that MBP exists in the form of multiple isomers distributed in different microdomains of myelin. About 90–95% of the MBPs present in myelin require detergent for their extraction. Only some MBP (about 5%) is extracted by salts, as occurs in the case of extrinsic membrane proteins. MBP is found also in detergent-resistant domains, a finding that is not expected for a water-soluble protein such as classic MBP. What determines the interaction of MBP with a specific myelin domain are the post-translational modifications and not the innate adaptation of MBP to a certain environment.

Another important aspect to be taken into consideration is that the detergent-extracted MBPs, purified with their original lipid environment, were found to have a noticeable amount of ordered secondary structure and to differ not only in functional but also in structural aspects from the corresponding lipid-free form. The data on the conformation of MBP in its native lipid environment in solution point toward a compact, but not spherical, C-shaped protein-lipid complex with regions of different electron density (Haas *et al.*, 2004). This is in accord with the idea that the isolation of specific protein-lipid complexes may be the more desirable goal for structural and functional studies of membrane proteins (Garavito, Ferguson-Miller, 2001), supposing they are structured and stable (Rosenbusch, 2001).

Since the presence of MBP in the different regions of myelin depends on its chemical modifications and these are certainly reproducible, it may be suggested that the extraction and the purification of MBP-lipid-detergent complexes may represent a better (more homogeneous with respect to the acid-extracted MBP) starting material for crystallization attempts. Moreover, removal of detergent and incorporation of MBP into liposomes may represent a further promising step in these studies.

Acknowledgments

This study is an outcome of the MARIE Network of the European Science Foundation on *Myelin Structure and Its Role in Autoimmunity*, 2004–2006.

The author gratefully acknowledges the work of Anna Fasano, Grazia Maria Liuzzi and Giulia Carlone, University of Bari, Italy; Antonella Bobba, IBBE, CNR, Bari, Italy; Eugenia Polverini, Paolo Cavatorta and

Marco P. Fontana, University of Parma, Italy; Ranieri Rolandi, University of Genova, Italy; Francesca Natali, OGG-INFM, Grenoble Cedex, France; Rocco Rossano, University of Basilicata, Potenza, Italy; Iris L. Torriani and Cristiano L. P. Oliveira, Campinas, Brazil; and Heinrich Haas, MediGene AG, Munich, Germany; Daniel Kirschner, Boston College, MA, USA.

Funding by the Italian Foundation for Multiple Sclerosis (FISM) is gratefully acknowledged.

References

Anthony JS, Moscarello MA. (1971) A conformation change induced in the basic encephalitogen by lipids. *Biochim Biophys Acta* **243**: 429–433.

Alvord EC, Kies MW, Suckling AL. (1984) *Experimental Allergic Encephalomyelitis: A Useful Model for Multiple Sclerosis*. Alan R. (ed). Liss Inc, New York.

Aquino DA, Peng D, Lopez C, Farooq M. (1998) The constitutive heat shock protein-70 is required for optimal expression of myelin basic protein during differentiation of oligodendrocytes. *Neurochem Res* **23**: 413–420.

Barylko B, Dobrowolski Z. (1984) Ca^{2+}-calmodulin-dependent regulation of F-actin-myelin basic protein interaction. *Eur J Cell Biol* **35**: 327–335.

Beniac DR, Luckevich MD, Czarnota GJ, Tompkins TA, Ridsdale RA, Ottensmeyer FP, Moscarello MA, Harauz G. (1997) Three-dimensional structure of myelin basic protein. I. Reconstruction via angular reconstitution of randomly oriented single particles. *J Biol Chem* **272**: 4261–4268.

BIOMED 2. High Resolution Structures of Myelin Proteins. (1996–1999). Cost-sharing contract. Participants: Guy Ourisson, Strasbourg, France Jean-Marie Ruysschaert, Bruxelles, Belgium; Dino Moras, Illkirch, France; Wilhelm Stoffel, Koeln, Germany; Jacques Gostelli, Budendorf, Switzerland; Paolo Riccio, Potenza, Italy; Jurg Rosenbusch, Basel Switzerland.

Bobba A, Munno I, Greco B, Pellegrino NM, Riccio P, Jirillo E, Quagliariello E. (1991) On the spontaneous adherence of myelin basic protein to T-lymphocytes. *Biochem Biophys Res Commun* **180**: 1125–1129.

Boggs JM, Moscarello MA, Papahadopoluos D. (1982) Structural organization of myelin — Role of lipid-protein interactions determined in model systems. In Jost PC, Griffith OH (eds). *Lipid-Protein Interactions*, Vol. 2. pp. 1–51. Wiley & Sons, New York.

Boggs JM, Rangaraj G, Hill CM, Bates IR, Heng YM, Harauz G. (2005) Effect of arginine loss in myelin basic protein, as occurs in its deiminated charge isoform, on mediation of actin polymerization and actin binding to a lipid membrane *in vitro*. *Biochemistry* **44**: 3524–3534.

Boggs JM. (2006) Myelin basic protein: A multifunctional protein. *Cell Mol Life Sci* **63**: 1945–1961.

Boggs JM, Wang H. (2004) Co-clustering of galactosylceramide and membrane proteins in oligodendrocyte membranes on interaction with polyvalent carbohydrate and prevention by an intact cytoskeleton. *J Neurosci Res* **76**: 342–355.

Brown DA, London E. (1998) Functions of lipid rafts in biological membranes. *Annu Rev Cell Dev Biol* **14**: 111–136.

Burns PF, Campagnoni CW, Chaiken IM, Campagnoni AT. (1981) Interactions of free and immobilized myelin basic protein with anionic detergents. *Biochemistry* **20**: 2463–2469.

Campagnoni AT, Skoff RP. (2001) The pathobiology of myelin mutants reveal novel biological functions of the MBP and PLP genes. *Brain Pathol* **11**: 74–91.

Carlone G, Fasano A, Rossano R, Riccio P. (2005) Raft and non-raft distribution of myelin proteins. In *Proceedings of the ESF MARIE Round Table on "Rafting and Misrafting Myelin with a Look to Autoimmunity"*, 28th September–1st October 2006, Giovinazzo, Bari, Italy.

Cavatorta P, Giovanelli S, Bobba A, Riccio P, Szabo AG, Quagliariello E. (1994) Myelin basic protein interaction with zinc and phosphate: Fluorescence studies on the water-soluble form of the protein. *Biophys J* **66**: 1174–1179.

Cristofolini L, Fontana MP, Serra F, Fasano A, Riccio P, Konovalov O. (2005) Microstructural analysis of the effects of incorporation of myelin basic protein in phospholipid layers. *Eur Biophys J* **34**: 1041–1048.

Cwiklinska H, Mycko MP, Luvsannorov O, Walkowiak B, Brosnan CF, Raine CS, Selmaj KW. (2003) Heat shock protein 70 associations with myelin basic protein and proteolipid protein in multiple sclerosis brains. *Int Immunol* **15**: 241–249.

Dawson R, Muller L, Dehner A, Klein C, Kessler H, Buchner J. (2003) The N-terminal domain of p53 is natively unfolded. *J Mol Bio* **332**: 1131–1141.

DeBruin LS, Haines JD, Wellhauser LA, Radeva G, Schonmann V, Bienzle D, Harauz G. (2005) Developmental partitioning of myelin basic protein into membrane microdomains. *J Neurosci Res* **80**: 211–225.

DeBruin LS, Harauz G. (2007) White matter rafting-membrane microdomains in myelin. *Neurochem Res* **32**: 213–228.

Deibler G, Martenson RE, Kies MW. (1972) Large-scale preparation of myelin basic protein from central nervous tissue of several mammalian species. *Prep Biochem* **2**: 139–165.

Deibler GE, Boyd LF, Kies MW. (1984) Proteolytic activity associated with purified myelin basic protein. In Alvord EC Jr, Kies MW, Suckling AJ (eds). *Experimental Allergic Encephalomyelitis: A Useful Model for Multiple Sclerosis*, pp. 249–256. Liss, New York.

Dyer CA. (1997) Myelin proteins as mediators of signal transduction. In Juurlink BHJ, Devon RM, Doucette JR, Nazarali AJ, Schreyer DJ, Verge VMK (eds). *Cell Biology and Pathology of Myelin: Evolving Biological Concepts and Therapeutic Approaches*, pp. 69–74. Plenum Press, New York.

Eylar EH, Thompson M. (1969) Allergic encephalomyelitis: The physiko-chemikal properties of the basic protein encephalitogen from bovine spinal cord. *Arch Biochem Biophys* **129**: 468–479.

Facci P, Cavatorta P, Cristofolini L, Fasano A, Fontana MP, Riccio P. (2000) Kinetic and structural study of the interaction of myelin basic protein with phospholipid layers. *Biophys J* **78**: 1413–1419.

Garavito RM, Ferguson-Miller S. (2001) Detergents as tools in membrane biochemistry. *J Biol Chem* **276**: 32403–32406.

Gendreau S, Schirmer, J, Schmalzing G. (2003) Identification of a tubulin binding motif on the P2X(2) receptor. *J. Chromatogr B Anal Technol Biomed Life Sci* **786**: 311–318.

Givogri MI, Bongarzone ER, Campagnoni AT. (2000) New insights on the biology of myelin basic protein gene: The neural-immune connection. *J Neurosci Res* **59**: 153–159.

Givogri MI, Bongarzone ER, Schonmann V, Campagnoni AT. (2001) Expression and regulation of golli products of myelin basic protein gene during *in vitro* development of oligodendrocytes. *J Neurosci Res* **66**: 679–690.

Gold R, Linington C, Lassmann H. (2006) Understanding pathogenesis and therapy of multiple sclerosis via animal models: 70 years of merits and culprits in experimental autoimmune encephalomyelitis research. *Brain* **129**: 1953–1971.

Golds EE, Braun PE. (1978) Protein associations and basic protein conformation in the myelin membrane. The use of difluorodinitrobenzene as a cross-linking reagent. *J Biol Chem* **253**: 8162–8170.

Gunnarsson M, Jensen PE. (1998) Binding of soluble myelin basic protein to various conformational forms of alpha2-macroglobulin. *Arch Biochem Biophys* **359**: 192–198.

Gunnarsson M, Sundstrom P, Stigbrand T, Jensen PE. (2003) Native and transformed alpha2-macroglobulin in plasma from patients with multiple sclerosis. *Acta Neurol Scand* **108**: 16–21.

Haas H, Torrielli M, Steitz R, Cavatorta P, Sorbi R, Fasano A, Riccio P, Gliozzi A. (1998) Myelin model membranes on solid substrates. *Thin Solid Films* **327–329**: 627–631.

Haas H, Oliveira CLP, Torriani IL, Polverini E, Fasano A, Carlone G, Cavatorta P, Riccio P. (2004) Small angle X-ray scattering from lipid-bound myelin basic protein in solution. *Biophys J* **86**: 455–460.

Haas H, Steitz R, Fasano A, Polverini E, Cavatorta P, Riccio P. (2007) Laminar order within Langmuir–Blodgett multilayers from phospholipids and myelin basic protein. A neutron reflectivity study. *Langmuir* **23**: 8491–8496.

Harauz G, Ishiyama N, Hill CMD, Bates IR, Libich DS, Farès C. (2004) Myelin basic protein-diverse conformational states of an intrinsically unstructured protein and its roles in myelin assembly and multiple sclerosis. *Micron* **35**: 503–542.

Hu Y, Doudevski I, Wood D, Moscarello M, Husted C, Genain C, Zasadzinski JA, Israelachvili J. (2004) Synergistic interactions of lipids and myelin basic protein. *Proc Nat Acad Sci* **101**: 13466–13471.

Kim H, Jo S, Song HJ, Park ZY, Park CS. (2007) Myelin basic protein as a binding partner and calmodulin adaptor for the BKCa channel. *Proteomics* **7**: 2591–2602.

Kirschner DA, Blaurock AE. (1992) Organization, phylogenetic variations and dynamic transitions of myelin. In Martenson RE (ed). *Myelin: Biology and Chemistry*, pp. 3–78. CRC Press, Boca Raton, FL.

Kolehmainen E, Sormunen R. (1998) Myelin basic protein induces morphological changes in the endocrine pancreas. *Pancreas* **16**: 176–188.

Libich DS, Harauz G. (2008) Backbone dynamics of the 18.5 kDa isoform of myelin basic protein reveals transient alpha-helices and a calmodulin-binding site. *Biophys J* [March, Epub ahead of print].

Liuzzi GM, Tamborra R, Ventola A, Bisaccia F, Quagliariello E, Riccio P. (1996) Different recognition by clostripain of myelin basic protein in the lipid-bound and lipid-free forms. *Biochem Biophys Res Commun* **226**: 566–571.

Lolli F, Liuzzi GM, Vergelli M, Massacesi L, Ballerini C, Amaducci L, Riccio P. (1993) Antibodies specific for the lipid-bound form of myelin basic protein during experimental autoimmune encephalomyelitis. *J Neuroimmunol* **44**: 69–76.

Massacesi L, Vergelli M, Zehetbauer B, Liuzzi GM, Olivotto J, Ballerini C, Uccelli A, Mancardi L, Riccio P, Amaducci L. (1993) Induction of the autoimmune encephalomyelitis in rats and immune response to myelin basic protein in lipid bound form. *J Neurol Sci* **119**: 91–98.

Mazzanti B, Vergelli M, Riccio P, Martin R, McFarland HF, Liuzzi GM, Amaducci L, Massacesi L. (1998) T-cell response to myelin basic protein and lipid-bound myelin basic protein in patients with multiple sclerosis and healthy donors. *J Neuroimmunol* **82**: 96–100.

Modesti NM, Barra HS. (1986) The interaction of myelin basic protein with tubulin and the inhibition of tubulin carboxypeptidase activity. *Biochem Biophys Res Commun* **136**: 482–489.

Moscarello MA, Wood DD, Ackerley C, Boulias C. (1994) Myelin in multiple sclerosis is developmentally immature. *J Clin Invest* **94**: 146–154.

Moscarello MA. (1996) Evolving biological concepts and therapeutic approaches. In Devon RM, Doucette R, Juurlink BHJ, Nazarali AJ, Schreyer DJ, Verge VMK (eds). *Cell Biology and Pathology of Myelin*. Plenum Publishing, New York.

Pedraza L, Fidler L, Staugaitis SM, Colman DR. (1997) The active transport of myelin basic protein into the nucleus suggests a regulatory role in myelination. *Neuron* **18**: 579–589.

Perczel A, Hollosi M, Tusnady G, Fasman GD. (1991) Convex constraint analysis: A natural deconvolution of circular dichroism curves of proteins. *Protein Eng* **4**: 669–679.

Perczel A, Park K, Fasman GD. (1992) Analysis of the circular dichroism spectrum of proteins using the convex constraint algorithm: A practical guide. *Anal Biochem* **203**: 83–93.

Polverini E, Fasano A, Zito F, Riccio P, Cavatorta P. (1999) Conformation of bovine myelin basic protein purified with bound lipids. *Eur Biophys J* **28**: 351–355.

Polverini E, Arisi S, Cavatorta P, Berzina T, Cristofolini L, Fasano A, Riccio P, Fontana MP. (2003) Interaction of myelin basic protein with phospholipid monolayers: Mechanism of protein penetration. *Langmuir* **19**: 872–877.

Polverini E, Boggs JM, Bates IR, Harauz G, Cavatorta P. (2004) Electron paramagnetic resonance spectroscopy and molecular modelling of the interaction of myelin basic protein (MBP) with calmodulin (CaM) — diversity and conformational adaptability of MBP CaM-targets. *J Struct Biol* **148**: 353–369.

Privat A, Jacque C, Bourre J-M, Dupouey P, Baumann NA. (1979) Absence of the major dense line in myelin of the mutant mouse "shiverer". *Neurosci Lett* **12**: 107–112.

Readhead C, Popko B, Takahashi N, Shine HD, Saavedra RA, Sidman RL, Hood L. (1987) Expression of a myelin basic protein gene in transgenic mice: Correlation of the dismyelinating phenotype. *Cell* **48**: 703–712.

Riccio P, Rosenbusch JP, Quagliarello E. (1984) A new procedure to isolate brain myelin basic protein in a lipid-bound form. *FEBS Lett* **177**: 236–240.

Riccio P, Masotti L, Cavatorta P, De Santis A, Juretic D, Bobba A, Pasquali-Ronchetti I, Quagliariello E. (1986) Myelin basic protein ability to organize lipid bilayers: Structural transitions in bilayers of lysophosphatidylcholine micelles. *Biochem Biophys Res Commun* **134**: 313–319.

Riccio P, Quagliariello E. (1993) Lipid-bound, native-like, myelin basic protein: A well-known protein in a new guise, or an unlikely story? *J Neurochem* **61**: 787–788.

Riccio P, Bobba A, Romito E, Minetola M, Quagliariello E. (1994) A new detergent to purify CNS myelin basic protein isoforms in lipid-bound form. *Neuroreport* **24**: 689–692.

Riccio P, Giovannelli S, Bobba A, Romito E, Fasano A, Bleve-Zacheo T, Favilla R, Quagliariello E, Cavatorta P. (1995) Specificity of zinc binding to myelin basic protein. *Neurochem Res* **20**: 1107–1113.

Riccio P, Fasano A, Borenshtein N, Bleve-Zacheo T, Kirschner DA. (2000) Multilamellar packing of myelin modeled by lipid-bound MBP. *J Neurosci Res* **59**: 513–521.

Riccio P, Fasano A, Polverini E, Relini A, Carlone G, Ranieri R, Gliozzi S, Cavatorta P. (2002) *MBP Back Home. Proceedings of the ESF Exploratory Workshop on Myelin Structure and Its Role in Autoimmunity*. 5–8 June 2002, Potenza, Italy. p. 15.

Richter-Landsberg C. (2001) Organization and functional roles of the cytoskeleton in oligodendrocytes. *Microsc Res Tech* **52**: 628–636.

Rispoli P, Carzino R, Svaldo-Lanero T, Relini A, Cavalleri O, Liuzzi GM, Carlone G, Fasano A, Riccio P, Gliozzi A, Rolandi R. (2007) A thermodynamic and structural study of myelin basic protein in lipid membrane models. *Biophys J* **93**: 1999–2010.

Ridsdale RA, Beniac DR, Tompkins TA, Moscarello MA, Harauz G. (1997) Three-dimensional structure of myelin basic protein. II. Molecular modeling and considerations of predicted structures in multiple sclerosis. *J Biol Chem* **272**: 4269–4275.

Rosenbusch JP. (2001) Stability of membrane proteins: Relevance for the selection of appropriate methods for high-resolution structure determinations. *J Struct Biol* **136**: 144–157.

Sedzik J, Kirschner DA. (1992) Is myelin basic protein crystallizable? *Neurochem Res* **17**: 157–166.

Simons K, Ikonen E. (1997) Functional rafts in cell membranes. *Nature* **387**: 569–572.

Simons K, Toomre D. (2000) Lipid rafts and signal transduction. *Nat Rev Mol Cell Biol* **1**: 31–39.

Smith R. (1982) Self-association of myelin basic protein: Enhancement by detergents and lipids. *Biochemistry* **12**: 2697–2701.

Smith R. (1992) The basic protein of CNS myelin: Its structure and ligand binding. *J Neurochem* **59**: 1589–1608.

Staugaitis SM, Colman DR, Pedraza L. (1996) Membrane adhesion and other functions for the myelin basic proteins. *Bioessays* **18**: 13–18.

Taylor CM, Coetzee T, Pfeiffer SE. (2002) Detergent-insoluble glycosphingolipid/cholesterol microdomains of the myelin membrane. *J Neurochem* **81**: 993–1004.

Tompa P. (2002) Intrinsically unstructured proteins. *Trends Biochem Sci* **27**: 527–533.

Uversky VN, Gillespie JR, Fink AL. (2000). Why are natively unfolded proteins unstructured under physiologic conditions? *Proteins* **41**: 415–427.

Uversky VN. (2002) What does it mean to be natively unfolded? *Eur J Biochem* **269**: 2–12.

Vacher M, Nicot C, Pflumm M, Luchins J, Beychok S, Waks M. (1984) A heme binding site on myelin basic protein: Characterization, location, and significance. *Arch Biochem Biophys* **231**: 86–94.

Chapter 11

The Use of a Magnetic Field and Magnetic Force as a Means to Improve the Quality of Protein Crystals

*Mitsuo Ataka**

Single protein crystals are indispensable in the determination of the three-dimensional structure of proteins and other biological macromolecules by X-ray crystallography. The quality of the crystals governs the resolution limit or accuracy of the atomic positions calculated. The purpose of this chapter is to describe the idea of using magnetic fields to grow better quality protein crystals. The possibilities as well as limitations will be discussed.

Keywords: Magnetic field; magnetic force; magnetization force; crystals; microgravity; sedimentation; convection.

Introduction

The need to obtain better quality protein crystals has long been recognized. Microgravity experiments have been planned by a number of countries since the beginning of the 1980's, which aimed at using weightless and convectionless environments available in space. This kind of experiment requires a large amount of funding, effort, time and manpower. Nevertheless, the investment has been widely justified by explaining that accurate protein structure is essential to the basic understanding of life

*National Institute of Advanced Industrial Science and Technology, Kansai Center, Japan. E-mail: m-ataka@aist.go.jp.

as well as to the development of drugs that suppress or enhance the functions of those proteins.

These historical facts, repeatedly advocated by the space agencies of various countries, tell us at least the following things. First, single protein crystals and the fine molecular structure obtained by using them are a key factor in the life sciences of the future. Second, alternative methods of improving the quality of crystals are not easy to list; space-based experiments have become the primary choice.

The expectations about such experiments may be right. However, even so, there is good reason to look for alternative, Earth-based methods of improving crystalline quality. The use of magnetic fields described below is one such trial. The crystals obtained within magnets may be used *per se* for structure determination. The experiments may also be useful in selecting the best crystals or best crystallization conditions that will benefit from the space-based experiments in the future. The experiments may elucidate the mechanism by which the microgravity environment in space leads to quality improvements, thus further justifying investment.

A few reviews are already available on the effects of static, strong magnetic fields on the quality of protein crystals (Ataka,Wakayama, 2002; Wakayama, 2003). The topics not covered therein will be dealt with here.

Theoretical Background

Two things must be emphasized when applying a magnetic field to protein crystal growth.

Magnetic field orientation

First, protein molecules and protein crystals belong magnetically to the category of feeble (weak but present) magnetic substances. Feeble magnetic substances have long been referred to as non-magnetic substances, meaning that these substances, containing both diamagnetic and paramagnetic materials, have been considered not to interact with magnetic fields. Instead, ferromagnetic materials, including iron, have been the main focus of scientific research on the use of magnetic fields. It is only somewhat more recently that the effects of magnetic fields on diamagnetic substances have attracted attention. This is due partly to the development of superconducting materials that have enabled strong and stable magnetic fields to be generated without much difficulty in many laboratories in addition to those specializing in low temperature physics with sophisticated techniques. Various effects and interactions have been recognized, and as a consequence of these effects the term "non-magnetic" began to be considered to be not entirely correct, "feebly magnetic" being a better description. Among protein molecules, some (perhaps 10% or a little more) are classified into metal proteins, which means that they require metal ions for stability and function, in addition to the usual atoms of C, N, O, H and S as in the majority of proteins. The metal ions can be Fe, Co, Ni, Mn, and so on, and of various valencies. Since a metal ion can be paramagnetic, and its absolute value is two to three orders of magnitude larger than the diamagnetism, a protein molecule can be paramagnetic if they contain (a) metal ion(s). In other cases, protein molecules are diamagnetic, like the majority of substances. In any case, protein molecules are feeble magnetic substances.

Our second point is that there are two independent actions of a magnetic field interacting with feeble magnetic substances. One is magnetic orientation. As long as feeble magnetic substances have magnetic anisotropy, they have an easy axis or axes (an easy axis means an axis that is magnetically easy to orient along the magnetic field) that tend to align along the magnetic field. In many cases, this orientation tendency can be disordered by thermal agitation or hindered by adherence or fixation to an environment. However, if the magnetic field is strong enough so as to overcome thermal fluctuation and if the substances can rotate freely without interference from adhesion to an environment, their easy axis will align along the magnetic field. This magnetic orientation can be observed both in homogeneous and inhomogeneous magnetic fields.

Magnetic force

The other action is a magnetic force that is observed only in an inhomogeneous magnetic field, both for diamagnetic and paramagnetic substances.

The Gibbs free energy representation requires that when the free energy changes in the presence of the magnetic field, some force can be generated to minimize the free energy. When the strength of the magnetic field 1H is chosen as an independent variable, the free energy representation is

$$d\boldsymbol{G} = -\boldsymbol{S}\, d\boldsymbol{T} + \boldsymbol{V}\, d\boldsymbol{P} - \nabla\, \mu_0 \boldsymbol{M}\, d\boldsymbol{H}\ (+ \cdots),$$

where $d\boldsymbol{G}$ is the free energy change, $\boldsymbol{S}$ is entropy, $\boldsymbol{T}$ is temperature, $\boldsymbol{V}$ is volume, $\boldsymbol{P}$ is pressure, and $\boldsymbol{M}$ is magnetization. μ_0 is the permeability of the vacuum, and "…" means the extra and unnecessary terms coming from electric field, gravitational field, stress field, addition/deletion of masses, and so on.

For all feeble magnetic substances, $\boldsymbol{M}$ can be expressed as

$$\boldsymbol{M} = \chi \boldsymbol{H},$$

where χ is the magnetic susceptibility of the corresponding substance. For paramagnetic substances, $\chi > 0$, and for diamagnetic substances, $\chi < 0$. The value χ is regarded as a constant as long as the substance is feebly magnetic. Then it comes that

$$d\boldsymbol{G} = \cdots - \nabla\, \mu_0\, \chi \boldsymbol{H}\, d\boldsymbol{H} \qquad \text{or} \qquad \boldsymbol{G} = \cdots - \nabla\, \mu_0 \chi\, (1/2) \boldsymbol{H}^2.$$

Since the negative of the spatial derivative of the free energy is a force, the magnetic force $\boldsymbol{F}$ that results is

$$\boldsymbol{F} = \mu_0 \chi \boldsymbol{H}\ \text{grad}\ \boldsymbol{H}.$$

This force $\boldsymbol{F}$ becomes zero if (grad $\boldsymbol{H}$) = 0, i.e. if the magnetic field is homogeneous and if it does not depend on position. However, if $\boldsymbol{H}$ is nonzero and inhomogeneous, then $\boldsymbol{F}$ will have a nonzero value.

An intuitive explanation of the above description is as follows. All the substances are more or less magnetized when placed in a magnetic field. The Gibbs free energy change coming from the magnetization is positive for a diamagnetic substance and negative for a paramagnetic substance. In other words, diamagnetic substances obtain energy if a magnetic field is applied, and to reduce this excess energy due to the presence of the magnetic field, they have a tendency to escape from a magnetic field. If the magnetic field is homogeneous, this tendency does not appear (intuitively, the substance does not know where to move to reduce energy if there is no gradient of the magnetic field). However, if the field is inhomogeneous, the diamagnetic substance tries to escape from the magnetic field into a direction that is expressed by (−grad $\boldsymbol{H}$), thereby reducing its free energy. The magnitude of this force is proportional to three parameters, χ, $\boldsymbol{H}$, and (grad $\boldsymbol{H}$). Therefore, to increase this force the product $\boldsymbol{H}$(grad $\boldsymbol{H}$) must be increased.

This force has been given several different names, including magnetic force, magnetization force, and Kelvin body force. Here we will call it the magnetic force.

Technical Background

Readers who are not interested in the instruments for applying magnetic fields can skip this section.

Generation of magnetic fields by various magnets

Superconducting magnets, electromagnets, and permanent magnets can all be used for the purpose of applying a magnetic field.

A superconducting magnet can provide magnetic fields up to about 15 T. Though the superconducting wire must be kept at a liquid helium temperature to pass or apply an electric current of tens of amperes, it is not accompanied by heat generation since electric resistance is absent. Usually, there is a cylindrical bore inside the coil that is kept at room temperature. Even if the coil is kept at an extremely low temperature, heat insulation allows the bore temperature to be little affected by the low temperature. Since the superconducting current is stable, such magnets offer static magnetic fields of a large magnitude without fluctuation. They are also energy efficient, since the zero electrical resistance does not require electric power to pass the current. Initially, the liquid helium, the coolant, had to be supplied continuously from outside, which required considerable experience and cost. However, this scheme has recently been replaced with an electrical cooling system which is cryocooler-free or liquid helium-free. The cylindrical bore diameter can be 100 mm or more. This allows a temperature regulating tool (e.g. water circulation) to be installed within the bore. When magnetic fields of more than about 20 T are needed, a hybrid magnet is used, in which additional fields are supplied by a normal metal wire, in addition to the superconducting wire. In this case, one needs an enormous amount of electric power to continuously apply electric currents. Only a limited number of facilities in the world have such hybrid magnets.

Electromagnets have traditionally been used to generate magnetic fields of up to about 2 T. Masses of iron are used, making these magnets extremely heavy. Also, electric current must be applied continuously by using electric power.

With permanent magnets, using alloys containing neodymium and other metals, increasingly large field strengths have been generated, up to nearly 2 T. However, the volume of the spatial region of high magnetic field strength is limited compared with superconducting magnets. The cost and weight of a permanent magnet is low, and it does not require electric power. It is challenging to consider to what extent the effects of magnetic field and magnetic force can be derived from permanent magnets.

The magnitude of the geomagnetic field by comparison is about 0.06 mT; field strengths of 2–10 T are energetically larger by about 10^{10}.

Superconducting magnets for supplying uniform magnetic force

Regarding the use of the magnetic force, a specified magnetic environment is required. This is because the magnetic force is proportional to grad($\boldsymbol{H}^2$). Often, a superconducting magnet is designed to provide a uniform $\boldsymbol{H}$ or $\boldsymbol{B}$, where $\boldsymbol{B}$ is the magnetic induction. That is, at a place where a sample (e.g. an NMR tube, a long, cylindrical tube) is placed, the magnitude and direction of $\boldsymbol{B}$ are both designed to be as constant as possible. This requirement is achieved by considering the best winding of the superconducting wire, plus the use of small assisting magnets (called shim coils). When we consider the use of a magnetic force, the usual magnets designed to provide a constant $\boldsymbol{H}$ are only of limited utility. This is because the constancy of $\boldsymbol{H}$ and that of $\boldsymbol{H}^2$ are mathematically different from each other. For $\boldsymbol{H}^2$ to be constant, the product of $\boldsymbol{H}$ and (grad $\boldsymbol{H}$) should be constant. However, a magnet specifically made for this purpose did not exist when we conceived the need for it in 1997. We then made some effort to show that the manufacture of such a magnet is in fact possible by presenting an actual design of the configuration of coils and by demonstrating the constancy of the magnetic force distribution inside the bore of the magnet (Kiyoshi *et al.*, 1999).

The same problem was formulated a little differently by Brooks *et al.* (2000). They pointed out that if we use a conventional superconducting magnet and if we want to make "a microgravity-equivalent environment" by counterbalancing the downward force of gravity and the upward magnetic force, then a spherical space of only 0.1 mm radius may be used; outside of this small volume, the counterbalancing condition will not be satisfied. This meant that, experimentally, identifying this volume and reproducibly using it would extremely be difficult. This was another demonstration that using a conventional magnet would not be easy if one wants to use an upward magnetic force that will in principle cancel gravity. This is because the magnetic force distribution in a conventional superconducting magnet is not uniform whereas gravity is highly uniform on Earth.

Modes of operation of various magnets

Many experiments (such as NMR) on the effects of a magnetic field on substances have been carried out at the center of a magnet where the field strength becomes maximal. Usually, the field homogeneity at the center of a magnet is quite good (i.e. grad $\boldsymbol{H} = 0$), which means that the magnetic force does not operate. On the other hand, there are positions off the center of the magnet where $\boldsymbol{H}$ is not small, and a change in $\boldsymbol{H}$ with position is also significant; hence, the product $\boldsymbol{H}$ (grad $\boldsymbol{H}$) has a considerable magnitude. In many commercial superconducting magnets, such a position is found about 100 mm separated from the magnet center. In many cases, there are two positions that are symmetrically positioned with respect to the center on both sides. It is at these positions that the effect of the magnetic force can best be observed.

Experiments to observe the effects of the magnetic force can be carried out, in the case of using a superconducting magnet, in either a vertically or a horizontally directed bore. If the bore is vertical, then the direction of the magnetic force is the same as that of gravity; the magnetic force is either downward or upward. In the former case, the magnetic force works to strengthen the gravitational pull on Earth. In the latter case, the magnetic force works to weaken gravity. In the special case where the magnetic force has the same magnitude as gravity but directed upward, the magnetic force and gravitational force will be balanced. It is in this case that feeble magnetic substances "float" within the magnet.

With regard to the direction of the magnetic field, an electromagnet usually generates a field in the horizontal direction, since a pair of large and heavy pole pieces are often placed at the same level, i.e. in the left-to-right direction. A superconducting magnet often has a bore that is vertical. In this case, vertical magnetic fields are generated at the center of the bore. The field gradually diminishes above and below the center. In such a configuration, the direction along which the field gradient is steepest becomes vertical. It is in this configuration that the upward magnetic force is generated and can be used to cancel out gravity. However, some superconducting magnets can be rotated by 90° around a horizontal axis. If this operation is carried out, the bore direction becomes horizontal. It is more difficult for mechanical reasons to direct the bore neither vertically nor horizontally.

Historical Background

As early as 1988, Ryuma Kuroda, who had just entered a graduate course at the University of Tokyo (Faculty of Engineering, Department of

Applied Physics), and his supervisor Prof. Koji Okano thought of applying a static magnetic field during protein crystal growth. They asked the present author how to grow protein crystals and learned about the conditions for growing hen egg-white lysozyme crystals, a protein biosynthesized by chickens. The protein's function is to catalyze the hydrolysis of polysaccharides comprising the cell walls of a number of bacteria. The egg-white contains a considerable amount of lysozyme; when bacteria invade the egg, the lysozyme can kill them by attacking the outer cell wall and protect the yolk. The crystallization conditions taught to Kuroda were basically the same as reported by Alderton *et al.* (1945): 30–50 mg/ml protein, 3–5% (w/w) NaCl as a crystallizing agent, bringing the pH to about 4.5, and keeping the temperature of the batch at around 15°C. Tetragonal crystals grow under these conditions. An electromagnet was

Fig. 1. A photomicrograph of hen egg-white lysozyme crystals grown without a magnetic field. A control to be compared with Fig. 2.

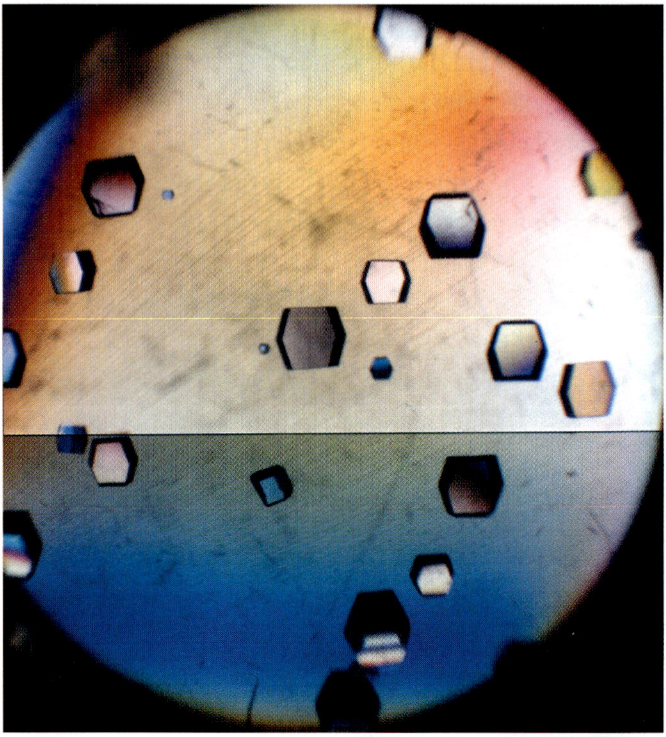

Fig. 2. A photomicrograph of the crystals grown under the same conditions as in Fig. 1, but in the presence of a magnetic field (direction is depicted by arrow); the intensity of the magnetic field is about 1 Tesla. (Taken from the Master's Degree Thesis of Ryuma Kuroda submitted in 1990 to the Department of Applied Physics, Faculty of Engineering, University of Tokyo.)

used to apply the magnetic field; a superconducting magnet requiring liquid helium for coolant was too sophisticated an instrument for normal laboratories at that time, prior to the situation changing as mentioned in the section on "Technical Background" (p. 5). Figures 1 and 2, taken from the Master's Thesis by Kuroda, clearly show that only the characteristic {110} faces are seen.

As Figs. 1 and 2 show, Kuroda found that, compared to the crystallization without applying the magnetic field, only one of the two typical crystalline faces that characterize the tetragonal hen egg-white lysozyme crystals is observed when the crystals are grown in a horizontal magnetic field provided by the electromagnet. The strength of the field was about 1 T and the vessel containing the supersaturated solution was observed from above. The only crystalline face observed was {110}, and not {101}. These facts can be explained if the *c*-axis was parallel with the magnetic field. It was very clear that the magnetic field orients the crystalline *c*-axis along the field direction.

In 1989, two undergraduate students, Takashi Imaeda and Yoshio Ohno participated in the research initiated by Kuroda. The new undergraduates calculated the magnetic moment of the tetragonal lysozyme crystal, using the atomic coordinates obtained from the Protein Data Bank (PDB). The calculation showed that the *c*-axis was in fact the easy axis. From the tensor representing the magnetic susceptibilities along the three principal axes, the magnetic anisotropy was calculated to be $1.45–6.75 \times 10^{-9}$ emu/cm^3. The range of values comes from a different PDB dataset as well as from the whole atomic coordinate versus polypeptide main chain atoms. The main chain atoms were used to exclude possible fluctuations in the side chain positions as evidenced from NMR measurements. The magnetic anisotropy in MKS units was calculated to be about 5×10^{-27} J/T^2. The prediction from calculation was further compared with an experiment, in which a single tetragonal crystal of a certain size, suspended with a thin thread, was rotated in the presence of a static magnetic field, and from the oscillatory period an experimental value comparable with the calculation was obtained. Thus, it was concluded that the magnetic orientation of the protein crystal used comes from diamagnetic anisotropy.

Unfortunately, the laboratory to which these three students belonged was a major research group of liquid crystals. For liquid crystals, i.e. rod-like anisotropic molecules, magnetic orientation is a well-known phenomenon — as is evident from the color change of a TV or PC monitor near a strong magnet. Since the magnetic orientation of the tetragonal lysozyme crystals was regarded as self-evident, the results were not reported further as a scientific paper. However, the Master's Degree Thesis of Kuroda as well as the Batchelor's Theses of Imaeda and Ohno, submitted in March of 1990, are kept in a library of the University of Tokyo.

Prior to the experiments described above, Rothgeb and Oldfield (1981) had used myoglobin (from sperm whale) crystals to study the effects of a static magnetic field of 0.3 T. In this case, already-grown and sedimented crystals were vigorously agitated to detach them from the vessel and then the vessel was placed in the magnetic field. With such experimental procedures, the suspended myoglobin crystals are considered to have sedimented to the vessel bottom in the presence of the magnetic field (the crystals do not grow in this case). Considerable orientation was observed, which was absent if the field was not present. The magnetic orientation was ascribed to the Fe ion contained in the heme in the myoglobin molecule, more specifically, to the anisotropy of the paramagnetism. Therefore, it was not clear from these analyses whether most proteins that do not contain Fe ions magnetically orient or not. Note also that the function of myoglobin in mammals including whales is to store oxygen in muscle and other parts of the body.

Our Studies on Magnetic Orientation

One of the main messages intended to be delivered in this chapter is the idea that magnetic orientation may in principle be used (at least in favorable cases) to reduce misalignment in protein crystals and to obtain better quality crystals. Microgravity in space has sometimes been suggested as a potential environment that can be used for similar purposes; however, the use of magnetic orientation may also be considered on some occasions.

Protein crystals grown in a magnetic field exhibit orientation

As shown in the section on "Historical Background" (p. 8), tetragonal lysozyme crystals exhibit orientation coming from diamagnetic anisotropy. We first used this fact to know when the crystals sediment to the bottom of the vessel (Ataka *et al.*, 1997). Usually, when we observe the crystals grown from a supersaturated solution of 1 ml or more using an optical microscope, nearly all of them are found to be at the bottom of the glass vessel. If the focus of the microscope is gradually raised to the inside of the solution, nothing is observed, except for some floating on top at the solution–air interface. There are two possibilities — either the crystals nucleated and started to grow at the vessel bottom surface, or they nucleated within the solution and then sedimented to the bottom later after reaching a certain size (usually, the density of crystals is greater than the density of surrounding solution). Distinguishing between the two cases is not easy, since seizing nucleation on site is difficult unless we know *a priori* where and when the nucleation starts. Therefore, we considered using the magnetic orientation phenomenon to solve this problem, which may otherwise be difficult to elucidate. The crystals at the bottom of the vessel do not respond to a magnetic field of 1 T or so and show orientation, since they adhere to the glass vessel. We carried out the crystal growth experiments in a magnet, applying the magnetic field all the time during the crystal growth. When the crystals were observed after they grew, almost all of them were at the vessel bottom, and nearly perfect orientation took place. By contrast, no clear orientation was observed if the magnetic field was not applied. On the other hand, when the magnetic field of 1.6 T (provided by an electromagnet) was applied some time (a few to ten hours) after supersaturating the solution, but before the crystals reached their final size, a considerable proportion of the crystals (particularly the larger ones that were considered to start growing earlier than the smaller ones) showed random orientation, in strong contrast with the situation where the field was present from the very beginning. The difference was attributed to the fact that only floating crystals within the solution which can freely rotate can respond to the magnetic field and show orientation, and thus this was taken as evidence that the crystals start growing within the solution. Also, the same observation indicates that when the crystals sediment, they are already oriented in one direction in the presence of a magnetic field of the order of 1 T. If the field strength is larger, then the magnetic orientation will be more perfect.

We next proceeded to show that other crystalline forms of hen egg-white lysozyme, as well as crystals of other proteins (without metal ions) can exhibit magnetic orientation (Sakurazawa *et al.*, 1999). The field strength necessary for the orientation to take place was found to be

0.2–0.6 T. These fields were provided with an electromagnet. These facts unambiguously showed that diamagnetic orientation is quite a general phenomenon.

Sedimenting crystals

We have mentioned above our studies on the emergence of the crystals within the solution, not at the surface of the vessel material. Another line of thought was provided from atomic force microscope (AFM) observations by the groups of McPherson and DeYoreo (McPherson, 1999). They noticed that when they were observing the surface of various protein and virus crystals with an AFM instrument, one of the surface probing microscopes, there are occasions where suddenly a microcrystalline mass appears on the surface and grows as a part of the existing, larger crystal. This kind of phenomenon was found to be rare for the usual inorganic and semiconductor crystals, whereas for protein and virus crystals it was not uncommon. For this reason, the investigators called this phenomenon "three-dimensional nucleation" and considered it as a more or less general mechanism of crystal growth for this group of substances. By comparison, other crystal growth mechanisms usually considered are: screw dislocation, two-dimensional nucleation, and kinetic roughening. Three-dimensional nucleation differs from these more common mechanisms in that an already existing crystal, though much smaller than a bulk crystal, seems to merge into the crystal being observed. The authors do not necessarily decide where the merging microcrystals originate from, but it is likely that they nucleated independently within the supersaturated solution and were transferred in the solution either by sedimentation or by convection flow. It must be recalled that protein and virus crystals need in general a much higher degree of supersaturation for spontaneous nucleation than smaller molecules, and for this reason there can be differences in nucleation behavior. Three-dimensional nucleation was independently observed via AFM by Astier *et al.* (2001) on porcine α-amylase crystals.

The above mechanism, merging microcrystals that nucleate separately and participate in growth, will certainly increases the chances of disorder in the crystalline lattice than the cases where the crystals grow mostly by the screw dislocation mechanism or by two-dimensional nucleation in which crystals basically grow by attachment of one molecule to the next molecule. If a magnetic field is applied so as to orient the direction of the merging and merged crystals, then the degree of disorder may be reduced. This is one of the effects expected from the application of the magnetic field. It must also be pointed out that for stronger magnetic fields, the smaller microcrystals can be aligned overcoming thermal agitation. Hence, the application of a strong magnetic field may be effective in increasing order. A similar line was taken by McPherson (1999), who considered that improvements in microgravity-grown protein crystals was achieved owing to the absence of deterioration due to the sedimentation of three-dimensional nuclei.

Other Studies on Magnetic Orientation of Protein Crystals

The understanding of diamagnetic orientation became clearer in the 1990's due to the efforts of various groups of investigators.

Magnetic orientation was independently investigated by another Japanese group (Sazaki *et al.*, 1997; Yanagiya *et al.*, 1999). The protein that exhibited orientation was again hen egg-white lysozyme, and a homogeneous magnetic field of 10 T was provided with a superconducting magnet. The crystal number (i.e. nucleation) and habit were also reported to be affected by the magnetic field. It should be noted that we have not confirmed the latter two kinds of variations in the presence of a (homogeneous) magnetic field as a general tendency. The other protein these authors studied, ferritin from horse spleen, is not expected to show magnetic orientation, since the crystal symmetry is cubic, lacking in magnetic anisotropy.

A French group also reported the magnetic orientation of porcine pancreatic amylase and bovine pancreatic tripsin inhibitor crystals in addition to lysozyme (Astier *et al.*, 1998). A magnetic field of 1.25 T was provided with a permanent magnet made of a Nd–Fe–B alloy.

Other Uses of Magnetic Orientation

Figures 1 and 2 show that the group of equivalent crystalline faces {110} is directed perpendicularly to the line of sight. It must be emphasized that a similar orientation is rather difficult to achieve by any other means. An advantage of the orientation of the crystalline faces being perpendicular to the line of sight is that the growth rate with time of this particular face can be directly measured with precision. If the face is not perpendicular, we need to know the angle of its inclination to the line of sight in order to calculate the growth rate. The perpendicular orientation allows one to find the growth rate directly from the images recorded, e.g. in a video tape or in a recording medium through a CCD camera. Pusey (1993) developed a computer-controlled microscopy system for following protein crystal face growth rates for automatic recording — potentially useful in microgravity experiments. If the magnetic orientation is combined, the system developed by Pusey could prove to be an excellent solution.

Other Possible Mechanisms Through Which Homogeneous Magnetic Fields May Contribute to Quality Improvement

The action of primary importance for a homogeneous magnetic field to improve the quality of protein crystals, we believe, is the magnetic orientation of sedimenting microcrystals as already discussed. However, there are other possible actions. A strong, homogeneous magnetic field can align the precrystalline structure, the formation of which we have studied previously (Ataka, 1998), as long as such structure has magnetic anisotropy. The postulated (but not yet fully proven) "gel-formation" (Zhong *et al.*, 2001) may be related to such precrystalline structure. If such precrystalline structure aligns in response to a magnetic field then suppression of convection and/or impurity diffusion take place, which may also lead to quality improvement.

There is evidence that the crystal growth rate (and the crystal dissolution rate in undersaturated solutions) reduces in a strong, homogeneous magnetic field. The evidence has been collected by observing the growth (and dissolution) of hen egg-white lysozyme crystals (Yanagiya *et al.*, 2000; Yin *et al.*, 2001, 2002). The reason for the growth (and dissolution) rate reduction may be complex, since it is not a simple mechanism such as solubility change or convection suppression. Yin *et al.* (2001) observed convection in a crystallizing aqueous lysozyme solution using tracer particles, but found no difference between their results with and without magnetic fields. An alternative reason could be a diffusion rate decrease in a solution placed in the magnetic field due, for example, to magnetic orientation of precrystalline structure or suspending microcrystals (Yin *et al.*, 2002; Wakayama, 2003). Whatever the reason may be, if crystals grow slower, it is conceivable that fewer defects are introduced into the crystals. In addition, if the advantage comes only from a slower growth rate, then it must be ascertained whether other means (e.g. merely decreasing the degree of supersaturation) also lead to the same quality improvement.

Studies on Crystal Perfection in Homogeneous Magnetic Fields

Enhancement in the perfection of the crystals grown in the presence of a homogeneous magnetic field of 10 T was demonstrated by Sato *et al.* (2000, 2001). They determined rocking widths using highly collimated synchrotron radiation on orthorhombic crystals of hen egg-white lysozyme. It was found that the resolution limit was 1.33 Å without the use of a magnetic field, whereas it extended to 1.13 Å when the crystals were grown in a magnetic field. These crystals showed magnetic orientation; in this case, a thin and long crystal "stood" on a narrower plane, instead of resting on the widest face as in the case of the absence of the vertical field. Only the oriented crystals were used for comparison. Also, fewer crystals grew when the magnetic field was applied. Sato *et al.* (2000) suggested that the orientation of mosaic blocks in addition to the suppression of convection were reasons for the improvement. Saijo *et al.* (2005) later added the *B* factor and mosaicity analysis of the same crystals. It is thus clear that the application of a strong magnetic field can in fact improve the quality of a protein crystal and hence contribute to determining more accurate structure. Similar improvement in the quality of the orthorhombic form of crystals of hen egg-white lysozyme in a 15 T homogeneous magnetic field was confirmed independently by Yin *et al.* (2004). However, it is not entirely clear why the orthorhombic form was used for study since among the structural information deposited in Protein Data Bank a different crystal form, the tetragonal one, gives better resolution. It may be that the quality enhancement was achieved for a low-resolution crystal form, that the choice was a mere accident, or that the results obtained for the tetragonal form were not reported for some reason.

Yin *et al.* (2003b) showed more uniform morphology in a horizontal as well as a vertical magnetic field by growing tetragonal crystals of hen egg-white lysozyme with a crystallizing agent of $NiCl_2$ containing a paramagnetic cation. With this crystallizing agent, the crystal symmetry becomes orthorhombic through a slight distortion between the *a*- and *b*-axes. Also, a Ni^{2+} cation was shown to be incorporated into the active cleft by X-ray crystallography (Yin *et al.* (2003a)). Yin *et al.* (2003a) found a difference in morphology when $NiCl_2$ and the magnetic field were used in combination. This is an example of the possibility that magnetic field application may be used to change or regulate crystalline morphology, although in this case it was used in combination with a less common crystallizing agent.

Lübbert *et al.* (2004) used X-ray rocking-curve measurements and statistical data analysis to evaluate tetragonal hen egg-white lysozyme crystals grown in the absence and presence of a homogeneous magnetic field of 2.4 T. Their results showed that the magnetic field (as well as gel) does not affect the growth of the main volume of the crystals but can reduce or partially prevent the formation or growth of separated and misoriented small outlier crystallites. This is consistent with the potential of a magnetic field to more easily orient sedimenting crystallites with respect to the main crystal, as we discussed above. However, in the study by Lübbert *et al.* the tetragonal crystals were already of high quality without any magnetic field, and the improvement was, at least with the 2.4 T field used, small.

The effects of gels and magnetic fields were also studied by Moreno *et al.* (2007). They grew lysozyme, thaumatin and ferritin crystals in gels in the presence of 7 T or 10 T fields. They concluded that improved resolution as well as reduced mosaicity was observed due to convection suppression and probably due to crystal orientation.

Tu *et al.* (2007) used a strong magnetic field of 18.7 T (placing the crystallization plates alongside the field) and considered the accomplishment of molecular orientation of UMP kinase from *Xanthomonas campestris*, a plant pathogen. Crystals of improved diffraction resolution (from 3.6 Å to 2.35 Å) grew in the apo-form without complexing with any substrate, thus enabling the first determination of the structure of a flexible, important loop of this enzyme. We should add that, in this case, it is difficult to judge to what extent the magnetic field was spatially homogeneous. We have emphasized above that the effects of a magnetic field can be two-fold: two independent effects, orientation and magnetic force, are expected. In addition to demonstrating the positive effects that a magnetic field can bring about, we anticipate that the two kinds of possible actions are reported in a way that can distinguish between the contributions.

Another line of evidence showing the beneficial effects of a homogeneous magnetic field of 10 T is described by Kinoshita *et al.* (2003), then affiliated to a major pharmaceutical company. Bovine adenosine deaminase crystals were known to suffer from inferior quality, even though it was found that crystals grew to a certain size. Only rarely (about 5% of crystals) could highly diffracting crystals be obtained, hampering efforts to search for possible compounds to interact with them for therapeutic purposes. Hence, crystallization in a magnetic field was attempted. It was found that a slightly more supersaturated condition could be used, under which crystals inappropriate for structure determination grow without a magnetic field, and the crystals thus grown often diffract X-rays to a high resolution. The success ratio of obtaining crystals that could be used to determine accurate three-dimensional structure was high. In turn, with such high success ratios, the crystallization step no longer hampers progress in the pharmaceutical efforts to find a compound complexed with the given protein. It is notable that the desirable "harvest period" for obtaining the crystals is less than two weeks in the case of structure-based drug design. If this requirement is fulfilled, then the crystal growth step does not interfere with the search for drugs based on structural information.

It may be worth noting that nearly all the benefits magnetic fields may bring about that have been discussed in this and the preceding sections are peculiar and inherent to the use of this field. Microgravity environments, for example, are not expected to provide such merits at least not by exactly the same mechanisms. It is highly desirable that more and more proteins are used to investigate whether the application of a magnetic field has some positive effects, especially in determining more accurate structure of protein molecules. Indeed, it is good news that proteins other than lysozyme have been used, especially in recent few years, to demonstrate the advantages of applying strong magnetic fields.

Magnetic Force

A magnet can be used to provide inhomogeneous magnetic fields (see the "Technical Background" section, p. 5), which generate magnetic forces (the "Theoretical Background" section p. 2). Here, we discuss their use in the quality improvement of protein crystals.

Our studies on the use of a magnetic force

Based on the possibility that the magnetic force can be applied to aqueous protein solutions and to protein crystals, we compared crystal growth at three different positions in a vertical bore of a superconducting magnet (Lin *et al.*, 2000), preceded by a preliminary study using a permanent magnet (Wakayama *et al.*, 1997). A commercial superconducting magnet with a cylindrical bore of 100 mm diameter was used, generating a 10 T magnetic field at the center. It follows that at an upper position, about 100 mm upward from the center, the magnetic field strength became 6 T, with a gradient of about 66.7 T/m, leading to a field–field gradient product of 400 T^2/m. The magnetic force for a diamagnetic substance corresponds to 30% of gravity, meaning that a diamagnetic substance experiences an upward magnetic force amounting to 30% of gravity. Thus, the net force becomes 70% of gravity. Similarly, at a lower position, 100 mm downward from the magnet center, the magnetic field becomes 6 T (symmetrical with respect to the center), with a field gradient of about 66.7 T/m but weaker in the lower direction, with a field–field gradient product of –400 T^2/m. The minus sign means that the force for a diamagnetic substance is downward. Its strength also amounts to 30% of gravity, but in this case the direction is the same as gravity, leading to a net force of 1.3 times gravity.

The protein used to compare the results of crystal growth was fructose 1,6-bisphosphatase, purified from a Chinese snake muscle. This is a protein involved in gluconeogenesis. The solution, supersaturated by the addition of $MgCl_2$ and polyethylene glycol 3350, was divided into batches, which were placed at the "middle" position at the magnet center, at the "upper" and "lower" positions defined above, and at a position outside of the magnet where both the magnetic field and the magnetic force were absent. The crystals grown after two to six days were evaluated with X-rays using a resolution that gave an S/N ratio of two on 1° oscillation photographs. Owing to magnetic orientation, many crystals grown in the magnetic fields (of trigonal symmetry) "stood" on a narrower face. The results and reproducibility are shown in Table 1.

It was found that the crystals that grew at the "upper" position showed the best resolution with least fluctuation among the crystals. The resolution was better than the crystals grown at the magnet center, and also than those grown outside the magnet. Note that the magnet center provided the strongest magnetic field, but the crystals were better when the upward magnetic force was superimposed on the (weaker than the center) magnetic field of 6 T. We can conclude that the magnetic force, an independent environment that a magnet can provide in addition to the magnetic field, can be used beneficially to grow better quality crystals.

The superconducting magnet that we used could not provide sufficient upward magnetic force to cancel gravity entirely. But Yin *et al.* (2004) grew hen egg-white lysozyme crystals (orthorhombic form) in a strong and upward magnetic force that did totally cancel gravity. The crystals

Table 1. Results of Applying Magnetic Force in Addition to Magnetic Field

Experiment No.	Position[a]			
	Upper	Middle	Lower	Outside
1 $\boldsymbol{H}$	6	10	6	0
$\boldsymbol{H}$ (grad $\boldsymbol{H}$)	+400[c]	0	−400	0
Resolution[b]	3.2	3.42	6.5	4.1
2 $\boldsymbol{H}$	6	10	6	0
$\boldsymbol{H}$ (grad $\boldsymbol{H}$)	+400	0	−400	0
Resolution	2.99	3.4	3.9	6
3 $\boldsymbol{H}$	4.8	8	4.8	0
$\boldsymbol{H}$ (grad $\boldsymbol{H}$)	+256	0	−256	0
Resolution	3.15	2.9	4.15	3.15
4 $\boldsymbol{H}$	4.8	8	4.8	0
$\boldsymbol{H}$ (grad $\boldsymbol{H}$)	+256	0	−256	0
Resolution	2.9	4.15	—	3.7

[a] Position means the three different positions (upper, middle, lower) where crystal growth took place in a vertical bore of a superconducting magnet. At these places, a magnetic field $\boldsymbol{H}$ and a field, gradient product $\boldsymbol{H}$ (grad $\boldsymbol{H}$) of the specified values are present. Outside means that the crystal growth took place outside of the superconducting magnet, where $\boldsymbol{H} = 0$.

[b] Resolution corresponds to the specific resolution values at which the S/N ratio = 2 using an oscillation photograph.

[c] The plus sign means that (grad $\boldsymbol{H}$) reduces on going upward; a diamagnetic substance receives an upward magnetic force if the sign of $\boldsymbol{H}$ (grad $\boldsymbol{H}$) > 0. See the "Theoretical Background" section, p. 2, for reasons and details.

thus obtained had a consistently better quality, as judged in this case by the "modified or relative Wilson plot." The reason was ascribed to convection damping. The quality improvement was even considerably better than in a case where crystals were grown in a homogeneous 15 T magnetic field (Yin *et al.*, 2004), a situation where better crystals grew, as already described, than without a magnetic field at all.

Yin and his colleagues (2008) recently grew the tetragonal form of crystals of hen egg-white lysozyme in a completely floating droplet without any support or container. Given the constant relative positions between suspended crystals, it can be considered that no convection flow is present in the levitated droplet. The quality of these crystals has not yet been reported, as long as the present author is aware. Clearly, there is further need for other proteins to be grown and the quality of their crystals to be evaluated precisely. Yin *et al.* (2004) reported that the cubic form of crystals of porcine insulin did not exhibit a quality improvement — an interesting observation for crystal symmetry lacking in magnetic anisotropy.

The levitated state has found interesting applications. Santesson *et al.* (2003) used acoustic levitation of protein droplets of up to 2.3 microliters surrounded only by air to screen conditions for amorphous precipitation. The information was used to determine regions of high supersaturation and thus to find suitable conditions likely to lead to nucleation and growth of single crystals. The idea behind this is that amorphous precipitation appears rapidly when sufficient supersaturation is provided, and can likely be used to obtain conditions for crystal growth. Levitation by other means than acoustic waves has been considered to be appropriate also, which suggests that magnetic levitation is a possibility, not widely known at the time when this acoustic levitation was applied in 2003 in Sweden.

An additional study of ours used the magneto-Archimedes effect to achieve levitation (Maki *et al.*, 2004). With a commercial superconducting magnet with a rather wide bore diameter of 100 mm conventionally used, only about 30% of gravity can be canceled (Lin *et al.*, 2000). To overcome this and achieve complete levitation — 100% cancellation of the gravity — we considered the use of the magneto-Archimedes effect, first proposed and coined by Ikezoe *et al.* (1997). This effect uses, in addition to the response of a substance to a magnetic field, that of the environment surrounding the substance. If the substance to be considered is diamagnetic, its surroundings are intentionally converted into the paramagnetic kind by addition of a paramagnetic ion. Then, the response toward the magnetic force will be opposite, enabling superposition of the opposing tendency in the presence of the magnetic field gradient. Thus, when the substance experiences an upward force, the environment is pulled downward, and the net upward force is much enhanced. By using this effect, we could totally levitate the protein crystals within a solution supersaturated with respect to aqueous lysozyme. The crystals thus obtained were evaluated by X-ray topography and the results demonstrated that crystals grown in the levitated state were free from internal stress.

Another recent development includes the use of a permanent magnet, instead of the much larger superconducting magnet, for magnetic levitation using the magneto-Archimedes effect (Maki, Ataka, 2007).

Convection Suppression or Damping in the Presence of Magnetic Force

To evaluate the potential of using magnetic force, and also to evaluate the desirable mode of using it, we carried out a numerical simulation to study how convection can be suppressed or damped in the presence of inhomogeneous magnetic fields. Such a study was systematically carried out and documented in a book by Ozoe (2005). Two of the initial works that led to systematic studies are Qi *et al.* (2001) and Ramachandran and Leslie (2005). Our simulation has added to the body of work described in Ozoe's Book.

It has been shown that when a diamagnetic liquid is placed in the presence of upward magnetic force, convection is generally suppressed. The particular magnetic force distributions we have studied include: a horizontally-directed bore of a superconducting magnet (Maki, Ataka, 2004), and off-centered positions in a vertically-directed bore (Maki, Ataka, 2005). It was shown that even the convection of water in a vessel placed at the center of a magnet can be influenced by the realistic magnetic force that results from the distribution of the magnetic field (Maki, Ataka, 2005).

Recent meaningful progress has been demonstrated by the elucidation of the difference in the conditions that lead to convection suppression and those that lead to levitation (Wakayama, 2005; Poodt *et al.*, 2005).

Summary

A general means to improve the quality of protein crystals is worth developing, especially as an alternative or a supplement to microgravity experiments in space. Superconducting as well as permanent magnets can be considered. The use of magnetic orientation is one possibility that has been shown to align the directions of a growing crystal and sedimenting microcrystals. The use of upward magnetic force is a second possibility that has proved to suppress convection as well as to levitate crystals without sedimenting. The way forward will be to distinguish between these two independent actions and carefully combine the possible beneficial effects. Further examples should be studied, especially new proteins, so as to ultimately determine whether crystals grown in the presence of a magnetic field can in fact be used to obtain more accurate structures of protein molecules.

References

Alderton G, Ward WH, Fevold HL. (1945) Isolation of lysozyme from egg white. *J Biol Chem* **157**: 43–58.

Astier JP, Veesler S, Boistelle R. (1998) Protein crystals orientation in a magnetic field. *Acta Crystallog D* **54**: 703–706.

Astier JP, Bokern D, Lapena I, Veesler S. (2001) *J Cryst Growth* **226**: 294–302.

Ataka M. (1998) Protein crystallization at the initial stage — Studies on supersaturated solutions. In Ohtaki H. (ed). *Crystallization Processes*, Wiley Series in Solution Chemistry, Vol. 3 Chap. 6, pp. 131–157. Wiley, Chichester.

Ataka M, Wakayama NI. (2002) Effects of a magnetic field and magnetization force on protein crystal growth. Why does a magnet improve the quality of some crystals? *Acta Crystallog D* **58**: 1708–1710.

Ataka M, Katoh E, Wakayama NI. (1997) Magnetic orientation as a tool to study the initial stage of crystallization of lysozyme. *J Cryst Growth* **173**: 592–596.

Brooks JS, Reavis JA, Medwood RA, Stalcup TF, Meisel MW, Steinberg E, Amowitz L, Stover CC, Perenboom JA. (2000) New opportunities in science, materials, and biological systems in the low-gravity (magnetic levitation) environment. *J Appl Phys* **87**: 6194–6199.

Ikezoe Y, Hirota N, Nakagawa J, Kitazawa K. (1998) Making water levitate (Scientific Correspondence). *Nature* **393**: 749–750.

Kinoshita T, Ataka M, Warizaya M, Neya M, Fujii T. (2003) Improving quality and harvest period of protein crystals for structure-based drug design: Effects of a gel and a magnetic field on bovine adenosine deaminase crystals. *Acta Crystallog D* **59**: 1333–1335.

Kiyoshi T, Ozaki O, Morita H, Nakayama H, Jin HB, Wada H, Wakayama NI, Ataka M. (1999) *IEEE Appl Supercond* **9**: 362–365.

Lin S-X, Zhou M, Azzi A, Xu G-J, Wakayama NI, Ataka M. (2000) Magnet used for protein crystallization: Novel attempts to improve the crystal quality. *Biochem Biophys Res Commun* **275**: 274–278.

Lübbert D, Meents A, Weckert E. (2004) Accurate rocking-curve measurements on protein crystals grown in a homogeneous magnetic field of 2.4 T. *Acta Crystallog D* **60**: 987–998.

McPherson A. (1999) in *Crystallization of Biological Macromolecules*, p. 453. Cold Spring Harbor Laboratory Press, New York.

Maki S, Ataka M. (2004) Suppression and promotion of convection in water by use of radial components of the magnetization force. *J Appl Phys* **96**: 1696–1703.

Maki S, Ataka M. (2005) Three-dimensional computation of convection of water at the center of a superconducting magnet. *Phys Fluids* **17**: 087107-1–087107-7.

Maki S, Ataka M. (2007) Magnetic levitation with permanent magnet: Application to three types of plant seed, *Jpn J Appl Phys* **46**: 2910–2911.

Maki S, Oda Y, Ataka M. (2004) High-quality crystallization of lysozyme by magneto-Archimedes levitation in a superconducting magnet. *J Cryst Growth* **261**: 557–565.

Moreno A, Quiaroz-Garcia B, Yokaichiya F, Stojanoff V, Rudolph P. (2007) Protein crystal growth in gels and stationary magnetic fields. *Cryst Res Technol* **42**: 231–236.

Ozoe H. (2005) *Magnetic Convection*. Imperial College Press, London.

Poodt PWG, Heijna MCR, Tsukamoto K, de Grip WJ, Christianen PCM, Maan JC, van Enckevort WJP, Vlieg E. (2005) Suppression of convection using gradient magnetic fields during crystal growth of $NiSO_4$ $6H_2O$. *Appl Phys Lett* **87**: 214105-1–214105-3.

Pusey ML. (1993) A computer-controlled microscopy system for following protein crystal face growth rates. *Rev Sci Instrum* **64**: 3121–3125.

Qi J, Wakayama NI, Ataka M. (2001) Magnetic suppression of convection in protein crystal growth. *J Cryst Growth* **232**: 132–137.

Ramachandran N, Leslie FW. (2005) Using magnetic fields to control convection during protein crystallization — Analysis and validation studies. *J Cryst Growth* **274**: 297–306.

Rothgeb TM, Oldfield E. (1981) Nuclear magnetic resonance of heme protein crystals. General aspects. *J Biol Chem* **256**: 1432–1446.

Saijo S, Yamada Y, Sato T, Tanaka N, Matsui T, Sazaki G, Nakajima K, Matsuura Y. (2005) Structural consequences of hen egg-white lysozyme orthorhombic crystal growth in a high magnetic field: Validation of X-ray diffraction intensity, conformational energy searching and quantitative analysis of B factors and mosaicity. *Acta Crystallog D* **61**: 207–217.

Sakurazawa S, Kubota K, Ataka M. (1999) Orientation of protein crystals grown in a magnetic field. *J Cryst Growth* **196**: 325–331.

Santesson S, Cedergren-Zeppezauer ES, Johansson T, Laurell T, Nilsson J, Nilsson S. (2003) Screening of nucleation conditions using levitated drops for protein crystallization. *Anal Chem* **75**: 1733–1740.

Sato T, Yamada Y, Saijo S, Hori T, Hirose R, Tanaka N, Sazaki G, Nakajima K, Igarashi N, Tanaka M, Matsuura Y. (2000) Enhancement in the perfection of orthorhombic lysozyme crystals grown in a high magnetic field (10 T). *Acta Crystallog D* **56**: 1079–1083.

Sato T, Yamada Y, Saijo S, Hori T, Hirose R, Tanaka N, Sazaki G, Nakajima K, Igarashi N, Tanaka M, Matsuura Y. (2001) Improvement in diffraction maxima in orthorhombic HEWL crystal grown under high magnetic field. *J Cryst Growth* **232**: 229–236.

Sazaki G, Yoshida E, Komatsu H, Nakada T, Miyashita S, Watanabe K. (1997) Effects of a magnetic field on the nucleation and growth of protein crystals. *J Cryst Growth* **173**: 231–234.

Tu J-L, Chin K-H, Wang AH-J, Chou S-H. (2007) The crystallization of apo-form UMP kinase from *Xanthomonas campestris* is significantly improved in a strong magnetic field. *Acta Crystallog F* **63**: 438–442.

Wakayama NI. (2003) Effects of a strong magnetic field on protein crystal growth. *Cryst Growth Design* **3**: 17–24.

Wakayama NI. (2005) Damping of solute convection during crystal growth by applying magnetic field gradients. *Jpn J Appl Phys* **44**: L833–L835.

Wakayama NI, Ataka M, Abe H. (1997) Effect of a magnetic field gradient on the crystallization of hen lysozyme. *J Cryst Growth* **178**: 653–656.

Yanagiya S-I, Sazaki G, Durbin SD, Miyashita S, Nakada T, Komatsu H, Watanabe K, Motokawa M. (1999) Effect of a magnetic field on the orientation of hen egg-white lysozyme crystals. *J Cryst Growth* **196**: 319–324.

Yanagiya S-I, Sazaki G, Durbin SD, Miyashita S, Nakajima K, Komatsu H, Watanabe K, Motokawa M. (2000) Effects of a magnetic field on the growth rate of tetragonal lysozyme crystals. *J Cryst Growth* **208**: 645–650.

Yin D, Inatomi Y, Kuribayashi K. (2001) Study of lysozyme crystal growth under a strong magnetic field using a Mach-Zehnder interferometer. *J Cryst Growth* **226**: 534–542.

Yin DC, Inatomi Y, Wakayama NI, Huang WD, Kuribayashi K. (2002) An investigation of magnetic field effect on lysozyme crystal dissolution and related phenomena. *Acta Crystallog D* **58**: 2024–2030.

Yin DC, Oda Y, Wakayama NI, Ataka M. (2003a) New morphology, symmetry, orientation and perfection of lysozyme crystals grown in a magnetic field when paramagnetic salts ($NiCl_2$, $CoCl_2$ and $MnCl_2$) are used as crystallizing agents. *J Cryst Growth* **252**: 618–625.

Yin D-C, Wakayama NI, Wada H, Huang W-D. (2003b) Significant effects of magnetic and gravitational fields on the morphology of protein crystals (orthorhombic lysozyme crystals grown using $NiCl_2$ as crystallizing agent). *J Phys Chem B* **107**: 14140–14144.

Yin D-C, Wakayama NI, Harata K, Fujiwara M, Kiyoshi T, Wada H, Niimura N, Arai S, Huang WD, Tanimoto Y. (2004) Formation of protein crystals (orthorhombic lysozyme) in quasi-microgravity environment obtained by superconducting magnet. *J Cryst Growth* **270**: 184–191.

Yin D-C, Lu H-M, Geng L-Q, Shi Z-H, Luo H-M, Li H-S, Ye Y-J, Guo W-H, Shang P, Wakayama NI. (2008) Growing and dissolving protein crystals in a levitated and containerless droplet. *J Cryst Growth* **310**: 1206–1212.

Zhong C, Wang L, Wakayama NI. (2001) Effect of a high magnetic field on protein crystal growth — Magnetic field induced order in aqueous protein solutions. *J Cryst Growth* **233**: 561–566.

Chapter 12

Interactive Crystallomic

*Jan Sedzik**

Biologically important molecules have to be crystallized from solution before their atomic structures can be determined. Once their structures have been determined, reliable and fine structural details, which are crucial for structure-based drug design, can be obtained. Crystallization is an important but very poorly understood process. Our inability to produce good quality crystals is a severe limitation of protein crystallography. In this chapter we present rational guidelines which aim to alter our know-how on crystallization from "black magic" into quantitative and affordable science. We have coined a new word, *crystallomic*, which can be applied to a diverse range of molecules and may intensify research related to crystallogenesis.

Keywords: Crystallization; factorial design; crystallomic; proteins; virus; crystals; crystallization screens.

Introduction

The most common procedure for uncovering structural information about biological molecules is producing crystals. (The NMR technique plays a minor role at this stage). Crystals are important for the production of structure-based drugs; however, accurate and reliable structures are only available when good quality crystals can be created. The acquisition of good quality crystals is followed by computer-intensive analysis, leading to high-resolution atomic details. Crystals are objects in which molecules are

*Protein Crystallization Facility, Department of Neurobiology, Care Sciences and Society, NOVUM, Karolinska Institutet, Stockholm, Sweden, and Department of Chemical Engineering and Technology, Protein Crystallization Facility, KTH, Teknikringen 28, 100 44 Stockholm, Sweden. E-mail: sedzik@swipnet.se.

arranged in a periodic and repeating fashion that extends in three (or two) dimensions. Crystals may comprise 10^{19} molecules. How many reliable three-dimensional structures are available? The current holding of accurate structures in the Protein Data Bank can be found online at www.rcsb.org/pdb. There is indeed a steady increase in solved protein structures: on February 15, 2005, 25 260 structures had been determined by X-ray diffraction (4376 determined by NMR); on August 1, 2006, there were 37 981 (4751 determined by NMR); on May 5, 2008, there were in total 42 342 structures determined by X-ray and 7150 by NMR. They are mainly water-soluble proteins, peptides, viruses, protein/nucleic acid complexes, nucleic acids or carbohydrates. In contrast, the output for crystallizing membrane proteins or viruses has been regrettably low (36 proteins, Aoyama *et al.*, 2004; 50 viruses, Sedzik *et al.*, 2001). The limiting factor, referred to by some researchers as "a bottleneck," is simply to find the right physico-chemical conditions in which dispersed, monomeric molecules can begin to nucleate (aggregate) forming nuclei. Following this, spontaneous and steady growth leads to a "large" crystal of size at least ~0.1 mm in all dimensions. Some proteins crystallize very easily and rapidly, for example lysozyme (McPherson, 1990), while for others successful crystallization was only achieved after 20 years (cytochrome c oxidase; Yoshikawa *et al.*, 1996).

In this chapter a new and rational approach to the process of crystallizing biological molecules, which we have called *interactive crystallomic*, is described. This strategy is affordable, efficient and economical, and can easily be adopted by any laboratory struggling to overcome the acute bottleneck of crystallization.

Experimental Procedure — An Example

Pre-crystallization assumptions

Before carrying out these crystallization trials of virus-like particles, two assumptions were made:

(i) use the same buffer for crystallization, in which virus-like particles are purified (50 mM Tris, pH 7.4), and

(ii) use precipitant polyethylene glycol, which has a molecular weight of 20 000 (PEG20 000) instead of commonly used salts, organic solutions or low molecular weight polyethylene glycols.

Chemicals and crystallization set-up

A stock solution of BK-3 virus-like particles at a concentration of 10 mg/ml (provided by Josefina Nilsson, KI, Sweden) was used. The BK-3 particles are spherical and have a diameter of about 400 Å. A database search on viruses (http://mmtsb.scripps.edu/viper/CRYSTALS/XTAL1/crystals.html) confirmed that crystallization of such virus-like particles has never previously been attempted. To check purity, SDS-PAGE was performed. The solution was not contaminated by other membranes or organelles since only one protein band was detected (data not shown here). The crystallization method used was the so-called "hanging drop" employing a reservoir of 1000 μl in an air-tight well (Costar plate, 4×6 wells, Hampton Research, Laguna Niguel, CA, USA). After mixing 3 μl of virus-like solution (dispersed in water) with 3 μl of reservoir, the total volume of the drop was 6 μl. The siliconized quartz cover slips — on which the crystallization drops were created — had a diameter of 18 mm.

Rationale and design of the crystallization trials

The strategy we implemented, which we call *crystallomic,* is an "interactive protocol for crystallization." In practical terms this means that the composition of the reservoir (mother liquid) for a subsequent crystallization trial (n), depends on what was noted immediately after the drop was created in trial ($n - 1$). When the drop is created by mixing one volume of solution of virus particles with the same volume of reservoir, there are most often two possibilities: (a) the drop is clear and transparent, or (b) the drop is not clear and an amorphous precipitate develops rapidly, the drop becomes cloudy because of phase separation, or the nearly instantaneous formation of micro-crystals has occurred. The idea of *crystallomic* is to set up several crystallization "hanging drops," and, as long as the drop is clear immediately after it was created, to simultaneously vary the virus (or protein) concentration *and* the concentration of precipitate.

Algorithm

The sequence of steps involved in setting up crystallization drops can be described as follows:

```
START
    CHOOSE precipitant
1   IF FOR EXPERIMENT n OUTCOME = Yp
    THEN
    APPLY X1 = [Cv(n + 1)] * [Cp(n + 1)] < [Cv(n)] * [Cp(n)] FOR
                EXPERIMENT n + 1
        IF OUTCOME = Yo THEN GOTO 1
    OTHERWISE
    APPLY X2 = [Cv(n + 1)] * [Cp(n + 1)] > [Cv(n)] * [Cp(n)] FOR
                EXPERIMENT n + 1
CONTINUE
END
```

The immediate outcomes after creating a drop n are classified as either Yp or Yo. When the drop is clear, it is classified as Yo; when the drop is not clear (because of precipitation or phase separation), it is classified as Yp. In the subsequent experiment $(n + 1)$, the condition X1 is applied such that the product of the multiplication of the concentrations of the virus-like particles and the precipitant for experiment $(n + 1)$ is smaller than the same product for experiment n. In other words, X1 satisfies the equation $[C_v(n + 1)] * [C_p(n + 1)] < [C_v(n)] * [C_p(n)]$, where $C_v(n)$ and $C_p(n)$ are the concentrations (mg/ml) of the virus and the precipitant (w/v, % or M) respectively. However, if the outcome of the drop from experiment $(n + 1)$ is Yo and the drop is clear, then the experiments are terminated. If the outcome of the drop is not Yo, then the condition X2 is applied such that the product of the concentrations (virus and precipitant) for experiment $(n + 1)$

is greater than that for experiment n. This can be expressed by the following equation: $[C_v(n + 1)] * [C_p(n + 1)] > [C_v(n)] * [C_p(n)]$. If this sequence is applied to all crystallization trials, it is expected that the vapor equilibration of the drop with the reservoir will most likely be terminated when the drop reaches the metastable state (Meyerson, 1999).

Results

A set of five crystallization drops was performed. The conditions were applied at room temperature (21–23°C) and are summarized in Table 1 and Fig. 1. The experiments were terminated after only one additional experiment, A6 (Table 1) for which the product of $[C_v] * [C_p] = 20.0$ was as for experiment A5. During the second (one month later) and third (three months later) evaluation of the drops, crystallization was observed (Fig. 1). In summary, only 10 μl of virus solution was used, which is equivalent to 0.1 mg of virus in total. The same rational protocol (previously termed "factorially designed trials/experiments") was applied to the crystallization of Semliki Forest virus (Sedzik *et al.*, 2001), rHEV (2002, unpublished data), bovine P2 protein in lipid-free and lipid-bound form (Sedzik, 2003), and pig P2 protein in lipid free-form (Sedzik *et al.*, 2005). In all cases, crystallization was achieved after performing no more than few dozen rationally designed experiments.

Discussion

To the best of our knowledge, a rational approach such as crystallomic has never been tested or applied to the crystallization of biological molecules. The name *crystallomic* refers to a procedure used to rationally and efficiently find initial conditions in which any biological molecule dispersed in aqueous solution can begin to nucleate and then (after refinement of the conditions) form visually pleasing 3D crystals that can be used for structural determination. The experiments (Table 1) were designed without prior knowledge of the precipitating agent or pH of the buffer. We have chosen a PEG of MW 20 000, since such a precipitant is very seldom used

Table 1. A Summary of the Crystallization Trials of BK Virus-like Particles[a]

A1	A2	A3	A4	**A5**	**A6**
Virus, 2 mg/ml	Virus, 4 mg/ml	Virus 6, mg/ml	Virus, 8 mg/ml	Virus, 8 mg/ml	Virus, 10 mg/ml
1% (w/v)	1.25% (w/v)	1.6% (w/v)	5.0% (w/v)	2.5% (w/v)	2% (w/v)
PEG20 000	PEG20 000	PEG20 000	PEG20 000	PEG20 000	PEG20 000
50 mM Tris, pH 7.5	50 mM Tris, pH 7.5	50 mM Tris, pH 7.5	50 mM Tris, pH 7.5	50 mM Tris, pH 7.5	50 mM Tris, pH 7.5
$[C_p] * [C_v] = 2$	$[C_p] * [C_v] = 5$	$[C_p] * [C_v] = 9.6$	$[C_p] * [C_v] = 40$	$[C_p] * [C_v] = 20$	$[C_p] * [C_v] = 20$
Outcome scores:	Outcome scores:	Outcome scores:	Outcome scores:	Outcome scores:	Outcome scores:
0 (0 day)	0	0	0.5	0	0
0 (10 days)	0	0	3	0	0
0 (1 month)	0	0	3	7	9
0 (3 months)	0	0	3	7	9

Note: The scoring of the outcomes of crystallization (Sedzik 1994) of each drop was as follows:

0 – drop was clear, without precipitate; 0.5 – very weak amorphous precipitate; 3 – very strong amorphous precipitate; 7 – several (shower) microcrystals of ~0.05 mm; 9 – single crystals of ~0.1 mm, and without any amorphous precipitate.

[a] The crystallization trials were performed in the following order: A1–A5. Experiment A4 gave an immediate precipitate when the virus solution (8 mg/ml) was mixed with the reservoir at a concentration of 5% (w/v) of polyethylene glycol, which has a molecular weight of 20000. Therefore, in experiment A5, the concentration of the precipitate was lowered to 2.5% (w/v). In all the drops, the buffer was 50 mM Tris, pH 7.5. The dishes were disturbed only three times during the three-month period of the experiments. Crystals were noticed in drop A5 during the second evaluation check, one month after the experiment was set up. The dishes were kept in darkness at room temperature (21–23°C). C_p — concentration of precipitate; C_v — concentration of virus.

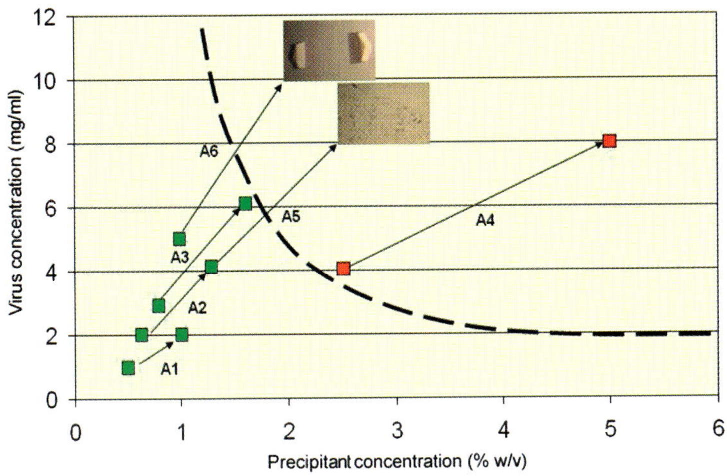

Fig. 1. Precipitation diagram after performing only five crystallization trials (A1–A5). The additional experiment, A6, resulted in single crystals and was performed after analyzing experiments A1–A5. Each experiment describes the conditions at the beginning of the experiment and again after two months. Red squares mean that precipitate had developed; green squares mean that the drop was clear, without detectable precipitate. Experiment A5 resulted in microcrystals, but A6 resulted in large single crystals (0.2 mm). The boundary between undersaturation and oversaturation is calculated using $c^*(\text{virus}) \times c(\text{precipitant}) = 9.6$ [$c^*(\text{virus})$ — concentration of virus in supersaturation]. Crystals (drop A5 and A6) were obtained after setting up the crystallization trials as described in Table 1 and according to the equation $c(\text{virus}) \times c(\text{precipitant}) = 20$ [8 (mg/ml) × 2.25 (% w/v) = 20]. Assuming that the relationship $c(\text{virus}) \times c(\text{protein}) = 20$ was valid, the next drop was set up with a concentration of 10 mg/ml of virus and 2% (w/v) concentration of precipitant so that $c(\text{virus}) \times c(\text{precipitant})$ still equals 20. How to plot the precipitation diagram is described in detail for membrane proteins in Sedzik *et al.* (2001, 2003), and Sedzik (2004).

when crystallizing viruses or proteins. The pH was the same as that in which the virus particles were purified.

It this short chapter, it is shown that — when searching for initial crystallization conditions — it is very beneficial to simultaneously vary the protein concentration and the concentration of the precipitant. From Table 1 and Fig. 1 it is evident that from each of the experiments performed, two very important pieces of information are available: namely, what occurred in the drop at the beginning of the experiment and the result after 1–3 months.

The same can be said about crystallization data performed by trial-and-error, or so-called "high-throughput crystallization." However, here making more or less reasonable decisions on how to continue with crystallization — in situations where crystals are not available — is impossible.

It is widely believed among crystal growers that crystallization by chance and trial-and-error methods is most successful when applied to biological macromolecules; consequently, thousands of randomly performed trials, generally known as "high-throughput crystallization" or "deliberate approaches to screening for initial crystallization conditions," are thought to guarantee further success in obtaining crystals (Luft *et al.*, 2003). The high-throughput crystallization offered, for example, by leading crystals provider Hauptman-Woodward Institute (Buffalo, USA) comprises the screening of experiments in 1536-well micro-assay plates and requires a minimum of 400 μl of protein solution at a concentration 10 mg/ml. Moreover, it costs US$1000 per plate, if the protein name is not released by the customer. (http://www.hwi.buffalo.edu/High-Through/High_Through.html).

The application of more rational approaches has gained little recognition among protein crystallographers. Indeed, it has been suggested to "get rid of all the crystals altogether" from structural biology (Patel, 2002). Instead of protein crystals, it would be sufficient to have only several molecules of protein in order to perform a full structural determination. Indeed, turning protein crystallization from an art into a science is still in the early stages of development (Chayen, 2004).

Optimizing protein concentration is very important when using commercially available screens (see www.hamptonresearch.com for references). When using these screens, the protein concentration is constant, usually >10 mg/ml. However, if the protein concentration is too high, there is nothing other than precipitate at the outset of all the experiments (sometimes phase separation may develop as well); however, when the concentration is too low the drops are usually clear in all trials. The remedy is pre-crystallization testing, requiring the adjustment of the stock protein concentration used in further trials. If after mixing the protein solution with the content of tube #6 of Crystal Screen kit formulation (30% PEG4000, pH 7.5 and 0.2 M magnesium chloride hexahydrate), precipitation in the drop appears too heavy, the protein concentration has to be reduced by half. If there is no immediate precipitation in drop #6 or even drop #4 (2 M ammonium sulfate, pH 8.5), protein concentration should be increased twofold for subsequent crystallization trials (www.hamptonresearch.com/stuff/tips99.html). It should be emphasized that protein concentration in this pre-crystallization test is not a variable factor, since it is kept constant from experiment to experiment. There has also been a report on growing crystals of R14 virus from ammonium sulfate at concentration x% saturation and y (mg/ml) concentration of R14 virus, but only if the product of x and y is numerically between 5–10 units (Erickson *et al.*, 1983).

Summary

The marketing tagline of Hampton Research Inc. (Laguna Hill, CA, USA) is: "behind every structure is a crystal." This small object (~0.1 mm) is the most important in determining the detailed arrangement of atoms. This knowledge is necessary for structure-based drug design, which in turn facilitates the production of effective drugs to treat disease. Traditional wisdom regarding crystallization indicates that there are no clearly defined rules governing how to find (or choose) the composition of the crystallization medium that will result in crystals suitable for the acquisition of high-resolution diffraction data. It is well known that protein and precipitant concentrations are two parameters that are crucial for crystallization. It is obvious that if two parameters are relevant to a result, their product will very likely be relevant too; the product of these two concentrations is a decisive parameter to vary. This approach has not yet been recognized or applied to crystallization. Instead, it has been recommended that varying protein concentration and keeping precipitant concentration constant or *vice versa* is a good way to find crystallization conditions.

The general approach of screening for crystallization conditions by simultaneously varying protein concentration and concentration of precipitate is defined in this paper as *crystallomic*. It is a cost effective and materials efficient approach. It requires the systematic collection of raw crystallization data in a computerized and logical form so that the analysis can always give more insights and more knowledge on how to implement subsequent trials. If crystallization fails, by plotting a precipitation diagram, new information can be used to focus further searches for nucleation and metastabile states. This is in contrast to other approaches (Luft *et al.*, 2003). Finally, the *crystallomic* approach is affordable for all laboratories, since it does not require costly instruments. Most importantly, each molecule being crystallized is treated individually, since it is not exposed to the same panel of screens.

Acknowledgments

This research was supported by the Research Board of the Karolinska Institute (Sweden). The author is grateful to Yvone Ishida for making corrections to the language in this chapter.

References

Aoyama H, Yamashita E, Sakurai K, Tsukihara T. (2004) Strategy to obtain high resolution structure of membrane protein by X-ray crystallography. In Cheng H, Hammar L (eds). *Conformational Proteomics of Macromolecular Architectures*, pp. 111–132. World Scientific Publishing, Singapore.

Chayen NE. (2004) Turning protein crystallization from an art into a science. *Curr Opinion Struct Biol* **14**: 577–583.

Erickson JW, Frankenberger EA, Rossman MG, Fout GS, Medappa KC. (1983) Virion orientation in cubic crystals of the human common cold virus HRV14. *Proc Nat Acad Sci (USA)* **80**: 931–924.

Giege R, Dock AC, Kern D, Lorber B, Thierry JC, Moras D. (1986) The role of purification in the crystallization of proteins and nucleic acids. *J Cryst Growth* **76**: 554–587.

Luft JR, Collins RJ, Fehrmas NA, Lauricella AN, Veatch CK, DeTitta GT. (2003) A deliberate approach to screening for initial crystallization conditions of biological macromolecules. *J Struct Biol* **142**: 170–179.

McPherson A. (2003) *Introduction to Macromolecular Crystallography*. John Wiley & Sons.

McPherson A. (1990) Current approaches to macromolecular crystallization. *Eur J Biochem* **189**: 1–23.

Meyerson AS. (1999) *Molecular Modeling: Applications in Crystallization.* Cambridge University Press, Cambridge.

Patel H. (2002) Shorter, brighter, better. *Nature* **415**: 110–111.

Sedzik J. (1994) DESIGN: A guide to protein crystallization experiments. *Arch Biochem Biophys* **308**: 342–348.

Sedzik J, Hammar L, Haag L, Skoging-Nyberg U, Tars K, Marko M, Cheng HR. (2001) Structural proteomics of enveloped viruses: Crystallization, crystallography, mutagenesis and cryo-electron microscopy. *Rec Res Dev Virol* **3**: 41–60.

Sedzik J, Kotake Y, Uyemura K, Ataka M. (2003) Factorially designed crystallization trials of the full-length P0 myelin glycoprotein. I. Precipitation diagram. *J Cryst Growth* **247**: 483–496.

Sedzik J, Carlone J, Fasano A, Liuzzi MG, Riccio P. (2003) Crystals of myelin P2 protein in lipid-bound form. *J Struct Biol* **142**: 292–300.

Sedzik J, Matsuura T, Kotake V, Hedman E, Tsukihara T. (2005) Structural proteomics and crystallomic of porcine P2 myelin protein. *J Neurochem* **94**(Suppl): 201–202.

Yoshikawa S, Shinzawa-Itoh K, Tsukihara T. (1996) Crystal structure and reaction mechanism of bovine heart cytochrom c oxidase. *Keio U Symp Life Sci Med* **1**: 13–26.

Chapter 13

Virtual Molecule: P0 Myelin Glycoprotein. I. Homology Modeling and Prediction of the Secondary and Tertiary Structure

Jan Pawel Jastrzebski[*,†] *and Jan Sedzik* [‡]

The P0 protein is a structural glycoprotein that accounts for more than 50% of the peripheral nervous system myelin membrane protein and is associated with human Charcot-Marie-Tooth disease. The P0 glycoprotein consists of three parts: the extracellular domain, the single helix transmembrane region, and the intracellular domain. The three-dimensional atomic structure of the full sequence P0 glyprotein has not yet been solved. Using homology modeling with secondary and tertiary structure prediction, it is possible to construct *in silico* a three-dimensional model of this protein embedded in the lipid bilayer. Further detailed analyses of such a model allow one to distinguish theoretical autoimmune epitopes and to visualize molecular effects and chemical changes in the molecule. For example, breaking a disulfide bridge can cause the opening of the P0_Ex part of the protein, leading to a decrease in migration speed of the "opened" protein in the SDS-PAGE.

Keywords: Homology modeling; comparative modeling; structure prediction; threading; myelin; P0; HNK-1; epitopes.

*Corresponding author.
[†]Department of Plant Physiology and Biotechnology, Faculty of Biology, University of Warmia and Mazury, ul. Oczapowskiego 1A/113, 10-719 Olsztyn, Poland. E mail: jan.jastrzebski@uwm.edu.pl.
[‡]Karolinska Institutet, Protein Crystallization Facility, NOVUM, Stockholm, Sweden. E-mail: sedzik@swipnet.se.

Introduction

P0 glycoprotein is the major structural protein of peripheral nerve myelin; it is thought to modulate inter-membrane adhesion by providing the stable link between apposed cytoplasmic membrane surfaces in peripheral myelin (Inouye *et al.*, 1985). Most of the studies on P0 protein — an abundant structural protein of peripheral myelin — have been focused on structure-function correlation in higher vertebrates: bovines and human (Luo *et al.*, 2006). It has even been observed that conformational changes of P0 protein are due to protonation-deprotonation of His^{52} at P0's putative adhesive interface (Luo *et al.*, 2006). Removal of the fatty acids that are attached to the single Cys^{153} residue in the cytoplasmic domain of P0 has not changed the PNS myelin structure of frog and mouse, suggesting that the P0-attached fatty acyl chain does not play a significant role in PNS myelin compaction and stability (Luo *et al.*, 2007; Luo *et al.*, 2008).

The P0 protein of peripheral myelin is very hydrophobic, and has one transmembrane segment. The soluble part, representing two–thirds of the total mass of the protein, has been cloned, expressed, and crystallized for three-dimensional structure determination (Shapiro *et al.*, 1996).

Methods and Tools for Homology Prediction

A virtual model of the complete P0 protein has been built using a homology modeling method (HMM), secondary structure prediction and threading. BLAST on NCBI, SWISS-MODEL (An Automated Comparative Protein Modelling Server: First approach mode) and Project (optimize) mode on the ExPASy Proteomics Server, Verify3D, WhatIf and the DeepView Swiss-PdbViewer off-line tool have all been used to perform homology modeling. ClustalW on EMBL-EBI on-line service, BioEdit and JalView have been used to analyze the protein sequences. SWISS-MODEL, JPred and SOSUI have been used to predict secondary structure and the transmembrane region. RCSB PDB, Pfam, ChemSketch, RasMol, Swiss-PdbViewer, PyMOL, VMD and NAMD and Cn3D have been used to visualize, verify, optimize and rearrange 3D models of the protein.

The 1NEU_A model from PDB database has been used as a template for P0 extracellular part (P0_Ex). This is a recombinant protein of rat (Shapiro *et al.*, 1996). The sequence similarity between 1NEU (rat) and the human P0 extracellular part is very high (Fig. 2), affirming that these two sequences are homologs and therefore must have identical 3D structure.

The theoretical pI and molecular weight of protein has been evaluated using the Compute pI/Mw tool (Bjellqvist, 1993), available on the ExPaSy (www.expasy.com) server. Electrostatic potential has been calculated using the Coulomb computation method of dielectric constant ($\varepsilon = 4.0$ for protein, $\varepsilon = 80$ for solvent), and using the SwissPDBViewer tool with ionic strength of solvent 5 mM.

Amino Acid Sequence — A Basic Input for Modeling

The sequence of the P25189 entry in the UniProtKB/Swiss-Prot database has been used as a full amino acidic sequence of human P0 protein (Sakamoto *et al.*, 1987), but the sequence denoted P25189.1 in the Swiss-Prot database has been used as a target; in homology modeling the "target" is the sequence whose structure is under modeling. This sequence contains two regions: short signal region-position 1–29, and mature chain region-position 30–248 (Fig. 1).

Multiple alignments have been calculated with the ClustalW online tool on the EBI server using the PAM matrix and defaults for other settings. The results of each pairwise alignment and the general multiple alignment of each group of pairwise alignments with the secondary prediction have been visualized using the JalView tool. The columns of conserved residues are colored blue in Fig. 1.

Construction and Analysis of a Three-Dimensional Atomic Model of P0 Glycoprotein

The protein part (excluding the HNK-1 epitope) of the P0 molecule (Table 1) consists of three elements called domains: extracellular (P0_Ex), transmembrane (P0_TM) and intracellular (P0_Int).

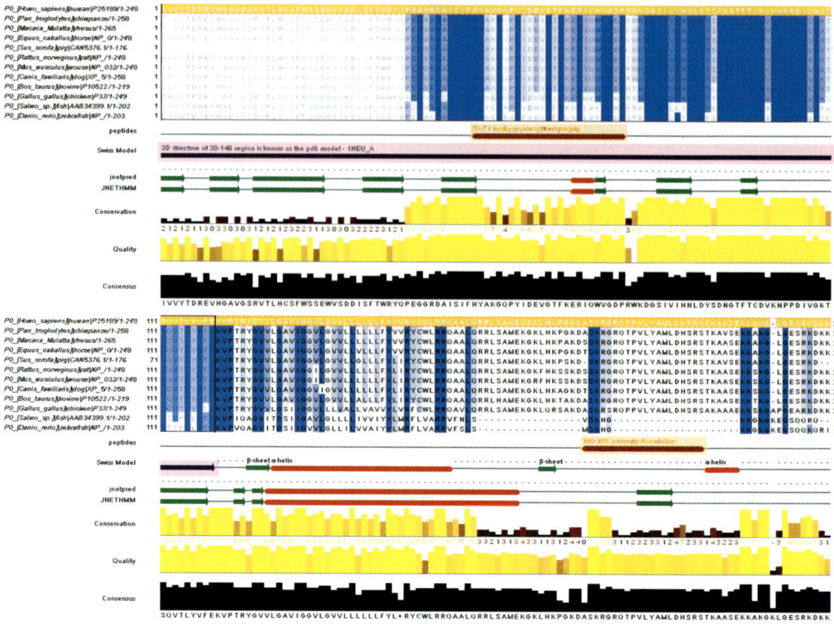

Fig. 1. The multiple alignments of a few sequences of P0 protein of vertebrates. The human P0 sequence is a reference (top) and is written on an orange background. The most conserved residues are marked in blue (columns) and the degree of conservation is quantitatified in the "Conservation" row. Yellow means residues are conserved; dark yellow means they are not. The "Swiss-model," "jnetpred" and "JNETHMM" rows present the results of the secondary structure predictions of each tool and method. The epitops causing experimental autoimmune neuritis (EAN) (Westall, 2006; Sedzik, 2008) are marked in the "peptides" row. The "Quality" row presents quality of alignment — the quality of the fit of each residue. The "Quality" row is adequate to the "Conservation" row. The diagram in the "Consensus" row describes the accuracy of the consensus sequence estimation and the level of variability of each residue. Higher bars in the "Consensus" row mean more conserved residue; lower bars mean more variable position. The region adequate to the 1NEU model is marked with a pale row (region 1–119). The multiple alignments were calculated using the ClustalW tool on the EBI online service. The presentation of the results of alignment was prepared using the JalView tool.

The P0 extracellular domain — Soluble part

The extracellular domain (positions 29–147) is defined in Pfam as the immunoglobulin V-set domain (acc: PF07686). It is the biggest part of the P0 protein: 126 animo acids, and a theoretical molecular weight of

Table 1. The Theoretical Properties of the Three Parts of P0 Protein[a]

Peptide position	P0 part	MW [Da]	pI
1–126	P0_Ex	14437.08	5.47
127–152	P0_TM	2758.51	8.56
153–219	P0_Int	7603.91	10.87

[a]The pI (isoelectric point) of the P0_Ex and P0_Int parts of the molecule is significantly different: very low for the extracellular part (5.47) and high for the intracellular part (10.87).

14.5 kDa (Table 1). The three-dimensional structure of this domain has been crystallized and the solved structure is deposited in RCSB (the Research Collaboratory for Structural Bioinformatics) and PDB (Protein Data Base) (www.rcsb.org) under 1NEU accession number (Shapiro *et al.*, 1996). This model has been used in research as a template for homology modeling of the extracellular domain. There is a 97% identity between the analyzed sequence of human P0 and the sequence of the 1NEU model (Fig. 2).

In the 1NEU model there is one gap in the structure: the lack of information about coordinates of segment ^{132}Pro-^{133}Pro-^{134}Asp-^{135}Ile (in mature protein without signal region: ^{103}Pro-^{104}Pro-^{105}Asp-^{106}Ile). This loop has been automatically generated by the SWISS-MODEL (Fig. 3). The spatial structure of each model after each step of modeling has been verified and optimized by running the Verify3D, WhatIf, Gromos and Anolea programs.

Spatial polypeptide chain conformation and the distance from the membrane of the P0_Ex domain have been calculated and established on hydrophobicity and hydrophilicity profiles (Figs. 5 and 6) (Choe, 1996; Sharpio *et al.*, 1996), as well as the spatial localization of peptides (Figs. 6B and 6C) known as autoimmune epitopes of myelin P0 protein (Westall, 2006) and the location of HNK-1[1] (Kirschner *et al.*, 1996; Sharpio *et al.*, 1996; Voshol *et al.*, 1996; Cebo *et al.*, 2002).

[1] HNK-1 stands for human natural killer-1, it is a protein carbohydrate epitope, also known as CD57 and LEU7.

```
> pdb|1NEU|A     Chain A, Structure Of Myelin Membrane Adhesion Molecule P0
Length=124

 Score =  258 bits (660),  Expect = 1e-69, Method: Compositional matrix adjust.
 Identities = 121/124 (97%), Positives = 123/124 (99%), Gaps = 0/124 (0%)

Query  30   IVVYTDREVHGAVGSRVTLHCSFWSSEWVSDDISFTWRYQPEGGRDAISIFHYAKGQPYI  89
            IVVYTDREV+GAVGS+VTLHCSFWSSEWVSDDISFTWRYQPEGGRDAISIFHYAKGQPYI
Sbjct  1    IVVYTDREVYGAVGSQVTLHCSFWSSEWVSDDISFTWRYQPEGGRDAISIFHYAKGQPYI  60

Query  90   DEVGTFKERIQWVGDPRWKDGSIVIHNLDYSDNGTFTCDVKNPPDIVGKTSQVTLYVFEK  149
            DEVGTFKERIQWVGDP WKDGSIVIHNLDYSDNGTFTCDVKNPPDIVGKTSQVTLYVFEK
Sbjct  61   DEVGTFKERIQWVGDPSWKDGSIVIHNLDYSDNGTFTCDVKNPPDIVGKTSQVTLYVFEK  120

Query  150  VPTR  153
            VPTR
Sbjct  121  VPTR  124
```

Fig. 2. BLAST results from searching the PDP database for the sequence template of the P0 protein. The best hit was 1NEU chain A. "Query" is an analyzed sequence of P0 protein; "Sbjct," subject sequence, is the sequence found in database by BLAST. Identity between target (query) and template (subject) was 97% and E value 2×10^{-71} (an E value close to zero means that both sequences are homologs). The sequence between "Query" and "Sbjct" means consensus sequence (full agreement) of this alignment; similar residues are depicted as "+" and different residues by "■" (space).

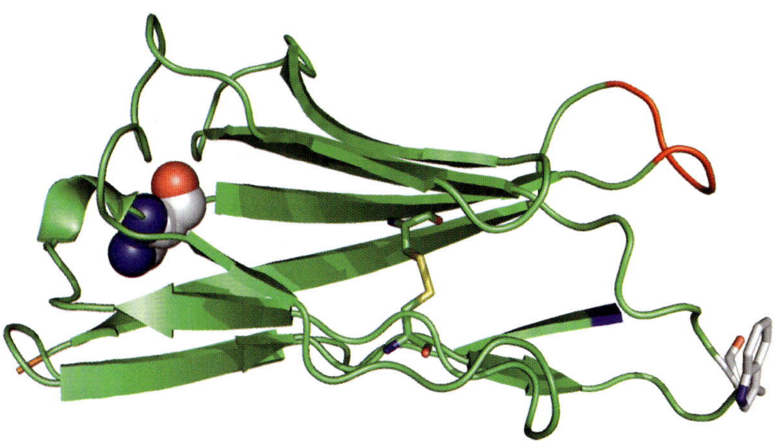

Fig. 3. The ribbon/wire model of the P0_Ex subunit of human P0 protein constructed using the homology modeling method. The structure of the ^{25}Ser···^{33}Ile loop (in red) is predicted using the loop database of the Swiss-Model — in the 1NEU model there is no information about the structure of this loop. The ending of the C-terminus is colored orange, and the ending of the N-terminus is colored blue. The ^{28}Trp residue is shown as a stick model (right). The disulfide bridge between ^{21}Cys···^{98}Cys is visible in the central part of the model.

The P0 transmembrane segment is 21 amino residues long

The transmembrane region has been estimated using SOSUI prediction (Fig. 4), JPred, JNetHMM and SWISS-MODEL predictions (Fig. 1), and hydrophobicity and hydrophilicity profiles (Fig. 5). Finally, the region from ^{130}Gly to ^{151}Arg has been selected as the most probable transmembrane region and has been modeled as a strongly hydrophobic transmembrane α-helix.

P0_Ex and P0_TM parts are linked by a 9-a.a.-long coil; ^{119}Glu···^{127}Val. This region has been generated using the Swiss-Model loops database. Modeling uncovered the ^{119}Glu···^{127}Val loop and three anchor positions (immovable residues): ^{119}Glu as a beginning of the loop, ^{127}Val as an end of the loop and ^{121}Val as a unique hydrophobic position in the loop. The ^{121}Val residue has been localized in the direct vicinity of the membrane surface (Figs. 5 and 8).

This amino acid sequence is of a MEMBRANE PROTEIN which have 1 transmembrane helix.

No.	N terminal	transmembrane region	C terminal	type	lengtl
1	125	YGVVLGAVIGGVLGVVLLLLLLF	147	PRIMARY	23

[Hydropathy profile]

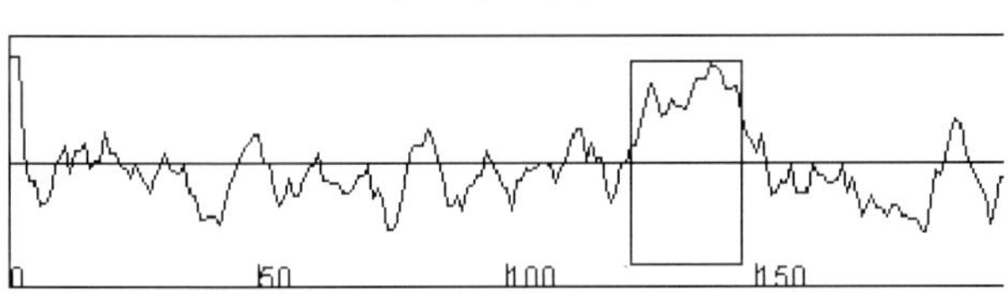

Fig. 4. The results of transmembrane region prediction using the SOSUI tool. The algorithm found only one 23-a.a.-long transmembrane region. It was the first step in the prediction of the secondary structure domains. The algorithms calculating the hydropathy profiles are accurate for this research.

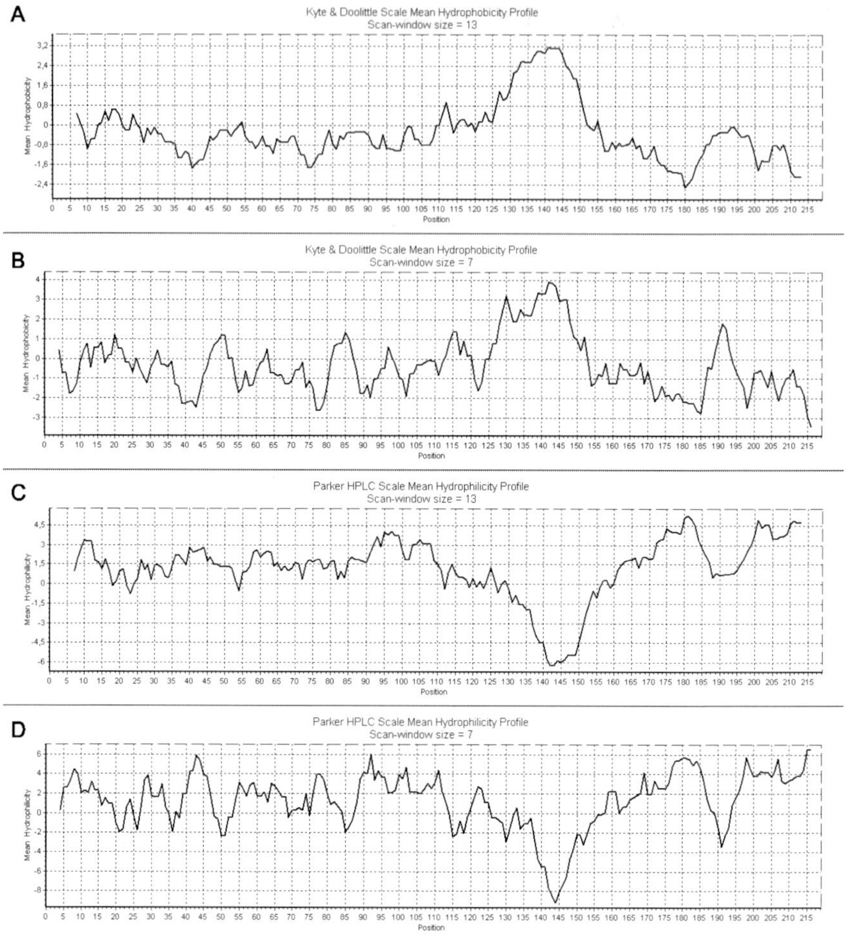

Fig. 5. Hydrophobicity and hydrophilicity profiles of P0 protein prepared using the BioEdit tool. A and B profiles are calculated based on the Kyte and Doolittle (1982) scale. C and D hydrophilicity profiles are calculated based on the hydrophilicity scale derived from high-performance liquid chromatography peptide retention data (Parker *et al.*, 1986). A and C profiles are prepared for window size = 13 (6 a.a. before analyzed position, 6 a.a. after analyzed position: 6 + 6 + 1 = 13). B and D profiles are for window size = 7 (3 + 3 + 1). The region between 138–148 positions discloses a strongly hydrophobic profile. This region belongs to the transmembrane helix (see Figs. 6 and 8). There is also a hydrophobic region between 190–195 residues.

The P0 intracellular part

This part is much smaller than P0_Int: it has a 67-a.a.-long region; MW is 7604 Da. This region is strongly hydrophilic (Figs. 4–6), of strong positive charge (Fig. 8), and has alkaline properties (Tables 1–3).

The intracellular part has been estimated and designed as a combination of two α-helixes and two ribbons of β-strands (Wong, Filbin, 1994; Luo *et al.*, 2007). One of the β-strands is strongly hydrophobic (in proximity to the 190–195 region; Figs. 1 and 5). There are two possibilities for modeling this

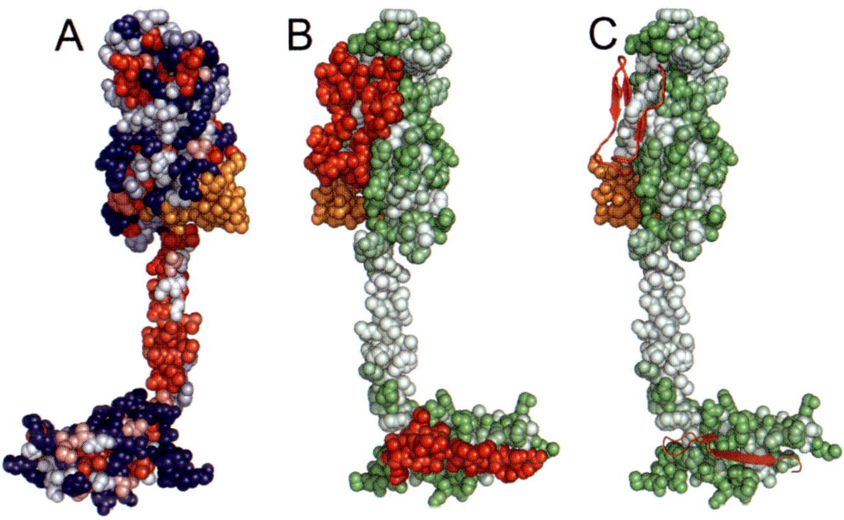

Fig. 6. The space filling models A, B, C of the whole sequence of P0 protein containing the HNK-1 epitope. (A) exhibits hydrophobic (red) and hydrophilic (blue) residues. Color intensity represents the strength of the property (for instance, light blue means faintly hydrophilicity; dark blue signifies strong hydrophilicity) in accordance with the Kyte, Doolittle (1982) scale data. A special PyMOL (company name) script was written to calculate the hydrophobicity profile. The strong hydrophobic transmembrane helix is colored dark red. (B) and (C) show similar types of hydrophobicity profiles, the intense green color meaning high hydrophilicity. The autoimmune epitops (Zou *et al.*, 2000; Westall, 2006) are colored red in these two figures. In addition, on (C), both peptides are shown as ribbons. The HNK-1 oligosaccharide is colored orange.

Table 2. The Theoretical Properties of Estimated Epitopes Assumed to be One-domain-long Epitopes[a]

Peptide position	Sequence	P0 part	MW [Da]	pI
52–76	HYAKGQPYIDEVGTFKERIQWVGDP	P0_Ex	2934.26	5.48
180–199	ASKRGRQTPVLYAMLDHSRS	P0_Int	2273.60	10.90
1–13	IVVYTDREVHGAV	P0_Ex	1457.65	5.32
14–26	GSRVTLHCSFWSS	P0_Ex	1466.63	8.26
102–123	NPPDIVGKTSQVTLYVFEKVPT	P0_Ex	2432.80	6.07
155–172	LRRQAALQRRLSAMEKGK	P0_Int	2112.53	12.01
168–179	MEKGKLHKPGKD	P0_Int	1367.63	9.52
200–219	TKAVSEKKAKGLGESRKDKK	P0_Int	2188.56	10.12

[a]The difference between overall pI (isoelectric point) of P0_Int epitopes and P0_Ex epitopes (52–76) is 5.48; the intracellular part (180–199) is 10.90. A very similar situation occurs for the short epitopes: the pI of extracellular epitopes ranges between ranges between 5.32–8.26, for intracellular epitopes it ranges between 5.52–12.01.

Table 3. The Theoretical Properties of Estimated Epitopes Assumed to be Two-domain-long Epitopes[a]

Peptide position	Sequence	P0 part	MW [Da]	pI
52–76	HYAKGQPYIDEVGTFKERIQWVGDP	P0_Ex	2934.26	5.48
180–199	ASKRGRQTPVLYAMLDHSRS	P0_Int	2273.60	10.90
1–26	IVVYTDREVHGAVGSRVTLHCSFWSS	P0_Ex	2906.27	6.91
155–179	LRRQAALQRRLSAMEKGKLHKPGKD	P0_Int	2888.43	11.58

[a]It is likely that two-domain epitopes are more specific for autoimmune reaction than one-domain epitopes. Physical properties of the long epitopes are more similar to adequate well-known epitopes.

part: first, a single hydrophobic β-strand located in the middle between hydrophilic domains oriented by its hydrophobic residues into the center of subunit; second, two antiparallel β-strands parallel to the two α-helixes. Our calculations have confirmed previously published data (Luo *et al.*, 2007) on the structure of the P0_Int part; the subunit is modeled as two β-strands parallel to two α-helixes. Since one of the β-strands is known as an autoimmune epitope of P0 protein (Westall, 2006), it should be on the outside surface of this subunit. This orientation is possible if the model consists of two β-strands parallel to two α-helixes. That is why this confirmation has been chosen to build the model of the whole P0 protein.

The P0_Int subunit is very flexible and movable (Fig. 7(b)) (Wong, Filbin, 1994; Luo *et al.*, 2007). in the presented model the hydrophobic domain of the P0 intracellular part is located close to the membrane, but the charge of the whole subunit (Fig. 8; Tables 1–3) allows one to change this orientation into the polar environment (in this case, water).

Theoretical model of P0

The final model of the whole P0 protein (Fig. 7) has been verified using the WhatIf, Verify3D, Anolea and Gromos tools. The three-dimensional structure of the elaborated model is correct but needs better optimization of the torsion angles for advanced simulations.

The oligosaccharide epitope has been designed using the ChemSketch tool and 3D models of oligosaccharide ligands of a number of models from the Protein Data Bank (for instance, 1AC0, 2H6O and 2V82). The composition of HNK-1, interactions with the protein and the position of the epitope have been constructed based on the data from numerous publications (Voshol *et al.*, 1996; Ong *et al.*, 1999; Cebo *et al.*, 2002; Kakuda *et al.*, 2004; Kleene, Schachner, 2004; Kakuda *et al.*, 2005).

Hydrophobicity and Electrostatic Potential of P0 Protein

The profiles of hydrophobicity and hydrophilicity of the P0 protein are shown in Figs. 4 and 5. There are clear sharp maxima (Figs. 5(B) and 5(D)) of the hydrophobic and hydrophilic zones. Strongly hydrophobic residues of P0_Ex and P0_Int are located in the central part of the molecule and are inaccessible for environment (red residues on Fig. 6(A), 8(a) and Tables 1–3). A very similar situation occurs in both P0_Ex and P0_Int parts; moreover, hydrophilic residues are oriented towards the outside of the molecule. The transmembrane part is highly hydrophobic (Figs. 4–7) and folded in the stable α-helix.

The general electrostatic potential of the P0_Ex part contains numerous positive and negative local charges of a short range (Fig. 8). The sources of charge are polar amino acids. The P0_TM part consists of non-polar and

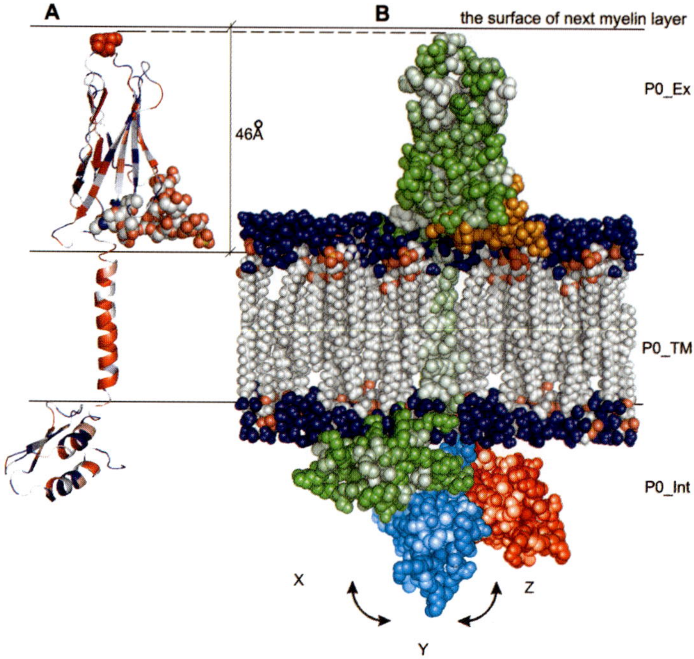

Fig. 7. (A) Ribbon model of P0 protein and possible arrangement relative to a lipid bilayer. (B) The space filling model of the whole P0 protein anchored in lipid bilayer. Higher intensity of green indicates more hydrophilicity. The HNK-1 epitop is colored orange. The surface of the membrane is covered by a narrow band of water (blue). (A) shows the ribbon/cartoon model of the protein part of P0 colored in accordance with the Kyte, Doolittle (1982) hydrophobicity scale (red for hydrophobic, blue for hydrophilic). The model of the HNK-1 epitop is shown as Van der Waals (VDW) spheres colored according to the type of atom (carbon is colored white; oxygen is colored red; sulfur is colored yellow; nitrogen is colored blue). The VDW model (a type of model where each atom is represented by a sphere at the center of the atom and the radius of sphere is equal to the Van der Waals radius, the range of Van der Waals interactions) illustrates the space occupied by the molecule in natural circumstances. The orientation of the P0_Ex head is estimated from published data (Westall, 2006; Jeremy, 2001; Shapiro *et al.*, 1996; Choe, 1996; Kirschner *et al.*, 1996; Luo *et al.*, 2007). The ^{28}Trp interacts with the surface of the next membrane and its aromatic rings are flat/parallel to the membrane surface. The distance between the next two lipid bilayers is about 46 Å (Shapiro *et al.*, 1996) and it is almost equal to the height of the P0_Ex subunit. The HNK-1 epitop stabilizes the spatial orientation of the extracellular head of protein linked by a single flexible coil to the transmembrane helix. The intracellular part of the P0 protein is also linked by the single long and dynamic loop to the membrane part of protein and this joint is not stabilized. In addition, the P0_Int part is highly charged (Fig. 8) and movable (Fig. 7B) (Wong, Filbin, 1994; Luo *et al.*, 2007). The movability of the P0_Int part is depicted as X, Y, Z positions (green, blue and red) with arrows.

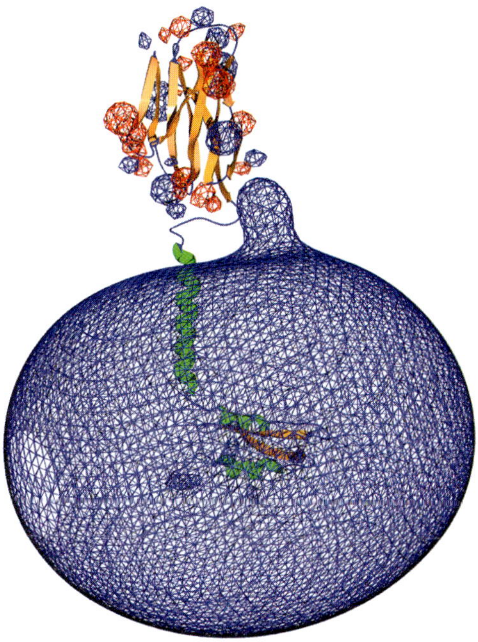

Fig. 8. The electrostatic potential of the electron density map of the whole P0 protein. The electrostatic potential is calculated using the Coulomb algorithm implemented in SwissPDBViewer. Adjusted variables for the algorithm are: dielectric constant of solvent equal $\varepsilon = 80.0$; dielectric constant of protein $\varepsilon = 4.0$; solvent ionic strength 5.0 mM. Blue signifies positive charge and red, negative charge. The range of the mesh depends on the strength of the electrostatic potential of the analyzed region.

hydrophobic residues. P0_Int is strongly hydrophilic and rich in basic amino acids: Arg and Lys; the average pI of P0_Int is more than 10 (Fig. 8, Table 1). This part contains a short, strongly hydrophobic region, but its hydrophobic nature is reduced by the charged surroundings. The spatial structure of the P0_Int part is sensitive to changes in physical-chemical solvent properties (Wong, Filbin, 1994; Luo *et al.*, 2007).

Structural Effects of Breaking the Disulfide Bridge

The P0 protein contains only one disulfide bridge, which is located in the central part of the P0_Ex subunit between ^{21}Cys-S—S-^{98}Cys. This disulfide

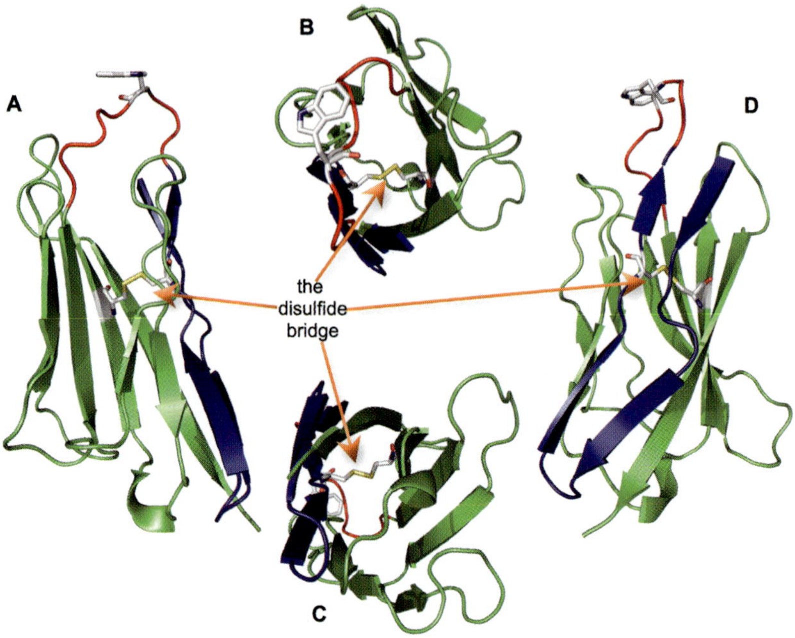

Fig. 9. Four projections of the extracellular part of the P0 protein: (B) from the top/from the next lipid bilayer side and (C) from the bottom/from the membrane side. The disulfide bridge ^{21}Cys-S—S-^{98}Cys is shown as a stick model and is visible in the central part of protein (A)–(D). Breaking the disulfide bridge (marked with orange arrows) can result in the opening of the molecule and uncover the strongly hydrophobic core of the protein. Three functional parts for the opening of the P0_Ex subunit are shown. The green colored part of this model is the stable and anchored lipid bilayer; this part does not move during the opening process. The blue colored part is movable after taking up the ^{21}Cys-S—S-^{98}Cys bond, because it is linked with the unmovable part by the single flexible and longest loop of the P0_Ex subunit (red).

bridge strongly stabilizes the 3D structure of the protein and its disruption can change the spatial conformation of the polypeptide chain causing dysfunction of the whole molecule. The disulfide bridge brings together the short part of the extracellular domain (24 a.a.: ^{1}Ile···^{24}Trp) with the biggest part of the P0_Ex subunit (Figs. 3 and 9). The long (^{25}Ser···^{33}Ile) and dynamic loop, is stabilized by the hydrophobic interaction between ^{28}Trp and the membrane (Shapiro *et al.*, 1996), and links the first 24-a.a.-long part with the large and stable remainder of the subunit anchored in the membrane (Figs. 7(A), 9 and 10).

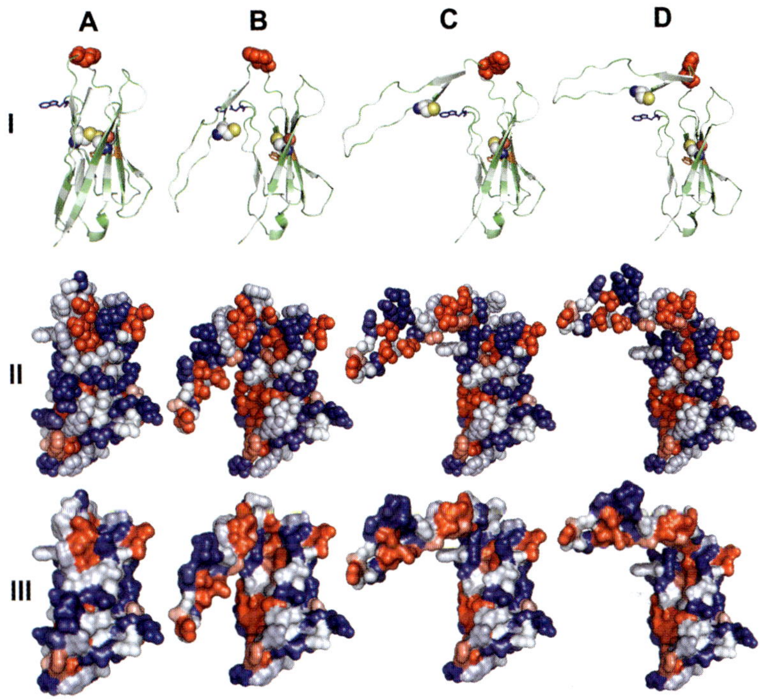

Fig. 10. Four steps of the P0_Ex protein opening (columns A, B C and D) caused by disulfide bridge breaking and interactions with detergent (Sedzik *et al.*, 1999). The cartoon models of P0_Ex presented in row I are colored green to represent hydrophilicity. The two cysteines creating the disulfide bond are represented as VDW spheres and are visible in the center of the molecule. The [28]Trp on the top of the model is colored red. The two tryptophans participating in dimer and tetramer creation by molecular associations observed in crystals of P0_Ex (Shapiro *et al.*, 1996) are shown as a stick model. The space-filling models colored in accordance with the Kyte, Doolittle (1982) hydrophobicity scale (hydrophobic residues — red spectrum; hydrophilic residues — blue spectrum) are shown in row II. Hydrophilic residues located on the external surface of molecule (blue) are clearly visible in IIA and IIIA. The hydrophobic core of the P0_Ex molecule is uncovered during opening (rows II and III, columns B, C and D). The hydrophobic pocket is constituted during this process (the red region in rows II and III).

This S-S bridge can be disrupted during SDS-PAGE electrophoresis causing P0 molecules to migrate in the electrostatic field more slowly, indicating the higher molecular weight of 3 kDa (Sedzik *et al.*, 1997).

```
/ Computations were done in vacuo with the GROMOS96 43B1 parameters set, without reaction field.
/ For more information about GROMOS96, refer to: W.F. van Gunsteren et al. (1996) in Biomolecular
/ simulation: the GROMOS96 manual and user guide. Vdf Hochschulverlag ETHZ (http://igc.ethz.ch/gromos).
/ When using those results, please mention that energy computations were done with the GROMOS96
/ implementation of Swiss-PdbViewer.
```

/ residue		bonds	angles	torsion	improper	nonBonded	electrostatic	constraint //		TOTAL
KJ/mol	P0_Ex_a	64.085	618.652	468.621	95.854	-3333.37	-2644.52	0.0000 //	E=	-4730.677
KJ/mol	P0_Ex_b	63.419	607.521	474.138	98.533	-2792.69	-2444.80	0.0000 //	E=	-3993.887
KJ/mol	P0_Ex_c	68.840	628.373	465.227	103.197	-2659.96	-2369.62	0.0000 //	E=	-3763.946
KJ/mol	P0_Ex_d	125.274	712.470	470.057	128.413	-2286.95	-2356.77	0.0000 //	E=	-3207.501

Fig. 11. A calculation of the total energy (the total energy of any molecule is a sum of chemical energy of bonds, torsion energy, electrostatic energy, and so on) of the P0_Ex subunit after breaking of the S–S bridge and during the "opening" process. The P0_Ex_a, P0_Ex_b, P0_Ex_c and P0_Ex_d models correspond with the A, B, C and D models in Fig. 10. More stable structures have lower total energy. The closed protein (model A) is the most stable structure. Wide opening causes more unfavorable interactions with the aquatic environment (particularly interactions between non-bonded atoms and atomic bonds). In this calculation all hydrogen atoms are neglected.

The ^{25}Ser···^{33}Ile loop is situated on the top of the P0_Ex subunit, close to the surface of the next myelin layer (Figs. 7A and 9) (Shapiro *et al.*, 1996). Breaking the ^{21}Cys-S—S-^{98}Cys disulfide bridge can cause the "opening" of the P0_Ex part of the protein, where the ^{1}Ile···^{24}Trp part is movable, the ^{34}Ser···^{19}Glu part is anchored in the membrane, and the ^{25}Ser···^{33}Ile loop functions like a hinge, connecting these two parts of the protein and allows the wide angle of rotation between them (Figs. 9 and 11). Strongly hydrophobic residues constitute the central part of the P0_Ex subunit (the hydrophobic core of the molecule). Spontaneous opening in a natural hydrophilic environment is difficult (Fig. 11). But strongly hydrophobic interactions in an opening pocket (Fig. 10(IIB), 10(IIC), 10(IID) and 10(IIIB), 10(IIIC), 10(IIID)) can be neutralized by molecules of detergent or phospholipids. Packing of phospholipids into the hydrophobic pocket and interactions with detergent can effectively stabilize the structure of the "opened" P0_Ex subunit (Sedzik *et al.*, 1999).

Three models of "opening" P0 protein have been built in the following research to ascertain possible phases of opening (Fig. 10) and calculate the theoretical energy of the models at each step of the opening of the dehiscent protein (Fig. 11).

It is well known that the molecule is stable when the interaction energy between atoms comprising the molecule is at the global minimum

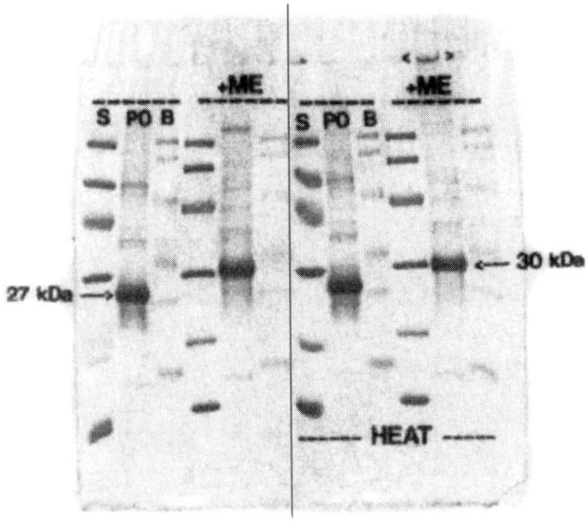

Fig. 12. Breaking of the S–S bridge of the P0 protein. The effect of heat and mercaptoethanol (5% v/v) on P0 protein in SDS-15%PAGE. Signifies the MW standards. P0 indicates the P0 protein eluted from the Cu2+-IMAC column. B stands for prestained SDS-PAGE Standards, Low Range (Bio-Rad, Hercules, Ca, USA). From top: phosphorylase B, bovine serum albumin, ovalbumin, carbonic anhydrase, soybean tyrosine inhibitor and lysozyme. The rationale for including prestained proteins was to check the influence of mercaptoethanol, heat, and prestaining on their electrophoretic behavior. The prestaining of proteins evidently inhibits electrophoretic migration of protein and abolishes staining with the Coomassie Brilliant Blue R-250. +ME, mercaptoethanol (5% v/v), was included to reduce the SDS sample buffer, HEAT — before electrophoresis samples were heated at 95°C, for 10 minutes. The MW of P0 increases approximately 3 kDa when mercaptoethanol is included in the SDS reducing buffer. Heating promotes excessive aggregation of myelin proteins (< >); they do not enter the 15% polyacrylamide gel. As the molecular weight of P0 solubilized in neutral detergents is similar to the MW of protein in the presence of mercaptoethanol, we conclude that neutral detergents may oxidize P0 as well. SDS seems to prevent the oxidizing of P0. Purified P0 seems to be folded and most likely resembles the native conformation in the myelin membrane. This is the most suitable state when crystallizing P0. (Reproduced with permission from *Neurochem Res*, **24**(6), (1999) 723–732.)

(Van Gunsteren *et al.*, 1996; Van Gunsteren, Karplus, 1980). The rule is that the molecule (or model) is energetically more stable when its energy of interaction is at a lower level. The P0_Ex_A model (Fig. 10(A)) is energetically the most stable because its energy is equal to –4730.677 kJ/mol (kilojoule per mol), the lowest of all the compared models (Fig. 11). Any structural reorganization of the protein backbone in one part of the molecule causes structural and energy changes in the whole molecule. Each step of the opening of the model causes an increase in the energy of the whole molecule (the total energy of the P0_Ex_B model is –3993.887 kJ/mol; for the P0_Ex_C model it is –3763.946 kJ/mol; and for P0_Ex_D it is 3207.501 kJ/mol).

Medical Implications — Autoimmune Epitopes of Myelin P0 Protein

The regions ^{52}His···^{76}Pro and ^{180}Ala···^{199}Ser of the P0 protein (Figs. 6(B) and 6(C) and 14(A) and 14(B)) are known to be autoimmune epitopes (Zou *et al.*, 2000; Westall, 2006; Sedzik, 2008). The study of the 3D structure of the model of the P0 protein allows one to estimate more potential autoimmune peptides. Here, two types of epitopes have been selected: short epitopes consisting of a single secondary structure domain, i.e. one strand of β-sheet (Table 2) and long epitopes containing two secondary structure domains, i.e. two anti-parallel strands of β-sheet (Table 3).

Two structural criteria of epitope estimation have been taken into account. First, each epitope has to be located on the external surface of the protein in order to be easily accessible for the receptor. Second, two structural domains of "long epitopes" should be neighbors (for example, two parallel or antiparallel strands). In accordance with the above criteria, six short and two long peptides are favored (Tables 2 and 3). Among the short peptides, three are located in the P0_Ex part and three others in the P0_Int part of the molecule (Table 2). P0_Ex contains one long epitope and the second is in the P0_Int subunit (Table 3). All the details of each peptide are summarized in Tables 2 and 3.

Summary and Discussion

The P0 protein is an abundant protein of peripheral myelin. A very hydrophobic membrane protein, it is not soluble in water and, therefore, poses a problem when attempting crystallization. Indeed, not many membrane proteins are crystallized, and their structures are known. In this chapter, we have presented a three-dimensional model of the P0 protein with linked sugar moiety of the NHK-1 epitope. The model needs to be tested experimentally; protein must therefore be purified and crystallized before the three-dimensional arrangement of atoms can uncover its overall structure and shape. The P0 protein has one transmembrane spanning segment, making the study of the protein very attractive.

For further calculations and analysis of the theoretical model of the P0 protein, predicted and estimated in this research, more work needs to be done on molecular dynamics in the membrane and simulation of local energy minimization. The whole model of the glycoprotein also requires precise docking of the HNK-1 epitope and structural optimization of oligosaccharide.

Finally, deep database mining and clinical research are required to establish whether the estimated autoimmune epitopes of the myelin P0 protein are able to be active autoimmune peptides in humans.

Acknowledgments

J.P. Jastrzebski is very grateful to Prof. Marcella Attimonelli (Bari, Italy) for help in learning and understanding of bioinformatics, and for constructive criticism and support. He is also grateful to Katarzyna Wiśniewska and Izabela Dorocka for technical assistance and to his wife Aneta Jastrzębska for her encouragement.

References

Bjellqvist B, Hughes GJ, Pasquali Ch, Paquet N, Ravier F, Sanchez J-Ch, Frutiger S, Hochstrasser DF. (1993) The focusing positions of polypeptides in immobilized pH gradients can be predicted from their amino acid sequences. *Electrophoresis* **14**: 1023–1031.

Bond JB, Saavedra RA, Kirschner DA. (2001) Expression and purification of the extracellular domain of human myelin Protein Zero. *Protein Exp Purif* **23**: 398–410.

Cebo C, Durier V, Lagant P, Maes E, Florea D, Lefebvre T, Strecker G, Vergoten G, Zanetta JP. (2002) Function and molecular modeling of the interaction between human interleukin 6 and its HNK-1 oligosaccharide ligands. *J Biol Chem* **277**: 12246–12252.

Choe S. (1996) Packing of myelin protein zero. *Neuron* **17**: 363–365.

Inouye H, Ganser AL, Kirschner DA. (1985) Shiverer and normal peripheral myelin compared: Basic protein localization, membrane interactions, and lipid composition. *J Neurochem* **45**: 1911–1922.

Kakuda S, Shiba T, Ishiguro M, Tagawa H, Oka S, Kajihara Y, Kawasaki T, Wakatsuki S, Kato R. (2004) Structural basis for acceptor substrate recognition of a human glucuronyltransferase, GlcAT-P, an enzyme critical in the biosynthesis of the carbohydrate epitope HNK-1. *J Biol Chem* **279**: 22693–22703.

Kakuda S, Sato Y, Tonoyama Y, Oka S, Kawasaki T. (2005) Different acceptor specificities of two glucuronyltransferases involved in the biosynthesis of HNK-1 carbohydrate. *Glycobiology* **15**: 203–210.

Kirschner DA, Inouye H, Saavedra RA. (1996) Membrane adhesion in peripheral myelin: Good and bad wraps with protein P0. *Structure* **4**: 1239–1244.

Kleene R, Schachner M. (2004) Glycans and neural cell interactions. *Nature Rev Neurosci* **5**: 195–208.

Kyte J, Doolittle RF. (1982) A simple method for displaying the hydrophobic character of a protein. *J Mol Biol* **157**: 105–142.

Luo XY, Cerullo J, Dawli T, Priest C, Haddadin Z, Kim A, Inouye H, Suffoletto BP, Avila RL, Lees JPB, Sharma D, Xie B, Costello CE, Kirschner DA. (2008) Peripheral myelin of *Xenopus laevis*: Role of electrostatic and hydrophobic interactions in membrane compaction. *J Struct Biol* **162**: 170–183.

Luo X, Inouye H, Gross AA, Hidalgo MM, Sharma D, Lee D, Avila RL, Salmona M, Kirschner DA. (2007) Cytoplasmic domain of zebrafish myelin protein zero: Adhesive role depends on beta-conformation. *Biophys J* **93**: 3515–3528.

Luo XY, Sharma D, Inouye H, Lee DA, Robin L, Salmona M, Kirschner DA. (2006) Cytoplasmic domain of human myelin p0 likely folded as beta-structure in compact myelin. *Biophys J* **92**: 1585–1597.

Ong E, Yeh JC, Ding Y, Hindsgaul O, Pedersen LC, Negishi M, Fukuda M. (1999) Structure and Function of HNK-1 Sulfotransferase. Identification of donor and acceptor binding sites by site-directed mutagenesis. *J Biol Chem* **274**: 25608–25612.

Parker JMR, Guo D, Hodges RS. (1986) New hydrophilicity scale derived from high-performance liquid chromatography peptide retention data: Correlation of predicted surface residues with antigenicity and X-ray-derived accessible sites. *Biochemistry* **25**: 5425–5432.

Sakamoto Y, Kitamura K, Yoshimura K, Nishijima T, Uyemura K. (1987) Complete amino acid sequence of P0 protein in bovine peripheral nerve myelin. *J Biol Chem* **262**: 4208–4214.

Sedzik J. (2008) Myelin sheaths and autoimmune response induced by myelin proteins and alphaviruses. I. Physicochemical background. *Current Med Chem* **15**: 1899–1910.

Sedzik J, Kotake Y, Uyemura K. (1999) Purification of P0 myelin glycoprotein by a Cu2+-immobilized metal affinity chromatography. *Neurochem Res* **24**: 723–732.

Shapiro L, Doyle J, Hensley P, Colman D, Hendrickson W. (1996) Crystal structure of the extracellular domain from P0, the major structural protein of peripheral nerve myelin. *Neuron* **17**: 435–449.

Van Gunsteren WF, Billeter SR, Eising AA, Hunenberger PH, Kruger P, Mark AE, Scott WRP, Tironi IG. (1996) *Biomolecular Simulation: The GROMOS96 Manual and User Guide*. VdF: Hochschulverlag AG an der ETH Zurich and BIOMOS b.v, Zurich, Gronigen.

Van Gunsteren WF, Karplus M, (1980) A method for constrained energy minimization of macromolecules *J Comput Chem* **3**: 266–274.

Voshol H, van Zuylen CWEM, Orberger G, Vliegenthart JFG, Schachner M. (1996) Structure of the HNK-1 carbohydrate epitope on bovine peripheral myelin glycoprotein P0. *J Biol Chem* **271**: 22957–22960.

Westall FC. (2006) Molecular mimicry revisited: Gut bacteria and multiple sclerosis. *J Clin Microbiol* **44**: 2099–2104.

Wong MH, Filbin MT. (1994) The cytoplasmic domain of the myelin P0 protein influences the adhesive interactions of its extracellular domain. *J Cell Biol* **126**: 1089–1097.

Zou LP, Ljunggren HG, Levi M, Nennesmo I, Wahren B, Mix E, Winblad B, Schalling M, Zhu J. (2000) P0 protein peptide 180–199 together with pertussis toxin induces experimental autoimmune neuritis in resistant C57BL/6 mice. *J Neurosci Res* **62**: 717–721.

Chapter 14

In meso Approaches to Membrane Protein Crystallization

Valentin I. Gordeliy,† and Ekaterina S. Moiseeva*†

Crystallization of membrane proteins remains a major challenge in modern structural biology. At issue is the amphiphathic character of these proteins and that the natural environment of these proteins is a lipid bilayer. This chapter describes different techniques of membrane protein crystallization focusing on a fascinating recent breakthrough in the field: methods of crystallization in the detergent-lipid environment (*in meso* methods).

Keywords: Membrane proteins; lipid bilayer; detergents; crystallization *in meso*; vesicles; bicelles.

Introduction

Structural information is the basis of our knowledge about matter, independently of whether we deal with elementary particles which are of interest to high energy physics, or atoms, or biological macromolecules of which living organisms are comprised.

Deciphering of the double helix structure of deoxyribonucleic acid (DNA) in 1953 revolutionized biology (Watson, Crick, 1953a; Watson, Crick, 1953b). Another important result was resolving the structure of the first proteins. High concentrations of myoglobin in muscle cells allow

*Corresponding author.
†Institute of Structural Biology J.P. Ebel, Grenoble, France; Institute of Neurosciences and Biophysics 2, Forschungszentrum Jülich GmbH, Jülich, Germany; Centre of Biophysics and Physical Chemistry of Supramolecular Structures of Moscow Institute of Physics and Technology, Moscow, Russia. E-mail: valentin.gordeliy@ibs.fr

organisms to hold their breaths longer. In 1958, Kendrew and associates successfully determined the structure of myoglobin by high-resolution X-ray crystallography (Kendrew *et al.*, 1958). For this discovery, John Kendrew shared the 1962 Nobel Prize in Chemistry with Max Perutz. Hemoglobin, the oxygen-carrying pigment found in red blood cells, was one of the first proteins to have its three-dimensional structure resolved by X-ray crystallography (by Max Perutz in the early 1960s (Perutz, 1962)).

Currently, the RCSB Data Bank contains over 20000 structures, but only about 30 are structures of different membrane proteins (Loll, 2003; Ruahani *et al.*, 2002), in spite of the fact that the topic of membranes occupies a central position in cell and molecular biology. It has become clear in recent years that the study of membranes at the molecular level is of great importance not only in the deciphering of all cellular processes but also in the understanding of the alterations leading to abnormal (transformed) cells and the action of drugs.

Membrane proteins are the main functional units of membranes and represent roughly one third of the proteins encoded in the genome, and yet 70% of drugs have membrane proteins as a target. However, they comprise only 1% of proteins of known structure.

High-throughput crystallography offers hope of correcting this imbalance. However, for large-scale membrane protein structural biology to realize its full promise, significant challenges must be overcome. One of the most substantial of them is the development of reliable methods for membrane protein crystal growth (Loll, 2003; Ruahani, *et al.*, 2002). Usually, to crystallize membrane proteins one first has to solubilize them with a detergent. However, it is not an easy problem to find a proper detergent to obtain a stable and functional protein. The difficulty is that the natural environment of these proteins is a lipid bilayer.

Membrane Protein Crystallization: A Standard Approach

Detergents are a relatively poor substitute for the lipid bilayer and membrane proteins are often unstable outside the membrane (Michel, 1991).

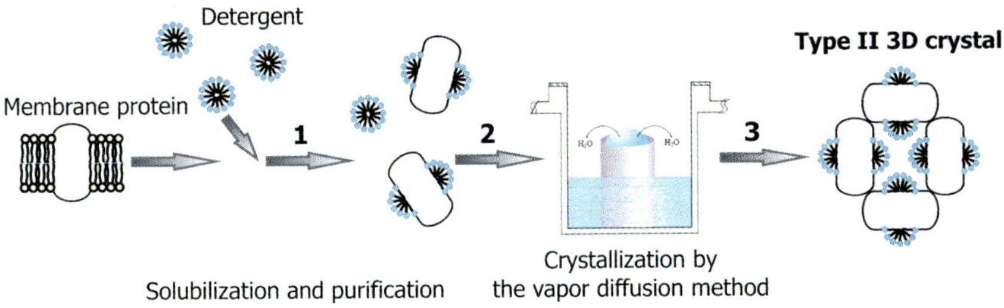

Fig. 1. A schematic representation of a standard approach to the crystallization of membrane protein: (1) solubilization and purification of the proteins, (2) crystallization by the vapor diffusion method, and (3) formation of crystals, usually of type II.

If the significant efforts to select the detergent are successful then one can use the quite well developed vapor diffusion methods for the crystallization of water soluble proteins (Fig. 1).

However, "standard" crystallization from detergent solutions may not be the optimal way of crystallizing membrane proteins. Indeed, success at the first step does not give a guarantee of obtaining crystals of the quality sufficient for X-ray crystallography. The next problem is characterized well by Michel: "what seems to be suited for stability of membrane proteins is less suited for their crystallization" (Michel, 1991). Indeed, detergents with short chains are not well suited for membrane protein stability. However, the use of longer chain detergents can lead to a steric conflict in the crystals. The reason for this is illustrated in Fig. 1. Nearly all the crystals obtained by the standard approach are those of type II. In this case the detergent layer ("belt") around the protein occupies a sufficiently large part of the crystal and close direct contacts between the polar parts of proteins, which are necessary for the creation of specific interactions, are not always possible. This is an explanation of why in spite of significant efforts the first X-ray crystallographic structure of a membrane protein — the photosynthetic reaction center from *Rhodopseudomonas viridis* — was obtained only in 1984. At that time it was quite widely accepted that it was not possible to obtain membrane protein crystals of sufficient quality to resolve their high resolution structure by means of X-ray crystallography. For their pioneering work

H. Michel, J. Deisenhofer and R. Huber received the Nobel Prize in Chemistry in 1988 (Deisenftofer *et al.*, 1984).

In spite of this success, obtaining membrane protein crystals was still a very difficult task. An illustration of this statement is the story of bacteriorhodopsin crystallization. Bacteriorhodopsin belongs to the family of seven alpha-helical retinal proteins and plays an important role in the bioenergetics of *Halobacterim salinarum*. Its chromophore retinal absorbs a photon, isomerizes and triggers a sequence of further molecular events leading to proton translocation against the electrochemical gradient of membranes. Further, the energy of the gradient is used by another membrane enzyme ATP synthase to catalyze the ADP-ATP reaction.

This protein is quite easy to solubilize and purify in large amounts, and it was used as a major model system to study the molecular mechanisms of proton transport (Landau, Rosenbusch, 1996). However, all attempts to obtain crystals of the protein diffracting to high resolution failed during a period of more than 20 years.

The success with bacteriorhodopsin (BR) crystallization came unexpectedly. In 1996, Rosenbusch and Landau obtained high quality BR crystals using a principally new approach: the protein was crystallized in the lipidic cubic phase (Landau, Rosenbusch, 1996). The major differences between the two approaches are illustrated in Fig. 2. A principal feature of this novel method is that after solubilization the protein

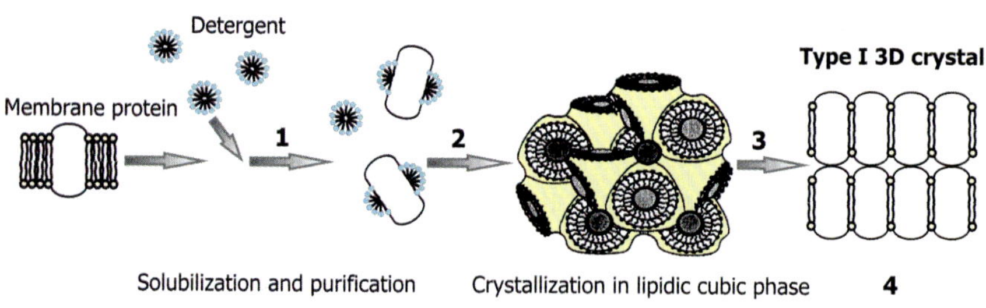

Fig. 2. A schematic representation of an *in meso* approach to crystallization of membrane protein: (1) solubilization and purification of proteins, (2) reconstitution of the protein into a lipid bilayer, (3) initiation of crystal growth by direct addition of the precipitant to the lipid/protein phase, and (4) formation of crystals, usually of type I.

is reconstituted back to a lipid bilayer and then precipitant is added directly to the lipid/protein phase to initiate membrane protein crystallization. This approach will be considered in more detail in what follows.

The most successful detergents for membrane protein solubilization and crystallization are presented the Table 1. Lists presenting the detergents that have been most efficiently used in structure determination are available from several Internet sites that track the known 3D structures of membrane proteins (for example:
http://blanco.biomol.uci.edu/Membrane_Proteins_xtal.html,
http://www.mpibp-frankfurt.mpg.de/michel/public/memprotstruct.html,
http://www.mpdb.ul.ie) (Prive, 2007).

In meso Membrane Protein Crystallization

Cubic phase crystallization

Recent success in the crystallization of membrane proteins in lipid cubic phase systems has provided the hope for a breakthrough in the field. Indeed, broad success has recently been achieved in using lipid bilayers to crystallize the *Halobacterial* family of rhodopsins. Among these proteins there is bacteriorhodopsin (Pebay-Peyroula *et al.*, 1997), which, as mentioned, has been one of the most difficult membrane proteins to crystallize in the detergent-soluble state.

Another example is the crystallization and elucidation of the structure and functional mechanisms of the first complex of two membrane proteins: sensory rhodopsin II and transducer II, which is mediating phototaxis in *Natronobacterium pharaonis* (Gordeliy *et al.*, 2002; Moukhametzianov *et al.*, 2006). It is worth mentioning that the complex has monomeric content in the detergent-soluble state and the native dimeric state only in lipid membranes. It is likely that the complex cannot be crystallized from detergent solutions.

Very recent success in cubic phase crystallization leading to the first high resolution structure of the first ligand binding GPCR, human

Table 1. Detergents which are often used for Solubilization and Crystallization of Membrane Proteins (www.anatrace.com)

Detergent		Structure Formula	FW	CMC (H_2O)	Aggregation Number (H_2O)
C8E4	Octyl tetraethylene glycol ether	$CH_3(CH_2)_7O(CH_2CH_2O)_4H$	306.45	~8 mM (0.25%) (0.1M NaCl)	~82
OG	n-octyl-b-D-glucopyranoside	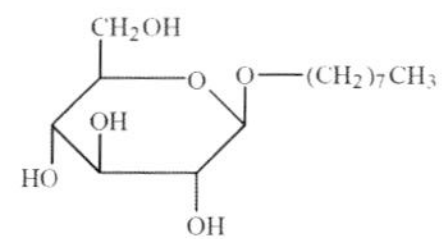	292.4	18–20 mM (0.53%)	~27–100
DM	n-decyl-b-D-maltopyranoside	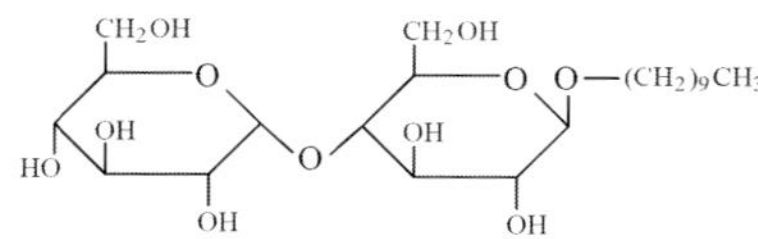	482.6	1.8 mM (0.087%)	~69
DDM	n-dodecyl-b-D-maltopyranoside	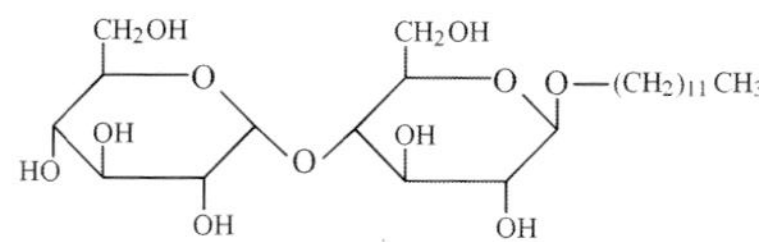	510.6	0.17 mM (0.0087%)	~78–149
FOS-12	n-Dodecylphosphocholine	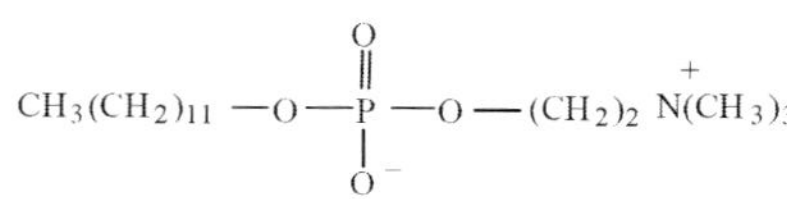	351.5	1.5 mM (0.047%)	~50–60
LDAO	N,N-dimethyldodecylamine-N-oxide	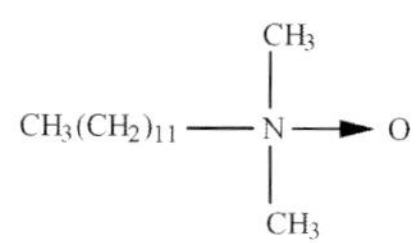	229.41	1–2 mM (0.023%)	~76

β_2-adrenergic G protein-coupled receptor (Cherezov *et al.*, 2007), is an important breakthrough in the field. Indeed, GPCRs comprise the largest integral membrane protein family in the human genome, with over one thousand members. These receptors participate in the transduction of signals across cellular membranes in response to a huge variety of extracellular stimuli, including light, proteins, peptides, small molecules, hormones, protons and ions. Activated GPCRs trigger a cascade of intracellular responses, primarily through interactions with their cognate heterotrimeric G proteins. In addition, GPCRs are associated with a multitude of diseases, which makes members of this family very important pharmacological targets.

These three examples show the power of the new approach. The method deserves more detailed consideration. First of all, most crystallizations were done using the monooleoyl (MO)/BR buffer system. A phase diagram of a simple MO/water system is shown in Fig. 3. The phase diagram is quite complex and comprises a wide range of phases: from two lamellar (*Lc* and *Lα*) to two cubic *bicontinuous* phases of different symmetries *Ia3d* and *Pn3m* (Briggs *et al.*, 1996). In this case the membrane divides the solvent into two interpenetrating continuous regions. Luzatti was the first to discover the lipid cubic phase and recognize that the midsurfaces of bilayers are close to cubic minimal surfaces, which have a zero mean curvature everywhere (Mariani *et al.*, 1988). It is interesting to note that bicontinuous cubic phases were observed in the cells (Landh, 1995), and they are used in the food industry (Fontell, 1990) as well as for drug delivery (Ericsson *et al.*, 1991).

Crystallization in the lipid cubic phase is very simple in practice. An example — crystallization of BR — can be described by the following procedure (Gordeliy *et al.*, 2003):

(i) Fill a 200 μl PCR tube with monooleoyl powder (~4 mg).
(ii) Melt the MO at 40°C and spin the lysolipid down for 10 minutes at 13 000 g at room temperature.

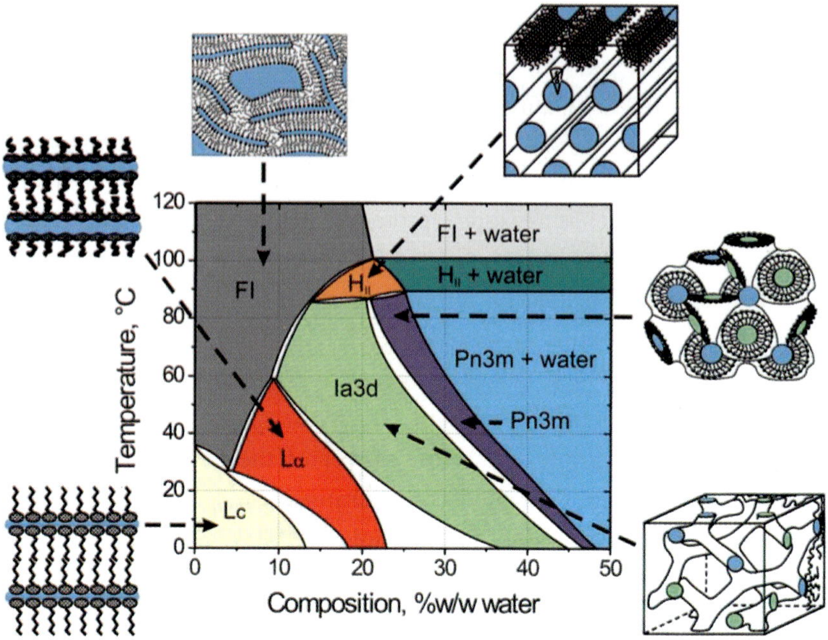

Fig. 3. A phase diagram of monooleoyl/water system (redrawn from Cherezov *et al.*, 2006).

(iii) To obtain the isotropic lipidic phase, the MO is kept for an additional 20 minutes at 40°C. Subsequently, let the lipidic phase cool to room temperature. The MO phase must remain transparent. If not, the melting procedure should be repeated.

(iv) Add 1 μl BR solution (10 mg/ml BR) with about 1.2% (w/w) of n-octyl-β-*D*-glucoside (OG) per mg of MO.

(v) A homogenous lipid/protein mixture as well as the formation of the cubic phase is achieved by the following centrifugation procedure. Spin the PCR tube with the sample at 22°C at 10000 rpm for 15 minutes. Rotate the tube within the rotor by 90° and spin again. Repeat this spinning procedure four times to obtain a homogenous mixture.

(vi) Leave the sample for 24 hours in the dark at 22°C.

(vii) Add a ground powder of KH_2PO_4 mixed with Na_2HPO_4 (95/5 w/w) to obtain a final concentration of 1–2.5 M phosphate (pH 5.6).

(viii) Homogenize the sample by repeated centrifugation as described in Step (v).

(ix) Leave the crystallization batch in the dark at 22°C. The first micro-crystals of BR (10–20 μm in diameter) usually appear within one week after the phosphate has been added (Fig. 4).

To summarize, BR in the buffer is mixed with MO in the *FI* phase. The following procedure of centrifugation facilitates cubic phase growth and reconstitution of BR in lipid bilayers. Crystal growth is initiated by the addition of a precipitant — ground powder of KH_2PO_4 mixed with Na_2HPO_4. The above-mentioned protocol for crystallization is quite close to that described in the original paper by Rosenbusch and Landau (Landau, Rosenbusch, 1996).

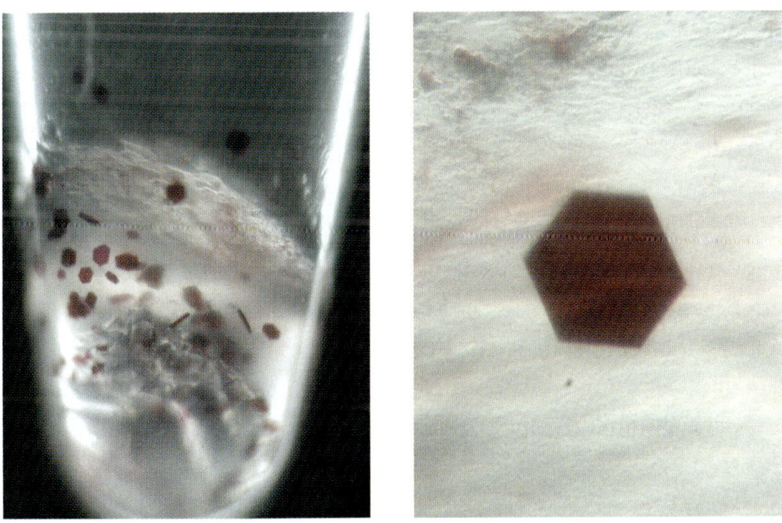

(a) (b)

Fig. 4. (a) A BR crystallization probe in a PCR tube, and (b) a single BR crystal in the probe.

Why do crystals grow in the lipidic cubic phase? There are several possible reasons for the efficiency of the method, which have been discussed in the literature. However, the first question is whether the system is really in this phase. In fact, the system for crystallization is different from the pure MO/water system. Indeed, it contains salts in the buffer and, most importantly, a certain amount of the detergent. *A priori* it was not evident that the phase diagram shown in Fig. 3 somehow describes the lipidic/membrane protein/detergent/salt/water system as well. Hence, the detailed kinetics of a proteolipidic cubic phase was studied by neutron scattering in the course of crystallization of the membrane protein bacteriorhodopsin. It was shown that the initiation of the crystallization process by salt addition leads to a dramatic decrease in the lattice constant, but no phase transition takes place. The cubic phase of *Pn3m* symmetry is observed during the entire crystallization process. No other phases are present in a macroscopic amount (Efremov *et al.*, 2005). However, this does not exclude the presence of a small amount of phases other than the cubic phase (Efremov *et al.*, 2005). There is evidence of the presence of a lamellar phase around the growing protein crystals but there is no definite proof of this (Cherezov *et al.*, 2007). What can one speculate about the mechanism of membrane protein crystal growth? There has been a theoretical attempt to understand the mechanism of membrane protein crystal growth in the lipidic cubic phase (Grabe *et al.*, 2003). Theoretical calculations showed that the elastic energy of the deformation of the curved bilayer due to embedded proteins may be the driving force for crystallization. Unfortunately, this theory is incomplete, and it is not yet known whether the model of the crystallization system used for the calculation is completely correct. Further experimental and theoretical studies are required to elucidate the mechanism of crystallization.

Crystallization in the sponge phase

The sponge (L_3) phase can be imagined as a disordered cubic phase. The ordered cubic phase structure is perturbed by thermally excited shape fluctuations of the membranes. The sponge phase, which is still bicontinuous, exhibits no long-range order. The transformation of the cubic to the sponge phase may occur when some small additional molecules like those of a polar solvent such as dymethyl sulfoxide, propylene glycol, or polyethylene glycol (MW ≈ 400) are added (Engstroem *et al.*, 1998). A major reason for such transformation is the reduced bending rigidity of the membranes. A schematic representation of the L_3 phase is shown in Fig. 5.

It is interesting to note that as in the case of the cubic phase there have been a number of studies exploring possible applications of the sponge phase for studying intracellular membranes like the endoplasmic reticulum (Lindblom *et al.*, 1988), the Golgi apparatus or as drug transport vesicles (Alfons *et al.*, 1998).

There are two reports in the literature on the crystallization of membrane proteins in the sponge phase (Cherezov *et al.*, 2006; Wadsten *et al.*, 2006). In one of the papers the appearance of the sponge phase was proven by "visual inspection, small-angle X-ray scattering and NMR

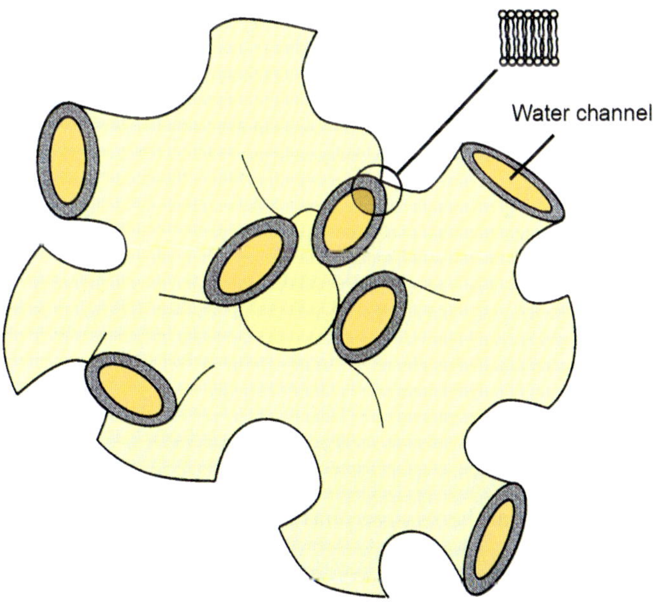

Fig. 5. A schematic representation of the sponge L_3 phase [modified from McGrath *et al.* (1997)].

spectroscopy." Crystals of the reaction center from *Rhodobacter sphaeroides* were obtained by a conventional hanging-drop experiment and harvested directly without the addition of lipase or cryoprotectant, and the structure was refined to 2.2 Å resolution. In contrast to the earlier lipidic cubic phase reaction center structure (Katona *et al.*, 2003), the mobile ubiquinone could be built and refined. In these experiments the components similar to those for cubic phase crystallization (structural resolution 2.35 Å) were used (Katona *et al.*, 2003): the MO/membrane protein/detergent/buffer. The only additional component was a small amphiphilic molecule, 1,2,3-heptanetriol or Jeffamine M600. The other work (Cherezov *et al.*, 2006) is similar to the paper mentioned above. The light harvesting II complex (LH2) was crystallized and a structural resolution of 2.45 Å was achieved. Several additives were used in this study: KSCN, butanediol, pentaerythritol propoxylate (PPO), t-butanol, Jeffamine, and 2-methyl-2,4-pentanediol (MPD). A 2.0 Å structure was available for LH2. It was obtained using vapor diffusion-grown crystals of the detergent-solubilized complex (Papiz *et al.*, 2003).

Is the "sponge phase" approach better than crystallization in the cubic phase? Unlike the former method, this approach has not led to a breakthrough in the structural biology of membrane proteins. The method has not resolved the structure of new membrane proteins nor has it achieved considerable improvements in structural resolution. In addition, the use of small additives to obtain the sponge phase can sometimes be harmful for a membrane protein. We have conducted a test of this approach using BR and the additives described in Cherezov *et al.* (2006). Some of the results are presented in Table 2 (Moisseva *et al.*, unpublished).

These experiments show that, indeed, most of the additives are harmful for bacteriorhodopsin and the approach does not allow one to obtain BR crystals of the same quality as in the cubic phase approach. Nevertheless, this does not mean that the sponge phase approach can be neglected; however, since for some proteins it will work well. Also, we cannot exclude the possibility that there will be further improvements of the method. As in the case of the previous approach, unfortunately, there is a lack of information about the behavior of *in meso* systems in the course of crystallization.

Table 2. Crystallization of BR in Swollen Lipidic L_3 Mesophases.[a]

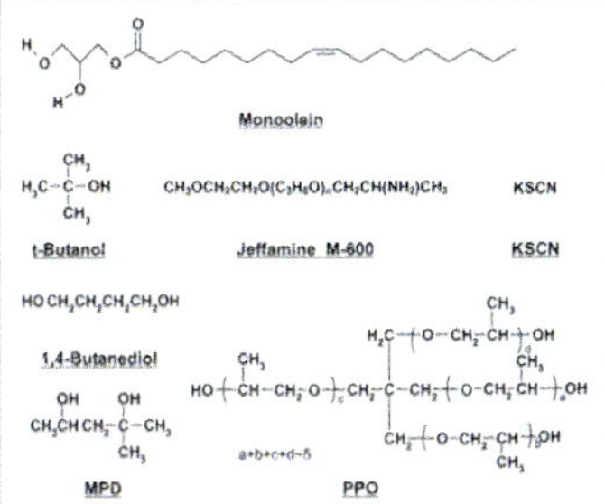

Additive		Bacteriorhodopsin				*crystals*
MPD	(%)	8	11	15	18	None
PPO	(%)	10	14	18	22	None
Jeffamine M-600	(%)	7	12	17	22	+
1,4-Butanediol	(%)	27	30	35	40	None
t-Butanol	(%)	18	22	26	30	None
KSCN	(M)	0.6	1.2	1.7	2.2	None

[a] The different colors of the cells correspond to the results of visual observation of the behavior of BR after the addition of the additives. Purple corresponds to stable BR and red, pink, yellow and white correspond to different levels of BR degradation.

Crystallization from vesicles

The third approach is crystallization from vesicles with a very large protein/lipid ratio (Takeda *et al.*, 1998). It was previously observed that purple membranes of *Halobacterim salinarum* (a two-dimensional BR crystal where the trimers of the protein are tightly packed in a perfect hexagonal lattice) treated with the detergent n-octyl-b-D-glucopyranoside under certain conditions lead to the creation of spherical protein clusters with a diameter of ~50 nm which are hexagonal close-packed (Kouyama *et al.*, 1994). This was revealed by atomic force microscopy images of the crystal surface. Electron micrographs of mechanically disintegrated crystals show that the inside of the protein cluster is filled with the mother liquor. The crystal is made up of hollow protein clusters. When disintegrated crystals are illuminated in the presence of a lipophilic anion, a significant alkalization of the external medium occurs. This result demonstrates that the protein cluster contains native lipids and that the cytoplasmic side of the protein faces the external medium. It has been claimed that X-ray diffraction patterns and the observed diameter of the spherical shell suggest that about 200 bacteriorhodopsin trimers are aligned on a polyhedral surface lattice. Vesicle formation from purple membranes was studied in detail in Denkov *et al.* (1998).

The authors of this approach to membrane protein crystallization (Takeda *et al.*, 1998) started with the preparation of BR vesicles as described in Kouyama *et al.* (1994) and Denkov *et al.* (1998) and then used a standard vapor diffusion method to obtain BR crystals.

Crystallization of BR from vesicles from PM can be described as follows:

(i) Resuspend the pellet of PM (preliminarily centrifuged 4000 g for 20 minutes) in a Tween 20 (0.3%) buffer (10 mM HEPES, pH 8.0, 160 mM NaCl) until a final concentration of 1 mg/ml.
(ii) Incubate for 20 minutes at 20°C.
(iii) Centrifuge immediately for 10 minutes at 16000 g.
(iv) Resuspend the pellet in a buffer for vesicle preparations (e.g. 10 mM HEPES, pH 8.0, 160 mM NaCl, 0.04% NaN_3).
(v) Repeat the washing procedure 3–4 times from Step (ii).
(vi) Dilute PM (10 mg/ml) with a double concentrated solution of detergent to a final concentration of 5 mg/ml of PM, 1–4 mg/ml of OG, 0.6–1.2 M of $(NH_4)_2SO_4$.
(vii) Incubate for 1–2 weeks at 32°C.
(viii) Centrifuge for 10 minutes at 4000 g.
(ix) Save supernatant and concentrate at 4°C.
(x) For crystallization, the protein sample is equilibrated against 80 mM of sodium citrate (pH 5.2) buffer containing 1.9–2.2 M of $(NH_4)_2SO_4$ in a reservoir by the standard vapor diffusion method.
(xi) Leave the crystallization batch at 4°C. The first crystals appear within a week.

A birefringent hexagonal crystal has been obtained diffracting X-rays beyond 2.5 Å resolution. This new crystal belongs to the space group P622 with unit cell dimensions of $a = b = 104.7$ Å and $c = 114.1$ Å. The highest announced structural resolution achieved by this method is 2.3 Å. Such an experiment is illustrated schematically in Fig. 6. A number of experiments have been carried out in our laboratories to optimize the method (Golubev *et al.*; Volkov *et al.*, unpublished). It has been shown

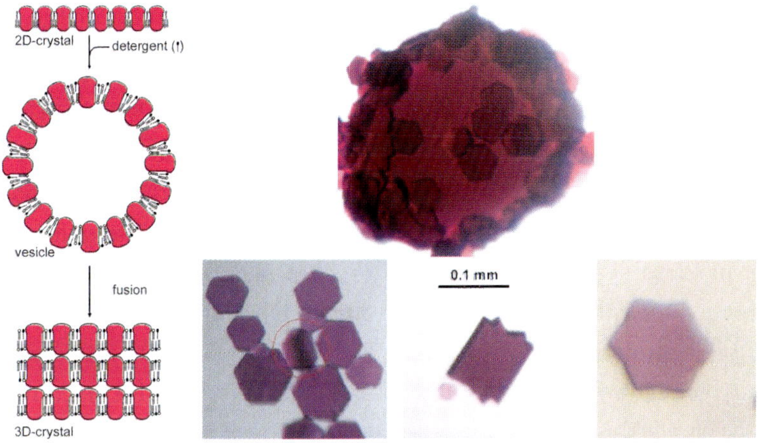

Fig. 6. A simplified representation of crystallization from vesicles obtained from 2-D crystalline native purple membranes. The authors of this method suggest that BR crystals are formed by fusion of the vesicles. (The schematic diagram on the left-hand side was modified from Takeda *et al.* (1998)). The crystals grown in our laboratories are also shown (Golubev *et al.*, unpublished).

that one can obtain different types of BR crystals (Fig. 6). However, at present, no other membrane proteins have been crystallized by this method. Nevertheless, it is not yet clear whether this approach is limited to some specific cases, like BR, or if it has a more general application. Unfortunately, as in the two other cases mentioned above, this system of crystallization has not yet been sufficiently characterized.

It is worth noting that there is an additional side of biological significance of these studies. For instance, the construction of artificial highly curved aggregates could help to understand more deeply the process of vesicle formation in living organisms. Indeed, well-known examples of the important roles of protein-containing spherical vesicles in eukariotic cells are: the clathrin coated vesicles that realize endocytosis of ligands bound to the cell surface; the coatomer coated vesicles that transport proteins to and from the Golgi complex; and synaptic vesicles in the nerve cells that transfer the neurotransmitters from the synaptic cell into the synaptic cleft. In addition, one should not forget proteins assembled in viral capsids.

Crystallization from bicelles

Another approach is crystallization from bicelles, which was first applied to obtain high quality BR crystals (Faham, Bowie, 2002; Faham *et al.*, 2005). It is known that mixtures of dimyristoylphosphatidylcholine (DMPC) with certain detergents form lipid-rich bilayer fragments that are edge-stabilized by the detergent component. The lipid-detergent aggregates in these systems were named *bicelles.* The most often employed detergents are either dihexanoylphosphatidylcholine (DHPC) or zwitterionic bile salt derivative, CHAPSO. The lipid to detergent ratios present in the bicellar systems are quite high (usually 4:1 to 1.5:1) compared to other micellar systems which have previously been used as model membranes in biophysical or biochemical studies (Sanders, Schwonek, 1992; Sanders, Prestegard, 1990). Moreover, the bicellar systems are unusual in that bicelles can be magnetically oriented to yield oriented solid state NMR spectra of high quality (Sanders, Landis, 1995).

It is interesting to mention that investigation of bicelle formation is of great biological importance. It may help us to understand better one of the fundamental processes in the cell: formation and transformation of biomembranes.

The procedure of crystallization of membrane proteins from bicelles can be described as follows:

(i) Prepare 40% (3:1) solution of DMPC/CHAPSO bicelles.
(ii) Mix solubilized BR (C = 10 mg/ml) with the bicelles (v/v ratio of 4:1). Thereafter, the homogenous mixture contains 8% of the bicelles and 0.8% of the protein. It is believed that at this stage the protein is reconstituted into bicelles. These steps are schematically illustrated in Fig. 7.
(iii) In the last step of the preparation of protein crystallization, keep all solutions on ice (higher temperature increases the viscosity of the probes). The protein sample is equilibrated against 3.2 M NaH_2PO_4 solution in a single reservoir by the standard vapor diffusion method.
(iv) Leave the crystallization batch at 37°C or room temperature. BR crystals appear within two weeks.

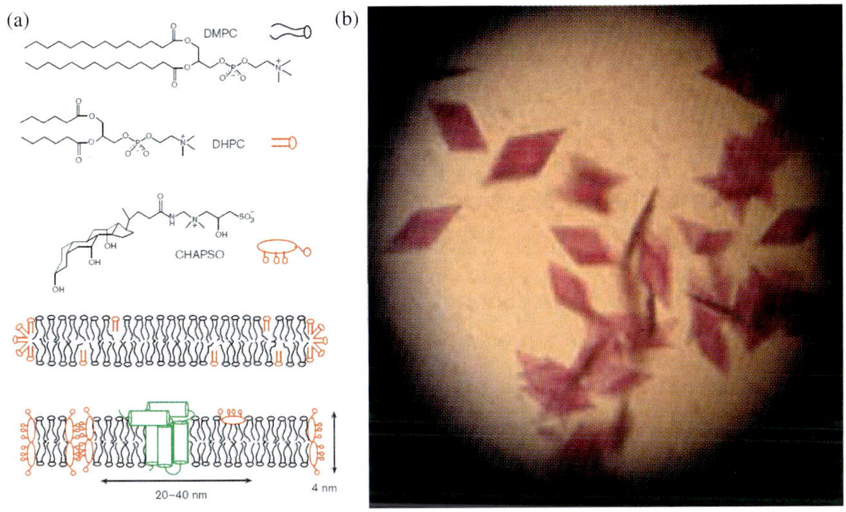

Fig. 7. A depiction of the preparation of a bicelle probe for crystallization [modified from Sanders *et al.* (1998)]. (a) A model CHAPSO–DMPC bicelle contains a reconstituted integral membrane protein (in green) [modified from Sanders *et al.* (1998)]. (b) The right-hand side of the figure shows crystals grown by this method (Ishchenko *et al.* unpublished).

BR crystals grown at room temperature are essentially identical to twinned crystals previously obtained at 37°C: space group $P2_1$ (2.0 Å resolution) with unit cell dimensions of $a = 44.7$ Å, $b = 108.7$ Å, $c = 55.8$ Å, $\beta = 113.6°$. The other room-temperature crystals are untwinned and belong to space group $C222_1$ (2.2 Å resolution) with the following unit cell dimensions: $a = 44.7$ Å, $b = 102.5$ Å, $c = 128.2$ Å.

After the first publication of this kind of crystallization, there has been no further success with other membrane proteins. However, very recently crystals of the human β_2-adrenergic G protein-coupled receptor were grown in DMPC bicelles (Rasmussen *et al.*, 2007). The structure was resolved to 3.5/3.7 Å resolution, which is considerably lower than that obtained by protein crystallization in the cubic phase (Cherezov *et al.*, 2007). However, taking into account the numerous failed attempts to crystallize a ligand-binding GPCR, one has to accept that this represents a considerable success of the method under discussion.

Towards a General Method of Membrane Protein Crystallization

New approaches to crystallization from lipid bilayers (or rather, from membranes) have led to a breakthrough in the field of structural biology of membrane proteins. Major progress has been due to the application of the lipidic cubic phase approach. Nevertheless, there is evidence that three other methods described in this chapter may have a more general application.

Unfortunately, the number of membrane proteins of known structure is still growing slowly and crystallization of membrane proteins remains a challenge. One of the major problems is that there have been no systematic experimental and theoretical studies of all crystallization events in the course of membrane protein crystallization. In other words we do not understand sufficiently well the systems that are used for crystallization; we cannot answer important questions. For instance, we do not know why all the above-mentioned *in meso* systems give rise to layer type (type I) membrane protein crystals. Does this mean that there is something important in common in all four cases?

A rational design of a crystallization experiment is not possible without in-depth knowledge about the systems used. A complimentary approach to the studies of complex fluids used for crystallization is of great importance for the development of more general and efficient methods of membrane protein crystallization. Physics with such powerful experimental methods as neutron and X-ray scattering as well as theoretical tools may play a key role in the further development of crystallization methods. Future success will have a high impact on our understanding of the molecular mechanisms underlying the function of living matter.

Summary

Crystallization of membrane proteins remains a major challenge in modern structural biology. The amphiphathic character of these proteins is significant as well as the natural environment of these proteins being a lipid bilayer. This chapter has described different techniques of membrane protein crystallization focusing on a recent breakthrough: methods of crystallization in the detergent-lipid environment (*in meso* methods). Major progress can be attributed to the application of the lipidic cubic phase approach, but there is evidence that three other methods (sponge phase, bicelles and vesicles) described in this chapter may also have a more general application. Further development of *in meso* approaches is needed in the future.

Acknowledgments

We are greatly indebted to A. Ishchenko, S. Golubev and O. Volkov for providing valuable information on their unpublished experiments. We would like to acknowledge Georg Bueld and Eva Pebay-Peyroula for continuous support of our work. Christian Baeken is gratefully acknowledged for his help with protein production.

This chapter was updated based on our lecture notes for the 39th IFF Spring School "Soft Matter — From Synthetic to Biological Materials," Forschungszentrum Jülich GmbH (www.fz-juelich.de/iff/fs2008_ prog_c).

References

Alfons K, Engstroem S. (1998) Drug compatibility with the sponge phases formed in monoolein, water, and propylene glycol or poly (ethylene glycol). *J Pharm Sci* **87**: 1527–1530.

Briggs J, Ching H, Caffrey M. (1996) The temperature-composition phase diagram and mesophase structure characterization of the monoolein/water system. *J Phys II* **6**: 723–751.

Cherezov V, Clogston J, Papiz MZ, Caffrey M. (2006) Room to move: Crystallizing membrane proteins in swollen lipidic mesophases. *J Mol Biol* **357**: 1605–1618.

Cherezov V, Caffrey M. (2007) Miniaturization and automation for high-throughput membrane protein crystallization in lipidic mesophases. *Faraday Discuss* **136**: 195–212.

Cherezov V, Rosenbaum MD, Hanson MA, Rasmussen SGF, Thian FS, Kobilka TS, Choi HJ, Kuhn P, Weis WI, Kobilka BK, Steven RC. (2007) GPCR engineering yields high-resolution structural insights into β_2-adrenergic receptor function. *Science* **318**: 1258–1265.

Deisenftofer J, Epp O, Miki K, Huber R, Michel H. (1984) X-ray structure analysis of a membrane protein complex. Electron density map at 3 Å resolution and a model of the chromophores of the photosynthetic reaction center from *Rhodopseudomonas viridis*. *J Mol Biol* **180**: 385–398.

Denkov ND, Yoshimura H, Kouyama T, Walz J, Nagayama K. (1998) Electron cryomicroscopy of bacteriorhodopsin vesicles: Mechanism of vesicle formation. *Biophys J* **74**: 1409–1420.

Efremov RG, Shiryaeva GN, Islamov A, Kuklin A, Yaguzhinsky L, Fragneto-Cusani G, Bueldt G, Gordeliy V. (2005) SANS investigations of the lipidic cubic phase behaviour in the course of bacteriorhodopsin crystallization. *J Cryst Growth* **275**: 1453–1459.

Engstroem S, Alfons K, Rasmusson M, Ljusberg-Wahren H. (1998) Solvent-induced sponge (L3) phases in the solvent-monoolein-water system. *Progr Colloid Polym Sci* **108**: 93–98.

Ericsson B, Eriksson PO, Loefroth JE, Engstroem S. (1991) ACS Symposium Series 469 (Polym. Drugs Drug Delivery Syst.), Chap. 22, pp. 251–265.

Faham S, Bowie JU. (2002) Bicelle crystallization: A new method for crystallizing membrane proteins yields a monomeric bacteriorhodopsin structure. *J Mol Biol* **316**: 1–6.

Faham S, Boulting GL, Massey EA, Yohannan S, Yang D, Bowie JU. (2005) Crystallization of bacteriorhodopsin from bicelle formulations at room temperature. *Prot Sci* **14**: 836–840.

Fontell K. (1990) Cubic phases in surfactant and surfactant-like lipid systems. *Colloid Polym Sci* **268**: 264–285.

Gordeliy VI, Labahn J, Moukhametzianov R, Efremov R, Granzin J, Schlesinger R, Bueldt G, Savopol T, Scheidig AJ, Klare J, Engelhard M. (2002) Molecular basis of transmembrane signalling by sensory rhodopsin II-transducer complex. *Nature* **419**: 484–487.

Gordeliy VI, Schlesinger R, Efremov R, Buldt G, Herbele J. (2003) Crystallization in lipidic cubic phase: A case study with bacteriorhodopsin. In Selinsky B (ed), *Membrane Protein Protocols: Expression, Purification and Crystallization, Methods in Molecular Biology Book Series*, pp. 305–316. The Humania Press Inc., USA.

Grabe M, Neu J, Oster G, Nollert P. (2003) Protein interactions and membrane geometry. *Biophys J* **84**: 854–868.

Katona G, Anderson U, Landau EM, Anderson EL, Neutze R. (2003) Lipidic cubic phase crystal structure of the photosynthetic reaction centre from *Rhodobacter sphaeroides* at 2.35 Å resolution. *J Mol Biol* **331**: 681–692.

Kendrew JC, Bodo G, Dintzis HM, Parrish RG, Wyckoff HW, Phillips DC. (1958) A three-dimensional model of the myoglobin molecule obtained by X-ray analysis. *Nature* **181**: 662–666.

Kouyama T, Yamamoto M, Kamiya N, Iwasaki H, Ueki T, Sakurai I. (1994) Polyhedral assembly of a membrane protein in its three-dimensional crystal. *J Mol Biol* **236**: 990–994.

Landau EM, Rosenbusch JP. (1996) Cubic phases: A novel concept for the crystallization of membrane proteins. *Proc Nat Acad Sci* **93**: 14532–14535.

Landh T. (1995) From entangled membranes to eclectic morphologies: Cubic membranes as subcellular space organizers. *FEBS Lett* **369**: 13–17.

Lindblom G, Rilfors L. (1989) Cubic phases and isotropic structures formed by membrane lipids — Possible biological relevance. *Biochim Biophys Acta* **988**: 221–256.

Loll PJ. (2003) Membrane protein crystallization: The high throughput challenge. *J Struct Biol* **142**: 144–153.

Mariani P, Luzatti V, Delacroix H. (1988) Cubic phases of lipid-containing systems: Structure analysis and biological implications. *J Mol Biol* **204**: 165–189.

McGrath KM, Dabbs DM, Yao N, Aksay IA, Gruner SM. (1997) Formation of a silicate *L3* phase with continuously adjustable pore sizes. *Science* **277**: 552–556.

Michel H. (1991) *Crystallization of Membrane Proteins*. CRC Press, Boca Raton, FL.

Moukhametzianov RE, Klare JP, Efremov RG, Baeken C, Göppner A, Labahn J, Engelhard M, Büldt G, Gordeliy VI. (2006) Development of the signal in sensory rhodopsin and its transfer to the cognate transducer. *Nature* **440**: 115–119.

Papiz MZ, Prince SM, Howard T, Cogdell RJ, Isaacs NW. (2003) The structure and thermal motion of the B800-850 LH2 complex from *Rps. acidophila* at 2.0 Å resolution and 100 K: New structural features and functionally relevant motions. *J Mol Biol* **326**: 1523–1538.

Pebay-Peyroula E, Rummel E, Rosenbusch JP, Landau EM. (1997) X-ray structure of bacteriorhodopsin at 2.5 Ångstroms from microcrystals grown in lipidic cubic phases. *Science* **277**: 1676–1681.

Perutz MF. (1962) Relation between structure and sequence of haemoglobin. *Nature* **194**: 914–917.

Prive GG. (2007) Detergents for the stabilization and crystallization of membrane proteins. *Methods* **41**: 388–397.

Rasmussen SGF, Choi H-J, Rosenbaum DM, Kobilka TS, Thian FS, Edwards PC, Burghammer M, Rratnala VRP, Sanishvili R, Fischetti RF, Schertler GFX, Weis WI, Kobilka BK. (2007) Crystal structure of the human β_2-adrenergic G-protein-coupled receptor. *Nature* **450**: 383–387.

Ruahani S, Facciotti MT, Woodcock G, Cheung V, Cunningham C, Nguyen D, Rad B, Lin C-T, Lunde CS, Glaeser RM. (2002) Crystallization of membrane proteins from media composed of connected-bilayer gels. *Biopolymers* **66**: 300–316.

Sanders CR, Prestegard JH. (1990) Magnetically orientable phospholipid bilayers containing small amounts of a bile salt analogue, CHAPSO. *Biophys J* **58**: 447–460.

Sanders CR, Schwonek JP. (1992) Characterization of magnetically orientable bilayers in mixtures of dihexanoylphosphatidylcholine and dimyristoylphosphatidylcholine by solid-state NMR. *Biochemistry* **31**: 8898–8905.

Sanders CR, Landis GC. (1995) Reconstitution of membrane proteins into lipid-rich bilayered mixed micelles for NMR studies. *Biochemistry* **34**: 4030–4040.

Sanders CR, Prosser RS. (1998) Bicelles: A model membrane system for all seasons? *Structure* **6**: 1227–1234.

Takeda K, Sato H, Hino T, Kono M, Fukuda K, Sakurai I, Okada T, Kouyama T. (1998) A novel three-dimensional crystal of bacteriorhodopsin obtained by successive fusion of the vesicular assemblies. *J Mol Biol* **283**: 463–474.

Wadsten P, Woehri AB, Anijder A, Katona G, Gardiner AT, Cogdell RJ, Neutze R, Engstroem S. (2006) Lipidic sponge phase crystallization of membrane proteins. *J Mol Biol* **364**: 44–53.

Watson JD, Crick FHC. (1953a) A structure for deoxyribose nucleic acid. *Nature* **171**: 737–738.

Watson JD, Crick FHC. (1953b) Genetical implications of the structure of deoxyribonucleic acid. *Nature* **171**: 964–967.

Chapter 15

Amyloid, The Amyloid β-Peptide and Alzheimer's Disease: Structural Considerations

*Lars O. Tjernberg**

Several proteins can form ordered, protease-resistant fibrillar deposits, called amyloid, in the human body. Amyloid is associated with several different diseases, amyloidoses, including Alzheimer's disease. This progressive neurodegenerative disease affects around 1% of the population of the Western world. The disease is pathologically characterized by amyloid deposits composed of fibrils formed by the amyloid β-peptide, Aβ, and it has been suggested that compounds that interfere with Aβ-polymerization could be used for treatment of Alzheimer's disease. Several studies aimed at revealing the structure of Aβ in solution and in amyloid fibrils have been performed. Structural studies of Aβ and other amyloid proteins are problematic since these proteins have poor solubility and do not form crystals. However, with the help of a combination of several different, mostly spectroscopic, techniques the structure of Aβ is being resolved. Here, we will give an introduction to amyloid, Alzheimer's disease, Aβ and different techniques used for studying Aβ structure. Finally, we will look at Aβ-polymerization and how this process can be inhibited.

Keywords: Amyloid; amyloid β-peptide; Alzheimer's disease; polymerization; β-sheet; spectroscopy.

Introduction

The term "amyloid" means starch-like, and was first used in the 1850s to describe histological deposits that were stained with iodine. These deposits

*Karolinska Institutet, Department of NVS, KI-Alzheimer's Disease Research Center, KASPAC — Phone: + 46 8 585 83620. E-mail: lars.tjernberg@ki.se.

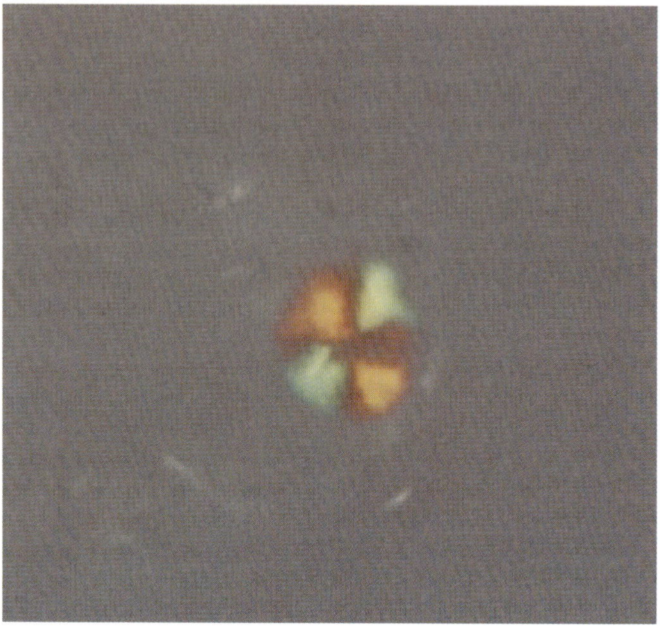

Fig. 1. A CR-stained plaque viewed under polarized light. One gram of human brain was homogenized and centrifuged at 4000 g for 2 min, and the pellet was washed and stained with CR for two days. The amyloid plaque cores were further purified by sucrose gradient centrifugation and filtration (40 μm mesh filter). The plaque cores were dispensed on a microscope slide and viewed under polarized light. The diameter of the plaque core is around 15 μm.

were at that time thought to be amorphous aggregates composed of starch, but were later found to be ordered fibrillar deposits made of specific proteins. At the beginning of the last century, the deposits were found to show red-green bi-refringence in polarized light after staining with the dye Congo red (CR) (Fig. 1). This phenomenon occurs when the CR molecules are arranged in an ordered fashion, and bi-refringence is one criterion for amyloid. Around 50 years ago, amyloid was subjected to electron microscopy (EM) examination and found to be composed of thin fibrils about 8 nm in diameter. When the amyloid fibrils were subjected to X-ray diffraction, two sets of arcs could be observed, corresponding to distances of 4.8 Å and 10–11 Å. These numbers reflect the distances between the peptide strands and between the β-sheets, respectively, in a cross-β fiber.

Thus, amyloid is composed of proteins arranged in ordered cross-β fibrils that show red-green bi-refringence after CR-staining and a fibrillar appearance in EM. In addition to these criteria, it has been suggested that the term "amyloid" should be reserved for extracellular aggregates. Interestingly, fibrils indistinguishable from *ex vivo* fibrils can be formed upon incubation of the corresponding synthetic peptide. For simplicity, "amyloid" will in this chapter be used also for intracellular fibrils and synthetic fibrils (Fig. 2).

Today, around 30 human amyloidoses have been described, each associated with a specific protein. These diseases can be systemic or organ specific, and are often associated with an increased production of the deposited protein or the expression of a more amyloidogenic variant of the protein. The different amyloidoses include a wide variety of proteins and peptides, and it is not obvious why they polymerize into amyloid. It is possible that most proteins could form amyloid under certain conditions, indicating that the polypeptide backbone has an inherited propensity to

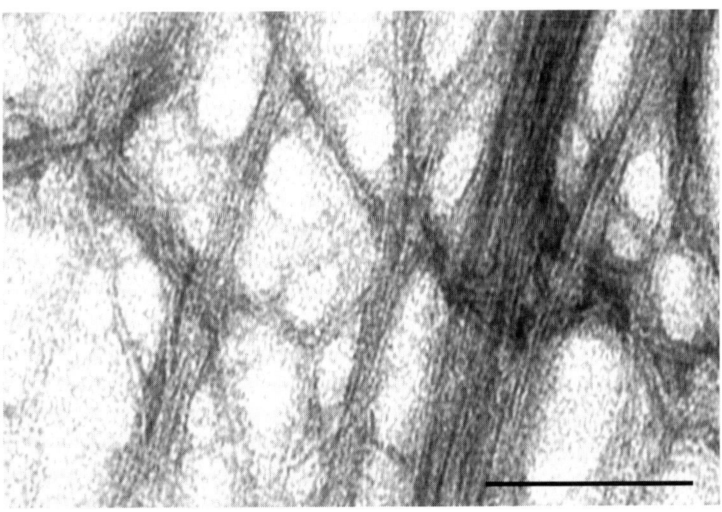

Fig. 2. Amyloid fibrils formed by Aβ1–40. Synthetic Aβ1–40 was incubated in buffer, pH 7.4, for three days. After a brief centrifugation, 8 μl of the sample was placed on an EM grid and stained with uranyl acetate and examined by EM. Scale bar = 100 nm. (EM by Johan Thyberg, Karolinska Institutet.)

polymerize into amyloid fibrils. An emerging view is that a low net charge, unstructured regions, exposed β-sheets and residues with a high β-sheet propensity (Thr, Ile, Tyr, Phe or Val) favors amyloid formation (Dobson *et al.*, 2003; Tjernberg *et al.*, 2002).

The structure of amyloid is very stable, and harsh conditions such as treatment with formic acid or high concentrations of guanidine are needed to dissolve the fibrils. Likewise, the fibrils are resistant to proteolytic degradation and can therefore accumulate over time. Although the fibrils are relatively well characterized, little is known about the processes leading to their formation. Here, we will focus on the amyloid found in Alzheimer's disease, since it is the most studied.

Amyloid in Alzheimer's Disease

Alzheimer's Disease (AD) is a progressive neurodegenerative disease, affecting around 24 million people worldwide. Age is the most important risk factor for AD, and thus, the number of affected individuals will rise rapidly as the mean life expectancy of the human population increases. The first sign of the disease is forgetfulness, followed by loss of cognitive functions and changes in personality. Eventually, the patient needs constant care and becomes bedridden. There is currently no cure for AD, but treatment with acetylcholine esterase inhibitors can in many cases have a beneficial effect and attenuate the progress of the disease for a limited time.

The disease was first described in 1906 by the German physician Alois Alzheimer. In silver stained sections from a demented patient's brain, he observed what today is known as the pathological hallmarks of AD: extracellular plaques and intraneuronal fibrils. Almost 80 years later, a breakthrough in AD research came with the discovery that the amyloid found in AD was composed of the amyloid β-peptide (Aβ) (Glenner *et al.*, 1984; Masters *et al.*, 1985). Extensive research followed this discovery, and the Aβ precursor protein (APP) was cloned a couple of years later, while the proteases responsible for processing APP into

Aβ were identified around the turn of the millennium. APP is a 695 to 770 amino acid type 1 transmembrane protein thought to be involved in cell adhesion, neurite growth and axonal transport. The first cleavage in the amyloidogenic processing of APP is mediated by a protease called BACE, which cleaves APP close to the extracellular/intralumenal side of the membrane, leaving a 99 residue fragment in the membrane. This fragment is, by a membrane protein complex called γ-secretase, further processed into Aβ and an intracellular domain suggested to have a role in gene transcription. There are two main variants of Aβ: Aβ40 and Aβ42 (peptides of 40 and 42 residues in length, respectively). The production of the shorter variant is around ten times higher, but the longer variant is more prone to polymerize into fibrils and is the main constituent of amyloid plaques. The physiological role of Aβ is still unclear.

Several lines of evidence suggest that Aβ is of importance in AD. Aβ plaques are by definition always present in AD, and elevated levels of Aβ in the brain correlate with cognitive decline (Näslund *et al.*, 2000). Numerous studies have shown that Aβ becomes neurotoxic upon incubation, and recent studies suggest that Aβ affects long-term potentiation (LTP), a process involved in memory formation. There are several examples of familial cases of AD, and all of these are associated with mutations in either APP or presenilin, which is a component of γ-secretase. Most of these mutations result in an increased Aβ42/Aβ40 ratio, leading to accumulation of amyloid and an early onset of the disease. Similarly, in Down's syndrome, an extra copy of the chromosome harboring APP leads to increased Aβ levels and early pathological signs of AD. Further evidence for the importance of Aβ comes from transgenic mice that express human Aβ. These mice develop AD-like lesions and memory impairments. Interestingly, these effects can be attenuated by vaccination against Aβ. Based on the evidence above, the "amyloid cascade hypothesis", which suggests that the aggregation of Aβ42 has a central role in AD, was formulated and modified (Hardy, 2006). A multitude of studies aimed at elucidating the structure of Aβ and understanding the polymerization process have been undertaken the last 20 years, and a brief review of these is given below.

Structure of Aβ in Solution

Several different techniques have been used to study the secondary structure of monomeric Aβ in solution. Since Aβ has a strong tendency to polymerize into fibrils, it is crucial to keep Aβ in a monomeric form during measurements. Moreover, there is a risk that the synthetic peptide contains small aggregates that can initiate the polymerization process, and it is important to use proper procedures for the preparation of a monomeric starting material.

Circular Dichroism

Circular dichroism (CD) spectroscopy is a frequently used technique for studying the secondary structure of proteins (Greenfield *et al.*, 1996). All common amino acids except for glycine are chiral, and therefore, a protein will absorb left and right circularly polarized light differently. The absorption is dependent on the conformation, and the technique is useful for the estimation of the secondary structure, i.e. the amount of β-sheet, α-helix, β-turn and random coil in proteins and peptides. The technique is relatively easy to use and a protein concentration around 10 μM is usually sufficient for analysis, but it is not possible to obtain detailed information. Earlier studies showed that Aβ to a large extent was present in a β-sheet conformation. However, if the peptide is carefully prepared, unordered structures dominate. It is clear that the structure is affected by the solvent, and in α-helix stabilizing solvents such as hexafluoroisopropanol (HFIP), a mostly α-helical structure dominates.

Nuclear magnetic resonance spectroscopy

Nuclear magnetic resonance (NMR) spectroscopy can provide information at the atomic level, but requires sophisticated equipment, higher concentrations (up to 1 mM) and longer analysis time than CD spectroscopy does. The high concentrations and the long analysis time result in an increased risk of polymerization of Aβ. Therefore, less amyloidogenic truncated Aβ-variants or conditions that promote the solubility of Aβ have often been used in these studies. In the first published study on full length Aβ40, 40% trifluoroethanol (TFE) in phosphate buffer at pH 2.8 was used (Sticht *et al.*, 1995). This study showed two helices, Gln15-Asp23 and Ile31-Met35, with the rest of the peptide being in random-coil conformation. TFE promotes the formation of ordered secondary structures, and can thus induce the formation of α-helices or β-sheets. Also the pH of the solution has a strong influence on secondary structure, and it is important to study the peptide under physiological conditions. When Aβ40 was studied at pH 5.1 in the presence of sodium dodecylsulfate (SDS) micelles, which are suggested to mimic the cell membrane to some extent, the peptide showed an α-helical structure from Gln15 to Val36 interrupted by a kink at Gly25-Asn27. Similar results were obtained when Aβ42 was studied at pH 7.2 in the presence of SDS micelles (Shao *et al.*, 1999). In this study it was concluded that the peptide binds to the surface of the micelles. The presence of an α-helix at the C-terminus of the peptide is in line with the notion that all residues after Lys28 in Aβ are part of the transmembrane region of APP. By combining data from NMR-analysis of Aβ40 (using short analysis time to avoid aggregation, resulting in less detailed data) and Aβ10–35 (which is more soluble and allows longer analysis times and/or higher concentrations), it was possible to obtain a structure of Aβ40 in water (Zhang *et al.*, 2000). In this case, the peptide showed a compact structure containing loops, strands and turns, but no α-helices or β-sheets. Later studies in water confirmed the absence of well-defined secondary structures, but a tendency for the hydrophobic region Leu17-Ala21 to adopt a β-strand structure. Moreover, the two extra residues in Aβ42, Ile and Ala, resulted in a more rigid C-terminus with a tendency to form a β-strand (Hou *et al.*, 2004). In summary, both NMR and CD spectroscopy support the notion that Aβ in a buffer free from detergents and organic solvents is mostly unstructured.

Structure of Aβ in Fibrils

X-ray diffraction shows that the peptide is present in a cross-β sheet conformation in the fibrils. Since amyloid is non-crystalline, it has not been possible to obtain a detailed structure by X-ray crystallography. By using cryo-EM in combination with advanced image processing, it was possible to visualize individual peptide strands in the fibril. In line with X-ray diffraction data, the strands were aligned perpendicular to the fibril axis with a repeat distance of 4.8 Å. In order to obtain more detailed information, solid state NMR (SS-NMR) has been used in the last decade. By SS-NMR analysis of fibrils formed from Aβ10–35 labeled with ^{13}C at Gln15, Lys16, Leu17 or Val18, it was possible to obtain a high resolution model of the central core of Aβ. This region showed a parallel β-sheet structure with the amino acids aligned in register, i.e. Gln15 interacts with Gln15 in adjacent strands (Benzinger *et al.*, 1998). These findings were confirmed and extended by labeling Aβ10–35 at several different positions, showing that the whole peptide was in a parallel β-sheet conformation (Benzinger *et al.*, 2000). Also Aβ1–40 as well as Aβ1–42 fibrils were shown to have an in-register parallel β-sheet conformation, but in these cases the N-terminus was unstructured (Antzutkin *et al.*, 2000, 2002). The latter study suggested that the peptide had two parallel β-sheet regions, 10–22 and 30–42 connected by a loop, which were folded over each other. Such folding results in a protofilament with one polar and one hydrophobic side. One strong driving force for peptide folding and aggregation is to exclude hydrophobic surfaces from water, and it is likely that the hydrophobic side of two such filaments would interact and form a fibril. A similar but more refined model of Aβ1–40 fibrils was suggested based on isotope dilution, restraints from SS-NMR measurements in combination with molecular dynamics and energy minimization calculations (Petkova *et al.*, 2002). Using hydrogen-exchange (see below) and SS-NMR, data from previous SS-NMR studies and pairwise substitutions of interacting residues, a detailed 3D structure of Aβ42 fibrils was presented (Lührs *et al.*, 2005) (Fig. 3).

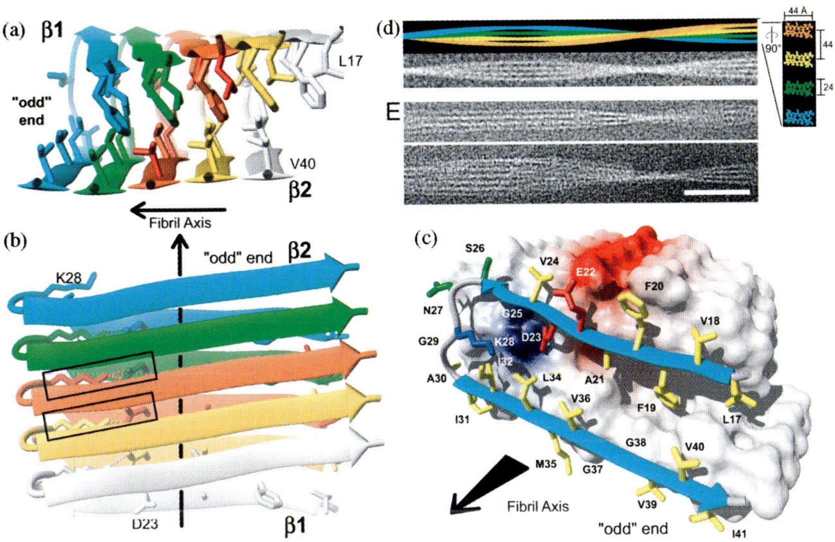

Fig. 3. The structure of an Aβ1–42 fibril. (a) and (b), Ribbon diagrams of the core structure of Aβ17–42. The β-strands are indicated by arrows, and intermolecular salt bridges are represented by dotted lines. The salt bridges formed by the central (orange) Aβ molecule are highlighted by boxes. (c) Van der Waals contact surface polarity/ribbon diagram. Color code for the residues: yellow = hydrophobic, green = polar, red = negatively charged, blue = positively charged. (d) Simulation of a fibril containing four protofilaments. (e) Cryo-EM of Aβ1–42 fibrils. Scale bar = 50 nm. From Lührs (2005).

Other Techniques

Several other techniques besides SS-NMR have been used for studying fibrils. Electron spin resonance was used for studying spin-labeled Aβ1–40 fibrils, and the results were in line with those from SS-NMR, supporting the presence of an unstructured N-terminus followed by a parallel in-register β-sheet region interrupted by a loop region around residues 23–26. Hydrogen exchange and limited proteolysis are other approaches that have been used to elucidate the structure of Aβ in fibrils. Proteolytic enzymes cannot degrade compact and rigid structure such as β-sheets, and thus, only the unstructured part of a fibril will be degraded by proteases. By using trypsin (which cleaves peptide bonds after Lys or Arg) and chymotrypsin (cleaves after large hydrophobic residues) it was concluded that fibrillar Aβ is unstructured up to residues around His13 to Lys16, the rest of the peptide being in a protease stable conformation (Wetzel *et al.*, 2001). Hydrogens that are involved in the hydrogen-bonding network of the peptide backbone in a fibril are resistant to exchange with hydrogens in the buffer. Aβ1–40 was allowed to form fibrils in H_2O, and the fibrils were thereafter transferred to D_2O and incubated for several days. In the next step the fibrils were rapidly dissolved and the resulting monomers were injected into a mass spectrometer. The authors found that approximately half of the backbone hydrogens had been exchanged, and concluded that half of the residues in the peptide are involved in β-sheet interactions in the fibril (Wetzel *et al.*, 2000).

In summary, several techniques have been used for the determination of Aβ structure in fibrils. Together, they support a cross-β-sheet model with the individual strands perpendicular to the fibril axis. An unstructured N-terminus is followed by a parallel in-register β-sheet, a loop around residues 24–29, and an in-register β-sheet C-terminal part folding back over the central part. The hydrophobic sides of two or more such protofilaments interact with each other and form a fibril.

Aβ Polymerization

A few years after the discovery of Aβ, several groups reported that freshly dissolved Aβ was harmless, while "aged" or "aggregated" Aβ was toxic to cells. The aggregated solutions showed fibrils upon examination by EM, and the conclusion was that fibrils were the toxic species (Lorenzo and Yankner, 1994). However, later studies showed that smaller species such as protofibrils or oligomeric variants were more toxic (Hartley *et al.*, 1999; Lambert *et al.*, 1998; Lesné *et al.*, 2006). Thus, it is of importance to study the polymerization process in order to learn more about what species are toxic, how they are formed, and how their formation can be inhibited.

Several different techniques have been used to study Aβ-polymerization, and a few of them will be described here. A classical technique for studying the size of particles in solution is light scattering. This technique is based on the fact that particles scatter light, and large particles scatter more than small. No labeling of the sample is required, but relatively high concentrations are necessary. Large aggregates are over-represented and it is difficult to determine the size of heterogeneous mixtures of aggregates. If no information on the size distribution of the particles is needed, the polymerization process can be followed in an ordinary spectrophotometer: increased light scattering will result in less light reaching the detector and an increased "absorbance" value. This technique was frequently used for the early studies on Aβ-polymerization (Jarrett *et al.*, 1993). It was shown that the polymerization could be divided into three major phases:

(i) *The lag phase or nucleation phase.* The process is dependent on a seed or nuclei to form before fibrils can grow. This phase is clearly concentration dependent; if the concentration is too low, no fibrils will form, and at high concentrations the lag phase is short and not detectable.
(ii) *The growth phase.* In this phase, the fibrils grow in proportion to the number of seeds formed and the concentration of Aβ.
(iii) *The equilibrium phase.* Finally, the concentration of free Aβ becomes low and the net growth of fibrils stops.

In studies aimed at following Aβ-polymerization, it is crucial to start from a monomeric solution since small amounts of seed will confound analysis. Depending on the manufacturer and batch-to-batch variations, different numbers of seeds are present from the beginning of the incubations. Therefore, many of the early polymerization studies were contradictory. Today, there are established protocols for obtaining a uniform, monomeric starting material. The most efficient solvents are formic acid, HFIP, high concentrations of urea or guanidine thiocyanate, and high pH buffers. It is important to keep in mind that many of these solvents can affect the primary structure of Aβ. Formic acid will oxidize Met35, guanidine will guanidinylate lysines, and a high pH increases the rate of hydrolysis. Thus, in these cases it is important to use short incubation times before lyophilizing (in the case of formic acid) or diluting the stock solutions. Treatment with HFIP is milder, and in this case the Aβ stock solution can be incubated for days. In order to remove potential seeds that have survived the initial treatment, centrifugation or filtration is frequently used. Each step in the preparation procedure leads to sample loss, and it is necessary to measure the concentration of Aβ in the final working solution. Since Aβ is a very "sticky peptide," it is important to select tubes, well plates and pipette tips with low affinity for the peptide. Several factors, such as temperature, buffer ions, pH, detergents, surface tension and shaking, affect the polymerization rate. Depending on the goal of the polymerization studies, different techniques are used for studying Aβ-polymerization. Here we will discuss the merits and drawbacks of the most frequently used.

Thioflavin T binding assay

Thioflavin T binding (ThT) is a fluorescent molecule that has a high affinity for amyloid fibrils. Upon binding, the fluorescence spectrum is shifted, and thus, the polymerization can be followed by monitoring the fluorescence of ThT (LeVine, 1993). The method is rapid, easy to use and suitable for well plate readings. However, it does not give information on the size or morphology of the aggregates, and ThT can bind to other aggregates than fibrils (Tjernberg *et al.*, 1999). Therefore, EM is usually used as a complement to ThT for showing the morphology of the aggregates. The ThT assay is frequently used for finding and studying polymerization inhibitors, but gives only limited information on at what stage the polymerization is inhibited, although it can be inferred that an increase in the lag phase would indicate that the inhibitor interferes with seed formation. In certain cases the binding sites for ThT can be blocked, for instance when fibril bundles are formed or when the studied molecule competes for the ThT binding site, resulting in a low signal (Söderberg *et al.*, 2005).

Gel Electrophoresis

Polyacrylamide gel electrophoresis (PAGE) has been used to visualize oligomers of different size. Using the strong ionic detergent SDS in the sample buffer, dimers and trimers could be observed. One problem with SDS-PAGE is that oligomers can be dissociated by SDS or during chromatography, and that SDS can induce oligomerization. In order to avoid dissociation, cross-linking of Aβ has been used. In this case it is important to use short cross-linking times to avoid diffusion-dependent cross-linking. Recently, it was suggested that photoinduced cross-linking of unmodified proteins (PICUP) could be used for studying Aβ oligomers (Bitan *et al.*, 2001). In this case, reagents that form radicals upon light exposure activate Aβ, and covalent bonds between Aβ molecules that are closely associated in an oligomer can form. The duration of the light pulse is usually one second or less, and the quenching of the reaction is achieved in a few seconds by the addition of sample buffer containing a reducing agent. After cross-linking, the oligomers are separated by PAGE and visualized by silver staining of the gel. Alternatively, western blotting can be used if high sensitivity is needed. By using urea in the sample buffer, artifactual interactions are minimized. One consideration regarding PICUP is that the reactive amino acids (e.g. Tyr and Lys) must be close in order to interact, and thus, it is likely that only a subset of oligomers will be cross-linked.

CD spectroscopy

As described above, CD spectroscopy can be used for studying the secondary structure of proteins. Since monomeric Aβ in a buffer free from organic solvents and detergents is in a mostly disordered conformation, while fibrillar Aβ is in β-sheet conformation, the polymerization can be followed as an increased amplitude in the spectrum at the wavelengths typical for β-sheet (e.g. a minimum around 217 nm). Unfortunately, this technique does not reveal minor changes in the sample, and it is not possible to follow the formation of oligomers since it is not clear what conformation they have. A comprehensive study including several different Aβ-variants showed that an α-helical conformation precedes the β-sheet conformation (Kirkitadze *et al.*, 2001).

Atomic force microscopy

Atomic force microscopy (AFM) can visualize fibrils and small aggregates by using a thin tip moving over the sample surface enabling structures in the nanometer range to be visualized (Harper *et al.*, 1997). The resolution is comparable to EM, but AFM has one clear advantage; fibril growth can be visualized in real time in solution. Thus, the growth of individual fibrils can be monitored. The technique is relatively slow, only species deposited on the surface are detected, and the surface affects polymerization.

Fluorescence correlation spectroscopy (FCS)

By using a confocal microscope setup, a sensitive detector, and an efficient correlator/computer, it is possible to monitor the diffusion of single fluorescent molecules through a tiny (femtoliter) volume element. Since single molecules are detected, the method is highly sensitive. The diffusion time is approximately proportional to the cubic root of the mass, i.e. an octamer will have a diffusion time twice that of a monomer. Therefore, it can be difficult to resolve heterogeneous samples with species of similar masses. However, single large aggregates can easily be detected in a mixture of low molecular weight species. This method was used to follow Aβ-polymerization, showing that large aggregates, undetected by CD spectroscopy, preceded the formation of fibrils (Tjernberg *et al.*, 1999). The formation of these aggregates was clearly concentration-dependent, and occurred only at Aβ concentrations above 40 μM. Thus, these aggregates are probably not relevant to AD, since the physiological levels of Aβ are around or below 1 nM. Their occurrence suggests that Aβ concentrations should be kept below this critical concentration to avoid non-relevant polymerization pathways. By using peptides with different fluorescent labels and two lasers, more information can be extracted from fluorescence correlation spectroscopy (FCS) experiments. For instance, in a 1:1 mixture of "red"-labeled and "green"-labeled Aβ, 50% of the dimers will carry both labels resulting in the simultaneous detection of red and green light when they pass the volume element.

Inhibition of Aβ Polymerization

The polymerization of Aβ leads to the formation of toxic species, and thus, interfering with this process is a potential strategy for the prevention and treatment of AD. Since oligomeric Aβ species are toxic, the inhibitors should stop the polymerization process as early as possible. As described above, it is difficult to study Aβ-polymerization, and finding polymerization inhibitors is therefore a complicated task. Moreover, the inhibitors should be stable, non-toxic, and pass the blood-brain barrier. Usually, a drug that acts as an inhibitor is directed to a well-defined site, e.g. the active site of a protein with a defined structure. Since monomeric Aβ is in a mostly random conformation, no such site exists, but there could still be regions in Aβ that are highly important for Aβ-Aβ interactions.

In order to investigate whether a specific region in Aβ could be crucial for Aβ-polymerization, peptide sequences corresponding to Aβ1–10, Aβ2–11, Aβ3–12, ..., Aβ31–40, were synthesized on a membrane. The membrane, containing 31 spots with different peptides, was then incubated with radiolabeled Aβ1–40. After washing, bound Aβ1–40 was detected by exposing an X-ray sensitive film to the membrane. It was found that peptides corresponding to the central region of Aβ mediated the strongest binding. By synthesizing a variety of shorter peptides, it was shown that Aβ16–20 (LysLeuValPhePhe or KLVFF in one letter code) was sufficient for mediating Aβ-Aβ binding (Tjernberg *et al.*, 1996). A peptide containing this sequence was synthesized and incubated together with Aβ1–40. Only a few fibrils were observed by EM in this sample while a massive network of fibril bundles were found after incubation of Aβ1–40 in the absence of the short peptide. Hence, a short peptide that binds to the KLVFF region in Aβ can inhibit fibril formation. Such a peptide is not useful for the treatment of AD, since it is rapidly degraded in the human body, but modifications making it more stable towards proteolytic degradation are possible. For instance, peptides composed of d-amino acids (natural amino acids are in the l-configuration) are relatively resistant to proteolysis. To find such peptides, a pentapeptide library was synthesized from d-amino acids on a membrane. The library was incubated with a radiolabeled peptide containing the KLVFF-motif and several strong binders were found with a sequence similar to a "reversed" KLVFF peptide, e.g. yfllr (d-amino acids are represented by lower case letters). Similarly to KLVFF, the yfllr peptide inhibited Aβ fibril formation as observed by EM (Tjernberg *et al.*, 1997). Proline is known to be incompatible with β-sheets, and a pentapeptide with the sequence LPFFD (compare with KLVFF) was found to inhibit fibril formation and to prevent Aβ-induced neuronal death in cell culture (Soto *et al.*, 1998). Moreover, this peptide was found to block fibril formation in a rat brain model of amyloidosis. A modified, endprotected variant of the peptide was tested in transgenic mice overexpressing Aβ, showing a clear reduction in amyloid burden and decreased neuronal damage (Permanne *et al.*, 2002). In this study, the peptide was administered by intraperitoneal injection or intra-cerebroventricular infusion, and further improvement regarding proteolytic stability in order to allow intravenous or oral administration was necessary. This was achieved by methylation of one nitrogen atom in the backbone of the peptide (Adessi *et al.*, 2003). The modified peptide showed good brain uptake, efficient inhibition of Aβ-polymerization and protected cells from Aβ-induced toxicity. Thus, it is possible to increase the stability of KLVFF-like peptides by different modifications and use such peptides, or peptidomimetics, for lowering of the amyloid burden and increasing neuronal survival in transgenic mice models of amyloid.

It is likely that small organic molecules would be more stable and better suited for pharmacological treatment of AD than peptide-derived molecules. Several different small organic molecules have been found to affect Aβ-polymerization, but only a few have entered clinical trials. One of these, 3-amino-1-propanesulfonic acid showed good results in animal studies and eventually entered phase III clinical trials. Unfortunately, the outcome of this trial was not positive and no clear effect on cognition could be observed. Another organic compound under investigation is curcumin, a polyaromatic yellow dye from the root of turmeric. Curcumin inhibits the polymerization of Aβ, and by oligomer specific antibodies it was shown that the formation of oligomers was inhibited (Yang *et al.*, 2005). Most of the methods that are used today to find Aβ-polymerization inhibitors are not suitable for screening large compound libraries. However, a better understanding of Aβ-polymerization together with methodological and technical development will hopefully enable such screens and increase the chances of finding "hits" that can be developed into drugs.

Summary

Amyloid are well ordered, protease resistant fibrillar deposits that show a cross-β sheet pattern upon X-ray diffraction analysis and red-green bi-refringence when viewed under polarized light in the microscope. There are around 30 diseases associated with amyloid, including Alzheimer's disease which affects around 1% of the population of the Western world. The amyloid is in this case formed by a 40–42-residues-long peptide called Aβ. Abundant evidence supports the notion that species formed during the polymerization of Aβ into amyloid deposits in the human brain are involved in the pathogenesis of Alzheimer's disease. Studies on the structure of Aβ in solution have been hampered by its poor solubility, but improved protocols for keeping the peptide in monomeric form, the use of more soluble Aβ variants, and a combination of spectroscopic techniques have shown that monomeric Aβ has a mostly unordered structure with a tendency to form a β-strand in the central region and at the C-terminus. Aβ does not form crystals, and therefore, it has not been possible to obtain the crystal structure of the peptide. However, several other techniques have been used for the determination of Aβ structure in fibrils. Solid state NMR has been important for obtaining a detailed model where the N-terminus is unordered while the rest of the peptide adopts an in-register parallel β-sheet interrupted by a turn around residues 24–29, allowing the two β-sheets to interact. Thus, a protofilament with one polar side and one hydrophobic side is formed. Two or more such filaments interact in order to exclude hydrophobic surfaces from water, forming an amyloid fibril.

Monomeric Aβ polymerizes into fibrils in a nucleation dependent reaction, involving several intermediate species, some of them being neurotoxic. Inhibitors that block the formation of such toxic intermediates may be used for the treatment of Alzheimer's disease. A central region of Aβ is crucial for Aβ-Aβ interactions, and compounds binding to this region can inhibit Aβ-polymerization. Such compounds have shown beneficial effects in transgenic mice and a couple of them have been tested in clinical trials, but the results are so far not convincing. Several different techniques have been used to study Aβ-polymerization. They all have their merits and drawbacks, and no single technique can give the full picture on its own. Hopefully, technical and methodological development will provide a more detailed understanding of the polymerization process and aid in the search for more efficient inhibitors that are therapeutically useful.

Acknowledgment

This work was supported by Stiftelsen Demensfonden, Stockholm, Sweden.

References

Adessi C, Frossard MJ, Boissard C, Fraga S, Bieler S, Ruckle T, Vilbois F, Robinson SM, Mutter M, Banks WA, Soto C. (2003) Pharmacological profiles of peptide drug candidates for the treatment of Alzheimer's disease. *J Biol Chem* **278**: 13905–13911.

Antzutkin ON, Balbach JJ, Leapman RD, Rizzo NW, Reed J, Tycko R. (2000) Multiple quantum solid-state NMR indicates a parallel, not antiparallel, organization of beta-sheets in Alzheimer's β-amyloid fibrils. *Proc Nat Acad Sci USA* **97**: 13045–13050.

Antzutkin ON, Leapman RD, Balbach JJ, Tycko R. (2002) Supramolecular structural constraints on Alzheimer's β-amyloid fibrils from electron microscopy and solid-state nuclear magnetic resonance. *Biochemistry* **41**: 15436–15450.

Benzinger TL, Gregory DM, Burkoth TS, Miller-Auer H, Lynn DG, Botto RE, Meredith SC. (1998) Propagating structure of Alzheimer's β-amyloid (10–35) is parallel beta-sheet with residues in exact register. *Proc Nat Acad Sci USA* **95**: 13407–13412.

Bitan G, Lomakin A, Teplow DB. (2001) Amyloid β-protein oligomerization: Prenucleation interactions revealed by photo-induced cross-linking of unmodified proteins. *J Biol Chem* **276**: 35176–35184.

Chiti F, Stefani M, Taddei N, Ramponi G, Dobson CM. (2003) Rationalization of the effects of mutations on peptide and protein aggregation rates. *Nature* **424**: 805–808.

Glenner GG, Wong CW. (1984) Alzheimer's disease: Initial report of the purification and characterization of a novel cerebrovascular amyloid protein. *Biochem Biophys Res Commun* **120**: 885–890.

Greenfield NJ. (1996) Methods to estimate the conformation of proteins and polypeptides from circular dichroism data. *Anal Biochem* **235**: 1–10.

Hardy J. (2006). Has the amyloid cascade hypothesis for Alzheimer's disease been proved? *Curr Alzheimer Res* **3**: 71–73.

Harper JD, Lieber CM, Lansbury PT Jr. (1997) Atomic force microscopic imaging of seeded fibril formation and fibril branching by the Alzheimer's disease amyloid-β protein. *Chem Biol* **4**: 951–959.

Hartley DM, Walsh DM, Ye CP, Diehl T, Vasquez S, Vassilev PM, Teplow DB, Selkoe DJ. (1999) Protofibrillar intermediates of amyloid β-protein induce acute electrophysiological changes and progressive neurotoxicity in cortical neurons. *J Neurosci* **19**: 8876–8884.

Hou L, Shao H, Zhang Y, Li H, Menon NK, Neuhaus EB, Brewer JM, Byeon IJ, Ray DG, Vitek MP, Iwashita T, Makula RA, Przybyla AB, Zagorski MG. (2004) Solution NMR studies of the Aβ (1–40) and Aβ (1–42) peptides establish that the Met35 oxidation state affects the mechanism of amyloid formation. *J Am Chem Soc* **126**: 1992–2005.

Jarrett JT, Berger EP, Lansbury PT Jr. (1993) The carboxy terminus of the β-amyloid protein is critical for the seeding of amyloid formation: Implications for the pathogenesis of Alzheimer's disease. *Biochemistry* **32**: 4693–4697.

Kheterpal I, Williams A, Murphy C, Bledsoe B, Wetzel R. (2001) Structural features of the Aβ amyloid fibril elucidated by limited proteolysis. *Biochemistry* **40**: 11757–11767.

Kheterpal I, Zhou S, Cook KD, Wetzel R. (2000) Aβ amyloid fibrils possess a core structure highly resistant to hydrogen exchange. *Proc Nat Acad Sci USA* **97**: 13597–13601.

Kirkitadze MD, Condron MM, Teplow DB. (2001) Identification and characterization of key kinetic intermediates in amyloid β-protein fibrillogenesis. *J Mol Biol* **312**: 1103–1119.

Lambert MP, Barlow AK, Chromy BA, Edwards C, Freed R, Liosatos M, Morgan TE, Rozovsky I, Trommer B, Viola KL, Zhang C, Finch CE, Krafft GA, Klein WL. (1998) Diffusible, nonfibrillar ligands derived from Aβ1–42 are potent central nervous system neurotoxins. *Proc Nat Acad Sci USA* **95**: 6448–6453.

Lesné S, Koh MT, Kotilinek L, Kayed R, Glabe CG, Yang A, Gallagher M, Ashe KH. (2006) A specific amyloid-β protein assembly in the brain impairs memory. *Nature* **440**: 352–357.

LeVine H 3rd. (1993) Thioflavine T interaction with synthetic Alzheimer's disease β-amyloid peptides: Detection of amyloid aggregation in solution. *Protein Sci* **2**: 404–410.

Lorenzo A, Yankner BA. (1994) β-amyloid neurotoxicity requires fibril formation and is inhibited by Congo red. *Proc Nat Acad Sci USA* **91**: 12243–12247.

Lührs T, Ritter C, Adrian M, Riek-Loher D, Bohrmann B, Dobeli H, Schubert D, Riek R. (2005) 3D structure of Alzheimer's amyloid-β (1–42) fibrils. *Proc Nat Acad Sci USA* **102**: 17342–17347.

Masters CL, Simms G, Weinman NA, Multhaup G, McDonald BL, Beyreuther K. (1985) Amyloid plaque core protein in Alzheimer disease and Down syndrome. *Proc Nat Acad Sci USA* **82**: 4245–4249.

Näslund J, Haroutunian V, Mohs R, Davis KL, Davies P, Greengard P, Buxbaum JD. (2000) Correlation between elevated levels of amyloid β-peptide in the brain and cognitive decline. *Jama* **283**: 1571–1577.

Permanne B, Adessi C, Saborio GP, Fraga S, Frossard MJ, Van Dorpe J, Dewachter I, Banks WA, Van Leuven F, Soto C. (2002) Reduction of amyloid load and cerebral damage in a transgenic mouse model of Alzheimer's disease by treatment with a beta-sheet breaker peptide. *FASEB J* **16**: 860–862.

Petkova AT, Ishii Y, Balbach JJ, Antzutkin ON, Leapman RD, Delaglio F, Tycko R. (2002) A structural model for Alzheimer's β-amyloid fibrils based on experimental constraints from solid state NMR. *Proc Nat Acad Sci USA* **99**: 16742–16747.

Shao H, Jao S, Ma K, Zagorski MG. (1999) Solution structures of micelle-bound amyloid β-(1–40) and β-(1–42) peptides of Alzheimer's disease. *J Mol Biol* **285**: 755–773.

Söderberg L, Dahlqvist C, Kakuyama H, Thyberg J, Ito A, Winblad B, Näslund J, Tjernberg LO. (2005) Collagenous Alzheimer amyloid plaque component assembles amyloid fibrils into protease resistant aggregates. *Febs J* **272**: 2231–2236.

Soto C, Sigurdsson EM, Morelli L, Kumar RA, Castano EM, Frangione B. (1998) Beta-sheet breaker peptides inhibit fibrillogenesis in a rat brain model of amyloidosis: Implications for Alzheimer's therapy. *Nat Med* **4**: 822–826.

Sticht H, Bayer P, Willbold D, Dames S, Hilbich C, Beyreuther K, Frank RW, Rosch P. (1995) Structure of amyloid A4-(1–40)-peptide of Alzheimer's disease. *Eur J Biochem* **233**: 293–298.

Tjernberg LO, Hosia W, Bark N, Thyberg J, Johansson J. (2002) Charge attraction and beta propensity are necessary for amyloid fibril formation from tetrapeptides. *J Biol Chem* **277**: 43243–43246.

Tjernberg LO, Callaway DJ, Tjernberg A, Hahne S, Lilliehök C, Terenius L, Thyberg J, Nordstedt C. (1999a) A molecular model of Alzheimer amyloid β-peptide fibril formation. *J Biol Chem* **274**: 12619–12625.

Tjernberg LO, Lilliehök C, Callaway DJ, Näslund J, Hahne S, Thyberg J, Terenius L, Nordstedt C. (1997) Controlling amyloid β-peptide fibril formation with protease-stable ligands. *J Biol Chem* **272**: 12601–12605.

Tjernberg LO, Näslund J, Lindqvist F, Johansson J, Karlström AR, Thyberg, J, Terenius L, Nordstedt C. (1996) Arrest of β-amyloid fibril formation by a pentapeptide ligand. *J Biol Chem* **271**: 8545–8548.

Tjernberg LO, Pramanik A, Björling S, Thyberg P, Thyberg J, Nordstedt C, Berndt KD, Terenius L, Rigler R. (1999b) Amyloid β-peptide polymerization studied using fluorescence correlation spectroscopy. *Chem Biol* **6**: 53–62.

Yang F, Lim GP, Begum AN, Ubeda OJ, Simmons MR, Ambegaokar SS, Chen PP, Kayed R, Glabe CG, Frautschy SA, Cole GM. (2005) Curcumin inhibits formation of amyloid beta oligomers and fibrils, binds plaques, and reduces amyloid *in vivo*. *J Biol Chem* **280**: 5892–5901.

Zhang S, Iwata K, Lachenmann MJ, Peng JW, Li S, Stimson ER, Lu Y, Felix AM, Maggio JE, Lee JP. (2000) The Alzheimer's peptide Aβ adopts a collapsed coil structure in water. *J Struct Biol* **130**: 130–141.

Chapter 16

Recent Advances in Structural Basis for Molecular Mimicry in Inflammatory Autoimmune Demyelinating Polyneuropathy

*Xin Yang**

Evidence is mounting to suggest a causal role of cellular or humoral mediated immune response arising from anti-myelin of peripheral nerve antibodies, for example anti-ganglioside antibodies, in a variety of neurological disorders. These disorders include the acute inflammatory autoimmune demyelinating and axonal forms of Guillain–Barré syndrome (GBS), Miller Fisher syndrome (MFS), and chronic inflammatory demyelinating polyneuropathy (CIDP). The origin of the auto-antibodies is discussed in light of the recent circumstantial evidence pointing to a molecular mimicry mechanism with infectious agents in GBS and MFS, but not in CIDP. Recent studies on the roles of anti-ganglioside and non-ganglioside in the pathogenesis of GBS, MFS and CIDP are summarized in this review. With a better understanding of the immunopathogenic mechanisms of these diseases, it will then be possible to devise rational and effective diagnostic and therapeutic strategies for the treatment of these neurological disorders.

Keywords: Guillain–Barré syndrome (GBS); Miller Fisher syndrome (MFS); chronic inflammatory demyelinating polyneuropathy (CIDP); molecular mimicry; pathogen.

*Department of Neurology, Ruijin Hospital, Medical School of Shanghai Jiao Tong University, Shanghai, China and Division of Neurodegeneration, Department of Neurobiology, Care Sciences and Society, Karolinska Institute, Karolinska University Hospital Huddinge, Stockholm, Sweden. E-mail: keading2006@yahoo.com.cn

Introduction

Molecular mimicry is one mechanism by which infectious agents may trigger an immune response against auto-antigens. Four criteria must be satisfied to conclude that a disease is triggered by molecular mimicry (Ang *et al.*, 2004): (i) the establishment of an epidemiological association between the infectious agent and the immune-mediated disease, (ii) the identification of T-cells or antibodies directed against the patient's target antigens, (iii) the identification of microbial mimics of the target antigen, and (iv) reproduction of the disease in an animal model. This review provides an update on the molecular mimicry of the acute inflammatory autoimmune demyelinating and axonal forms of Guillain–Barré syndrome (GBS), Miller Fisher syndrome (MFS), and chronic inflammatory demyelinating polyneuropathy (CIDP).

Guillain–Barré syndrome

GBS is a representative paralytic syndrome characterized by autoimmune response induced inflammation in the peripheral nervous system (PNS), with an annual incidence of two cases per 1 000 000 of the population. GBS is defined as a recognizable clinical entity characterized by rapidly evolving symmetrical limb weakness, loss of tendon reflexes, the absence of mild sensory signs and autonomic dysfunctions. Being the most frequent cause of acute flaccid paralysis initiated by a range of quite distinct immunopathological events, GBS has been identified as being triggered by many kinds of agents: *Campylobacter jejuni* (*C. jejuni*, in 13–39% of cases), cytomegalovirus (CMV, 5–22%), *Haemophilus influenzae* (*H. influenzae*, 1–8%), Epstein–Barr virus (EBV, 1–13%), and *Mycoplasma pneumoniae* (*M. pneumoniae*, 5%) (Godschalk *et al.*, 2007; Jacobs *et al.*, 1998; Koga *et al.*, 2005). All of these pathogens have carbohydrate epitopes which are common with glycoproteins of peripheral nerve tissue.

The most frequent pattern of GBS encountered in Europe and North America is that originally described as acute inflammatory demyelinating polyradiculoneuropathy (AIDP), with demyelination and a variable degree of lymphocytic infiltration. In some other developing areas in the world, in China and in Japan (Kuwabara *et al.*, 2003), the axonal pattern of GBS, acute motor axonal neuropathy (AMAN) and acute motor and sensory axonal neuropathy (AMSAN) has been found along with linkages to *C. jejuni* infection.

Much of the research into GBS over the last decade has focused on the forms mediated by anti-ganglioside antibodies (Willison, 2005). The identification of molecular mimicry between GBS associated pathogens, particularly *C. jejuni* lipopolysaccharides (LPSs) and GM1, GD1a and GT1a ganglioside in peripheral nerve tissue, has been investigated from mimic structural characterization to genetic polymorphism (Godschalk *et al.*, 2007). Besides the emerging correlations between anti-ganglioside antibodies and specific clinical phenotypes, notably between anti-GM1/anti-GD1a antibodies and AMAN, the complexes of different antibodies have also been found and shown to be reactive in "antibody-negative" GBS sera, i.e. GD1a-GD1b, GD1b-GT1b (Godschalk *et al.*, 2007). Despite this rapidly advancing progress in molecular mimicry of GBS, considerable gaps in our knowledge persist.

MFS is characterized by the acute onset of ophthalmoplegia, ataxia and flexia, and is the most common variant of GBS with which it often overlaps clinically, accounting for 5–10% of cases. It is now widely accepted that well over 90% of patients with MFS have the anti-GQ1b IgG antibody, which is completely absent in normal and other disease control groups (Willison, Yuki, 2002).

A true case of molecular mimicry

Molecular mimicry is one mechanism of GBS by which infectious agents, most frequently *C. jejuni,* may trigger an autoimmune response against myelin or axons, the nerve conduits for signals to and from the brain (Fig. 1).

The mistaken attach is considered to originate from the resemblance between the microbial LPSs and the ganglioside of human nerve tissues. It used to be thought that in AIDP the attack appears to be directed against a component of Schwann cells, while in AMAN the attack appears to be against the axonal and nodes of Ranvier. However, recent studies have shown that anti-ganglioside antibodies can: (i) destroy the nerve terminal (i.e. synaptic necrosis), (ii) kill the perisynaptic Schwann cell (pSC) that envelops the pre-synaptic region, or (iii) destroy both nerve terminal and pSC (Halstead *et al.*, 2005).

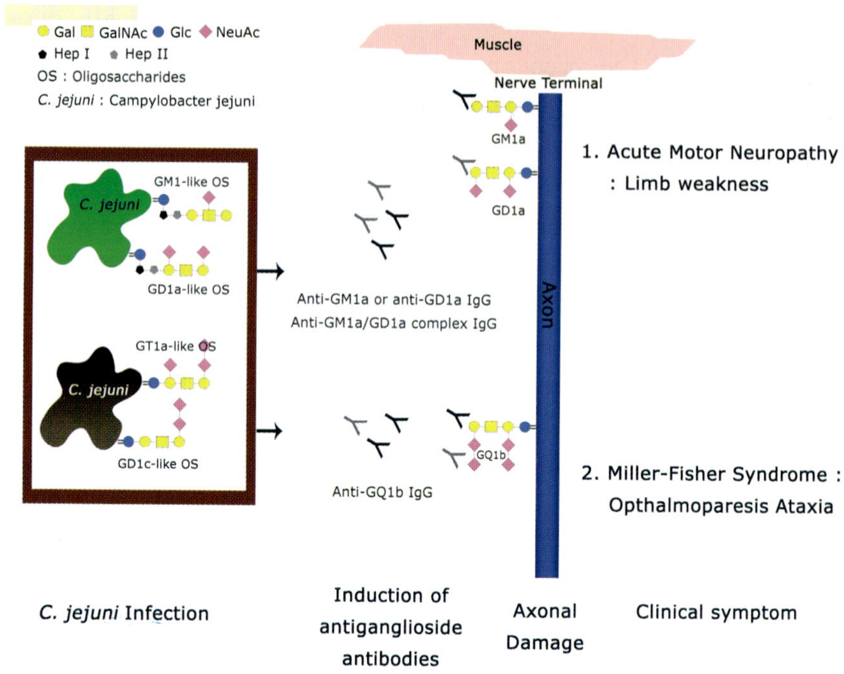

Fig. 1. Molecular mimicry hypotheses for Guillain–Barré syndrome. Infection by *Campylobacter jejuni* (*C. jejuni*) with a ganglioside GM1-like and GD1a-like structure on its cell surface may induce anti-GM1, anti-GD1a, or anti-GM1/GD1a complex IgG in some patients. Subsequently, with these antibodies binding to the GM1 or GD1a which are expressed on motor nerves, the patients suffer from limb weakness. On the other hand, *C. jejuni* that carries GT1a-like or GD1c-like oligosaccharides may raise the anti-GQ1b antibody production in GBS patients. Due to the distribution of GQ1b in oculomotor nerves and primary sensory neurons, Fisher syndrome and related conditions could be found.

Antecedent Pathogens

GBS occurs subsequent to the infections of these pathogens

Campylobacter jejuni

The epidemiological association of antecedent *C. jejuni* infection with GBS has been established for over ten years. However, this relationship has been dubitable owing to the fewer false-positives in cultured-confirmed

GBS compared to serology-confirmed GBS and a lack of serological assessments. In 2004, Kuwabara and his colleagues (Kuwabara *et al.*, 2004) determined antecedent *C. jejuni* infection by the strict criteria of positive *C. jejuni* serology and a history of diarrheal illness within the previous 3 weeks. For the first time, *C. jejuni* infection was identified as having no relationship with the AIDP pattern. More epidemiological features (10–30 year-old peak and the ratio of male to female as 1.7:1) were described in a later study of more than 100 Japanese patients with GBS in whom *C. jejuni* had been isolated (Takahashi *et al.*, 2005). A comprehensive study of antecedent infectious agents in MFS showed that 18% of patients with this syndrome were seropositive for recent *C. jejuni* infection, which was later shown to be related to MFS development and to induce anti-GQ1b auto-antibodies by molecular mimicking LPSs on those bacteria (Koga *et al.*, 2005a).

Cytomegalovirus

CMV infection, as a candidate for the pathogenesis of GBS, has been reported and confirmed in many studies. The frequency of this infection in GBS ranges from 13% to 15% (Jacobs *et al.*, 1998), and primary CMV infection was identified in almost 25% of patients with detectable CMV-specific antibodies in Steininger *et al.*'s study (2007). Compared with *C. jejuni*-related patients with GBS, CMV-related patients seem significantly younger, initially have a severe course as indicated by a high frequency of respiratory insufficiency, and often develop cranial nerve involvement and severe sensory loss (Visser *et al.*, 1996).

Hemophilus influenzae

Hemophilus influenzae (*H. influenzae*), a major pathogen of community-acquired respiratory infection, is considered a causative agent of GBS and FMS, but the frequency of this infection in GBS is controversial (Koga *et al.*, 2005a). Koga and his collogues reported 7% of patients with MFS and 2% of patients with GBS positive serology for *H. influenzae*

(Koga *et al.*, 2001) and showed that the production of anti-GT1a/anti-GQ1b auto-antibodies are mediated by the GT1a-mimicking and Gq1b-mimicking lipooligosaccharides (LOSs) on this bacteria (Koga *et al.*, 2005a). More importantly, Houliston *et al.* (2007) proved that the display by DH1 (a non-typical *H. influenzae* strain) of a surface glycan that mimics the terminal trisaccharide portion of disialosyl-containing gangliosides provided strong evidence for its involvement in the development of Fisher syndrome.

Other pathogens

In addition, *Mycoplasma pneumoniae* infection may be another candidate for the pathogenesis of chronic polyneuropathy in GBS. Recently, Susuki *et al.* (2004) demonstrated that, in certain cases, anti-GM1 antibodies induced by molecular mimicry with *M. pneumoniae* may cause acute motor axonal neuropathy.

The O-chains of a number of *Helicobacter pylori* strains exhibit mimicry of Lewis^x and Lewis^y blood group antigens (Moran *et al.*, 1996). By contrast, EBV and *M. pneumoniae* infection had no positive correlation with MFS in a recent prospective case-control serologic study (Koga *et al.*, 2005a). However, the details of these mimicries remain to be investigated.

Molecular Pathogenesis

Oligosaccharides of LPSs that mimic ganglioside structure

The presence of anti-GM1 antibody has been shown to be significantly associated with preceding *C. jejuni* infection by numerous reports (Yu *et al.*, 2006), particularly with the Penner 19 strain (Table 1). Using gas-liquid chromatography mass spectrometric (MS) analysis, Yuki and his colleagues firstly identified the purified LPSs of *C. jejuni* from a patient with GBS who had anti-GM1 IgG antibody. MS and nuclear magnetic resonance analyses have demonstrated many ganglioside-like structures in

Table 1. Molecular Mimicry between Bacterial OS of LPSs and Gangliosides Expressed by GBS- and MFS-associated Bacteria

Species Penner Serotype[a]	References Reporting Mimicry with Ganglioside Antigen									Other Antigen with Mimicry (ref. no.)
	GM1	GM2/ AIDP	GD1a	GD1c	GD2	GD3	GQ1b/ AMAN	GT1a	GA1	
C. jejuni serotypes										
HS:1	Aspinall *et al.* (1993) Moran *et al.* (2005) Prendergast *et al.* (2000) GM1b, Godschalk *et al.* (2007)	Moran *et al.* (2005) Prendergast *et al.* (2000)	Aspinall *et al.* (1993) Moran *et al.* (2005) Prendergast *et al.* (2000)		Moran *et al.* (2005) Prendergast *et al.* (2000)	Moran *et al.* (2005) Prendergast *et al.* (2000) Ritter *et al.* (1996)			Godschalk *et al.* (2007)	
HS:2	Moran *et al.* (2005) Ritter *et al.* (1996) Godschalk *et al.* (2007)	Moran *et al.* (2005) Prendergast *et al.* (2000) Godschalk *et al.* (2007)	Moran *et al.* (2005) Prendergast *et al.* (2000)	Godschalk *et al.* (2007)	Moran *et al.* (2005) Prendergast *et al.* (2000)	Moran *et al.* (2005) Prendergast *et al.* (2000)	Neisser *et al.* (1997)	Neisser *et al.* (1997)	Godschalk *et al.* (2007)	

(*Continued*)

Table 1. (*Continued*)

Species Penner Serotype[a]	References Reporting Mimicry with Ganglioside Antigen									Other Antigen with Mimicry (ref. no.)
	GM1	**GM2/ AIDP**	**GD1a**	**GD1c**	**GD2**	**GD3**	**GQ1b/ AMAN**	**GT1a**	**GA1**	
HS:2 (CF93-6)							Yuki *et al.* (1994)	Koga *et al.* (2005)		
HS:4	Aspinall *et al.* (1993) Moran *et al.* (2005) Prendergast *et al.* (2000) GM1b, Godschalk *et al.* (2007)	Moran *et al.* (2005) Prendergast *et al.* (2000) Godschalk *et al.* (2007)	Aspinall *et al.* (1993) Moran *et al.* (2005) Prendergast *et al.* (2000) Yuki *et al.* (1994)	Godschalk *et al.* (2007)	Moran *et al.* (2005) Prendergast *et al.* (2000)	Moran *et al.* (2005) Prendergast *et al.* (2000)			Godschalk *et al.* (2007)	GA2
HS:10	Moran *et al.* (2005) Prendergast *et al.* (2000)	Moran *et al.* (2005) Prendergast *et al.* (2000)	Moran *et al.* (2005) Prendergast *et al.* (2000)			Moran *et al.* (2005) Prendergast *et al.* (2000) Salloway *et al.* (1996)				

(*Continued*)

Table 1. (*Continued*)

Species Penner Serotype[a]	References Reporting Mimicry with Ganglioside Antigen									Other Antigen with Mimicry (ref. no.)
	GM1	GM2/ AIDP	GD1a	GD1c	GD2	GD3	GQ1b/ AMAN	GT1a	GA1	
HS:10 (PG836)					Moran *et al.* (2005) Prendergast *et al.* (2000)	Goodyear *et al.* (1999)				
HS:13	Godschalk *et al.* (2007)	Godschalk *et al.* (2007)	Godschalk *et al.* (2007)	Godschalk *et al.* (2007)					Godschalk *et al.* (2007)	
HS:19	Aspinall *et al.* (1993) Moran *et al.* (2005) Prendergast *et al.* (2000) Ang *et al.* (2002) Lee *et al.* (2002)	Moran *et al.* (2005) Prendergast *et al.* (2000)	Aspinall *et al.* (1993) Moran *et al.* (2005) Prendergast *et al.* (2000) Aspinall *et al.* (1994)			Moran *et al.* (2005) Prendergast *et al.* (2000) Usuki *et al.* (2006)				

(*Continued*)

Table 1. (*Continued*)

Species Penner Serotype[a]	References Reporting Mimicry with Ganglioside Antigen									Other Antigen with Mimicry (ref. no.)
	GM1	GM2/ AIDP	GD1a	GD1c	GD2	GD3	GQ1b/ AMAN	GT1a	GA1	
	Usuki *et al.* (2006) Yuki (2001) GM1a, Godschalk *et al.* (2007)		Godschalk *et al.* (2007)							
HS:19 (OH4382)						Aspinall *et al.* (1994)				
HS:19 (OH4384)			Aspinall *et al.* (1994)			Yuki *et al.* (2000)	Yuki (2001)	Aspinall *et al.* (1994) Yuki (2001)		

(*Continued*)

Table 1. (*Continued*)

Species Penner Serotype[a]	References Reporting Mimicry with Ganglioside Antigen									Other Antigen with Mimicry (ref. no.)
	GM1	GM2/ AIDP	GD1a	GD1c	GD2	GD3	GQ1b/ AMAN	GT1a	GA1	
HS:23	Moran *et al.* (2005) Prendergast *et al.* (2000)	Moran *et al.* (2005) Prendergast *et al.* (2000) Ritter *et al.* (1996) Godschalk *et al.* (2007)	Moran *et al.* (2005) Prendergast *et al.* (2000)		Moran *et al.* (2005) Prendergast *et al.* (2000)	Moran *et al.* (2005) Prendergast *et al.* (2000) Godschalk *et al.* (2007)	Neisser *et al.* (1997)	Neisser *et al.* (1997)		
HS:23/ HS:36 (81–176)		Guerry *et al.* (2002)								GM3, GD1b, Guerry *et al.* (2002)
HS:35		Godschalk *et al.* (2007)			Godschalk *et al.* (2007)	Godschalk *et al.* (2007)				

(*Continued*)

Table 1. (*Continued*)

Species Penner Serotype[a]	References Reporting Mimicry with Ganglioside Antigen									Other Antigen with Mimicry (ref. no.)
	GM1	GM2/ AIDP	GD1a	GD1c	GD2	GD3	GQ1b/ AMAN	GT1a	GA1	
HS:36	Moran *et al.* (2005) Prendergast *et al.* (2000)	Moran *et al.* (2005) Prendergast *et al.* (2000) Ritter *et al.* (1996)	Moran *et al.* (2005) Prendergast *et al.* (2000)		Moran *et al.* (2005) Prendergast *et al.* (2000)	Moran *et al.* (2005) Prendergast *et al.* (2000)				
HS:38	GM1a, Godschalk *et al.* (2007)		Godschalk *et al.* (2007)							
HS:41	Moran *et al.* (2005) Prendergast *et al.* (2000) Ritter *et al.* (1996)	Moran *et al.* (2005) Prendergast *et al.* (2000)	Moran *et al.* (2005) Prendergast *et al.* (2000)		Moran *et al.* (2005) Prendergast *et al.* (2000)	Moran *et al.* (2005) Prendergast *et al.* (2000)			Ritter *et al.* (1996)	
HS:43		Godschalk *et al.* (2007)								

(*Continued*)

Table 1. (*Continued*)

Species Penner Serotype[a]	References Reporting Mimicry with Ganglioside Antigen									Other Antigen with Mimicry (ref. no.)
	GM1	GM2/ AIDP	GD1a	GD1c	GD2	GD3	GQ1b/ AMAN	GT1a	GA1	
HS:44	GM1b, Godschalk *et al.* (2007)								Godschalk *et al.* (2007)	
HS:50	GM1a, Godschalk *et al.* (2007)		Godschalk *et al.* (2007)	Godschalk *et al.* (2007)						
HS:64	Godschalk *et al.* (2007)		Godschalk *et al.* (2007)	Godschalk *et al.* (2007)					Godschalk *et al.* (2007)	GA2, Godschalk *et al.* (2007)
HS:65	GM1a, Godschalk *et al.* (2007)		Godschalk *et al.* (2007)							

(*Continued*)

Table 1. (*Continued*)

Species Penner Serotype[a]	References Reporting Mimicry with Ganglioside Antigen									Other Antigen with Mimicry (ref. no.)
	GM1	GM2/ AIDP	GD1a	GD1c	GD2	GD3	GQ1b/ AMAN	GT1a	GA1	
HS:66				Godschalk *et al.* (2007)						
HS:UT	GM1a, Godschalk *et al.* (2007)		Godschalk *et al.* (2007)							
CMV		Ang *et al.* (2000)								
H. influenzae	Mori *et al.* (2000)						Houliston *et al.* (2007)	Koga *et al.* (2001)		
B. melitensis	Watanabe *et al.* (2005)									

[a] The serotypes of *C. jejuni* were determined via Penner's method with heat-stable (HS; O-antigen) serotyping. Modified from Yu *et al.* (2006) with permission from the publisher. Key: LPSs = lipopolysaccharides; GBS = Guillain–Barré syndrome; MFS = Miller Fisher syndrome.

oligosaccharides (OS) of bacteria isolated from patients with GBS and MFS (Table 2).

GM1a, GD1a and GD3 mimics on *C. jejuni* strains were associated with AMAN; GM2 was a possible target for IgM antibody in AIDP subsequent to CMV infection; and anti-GQ1b antibody was found in patients with MFS with antecedent infection (Koga *et al.*, 2005a). Godschalk *et al.* (2007) used high resolution magic angle spinning NMR spectroscopy to realize the prediction of OS structures for the outer core structures of 26 GBS- and MFS-associated *C. jejuni* strains (Table 2).

Their data showed that molecular mimicry between gangliosides and *C. jejuni* LPSs is the presumable pathogenic mechanism in most cases of *C. jejuni*-related GBS. However, in some cases, other mechanisms may play a role.

Since the GM1 epitope has been found on the LPSs of *H. influenzae* isolated from a patient with AMAN (Mori *et al.*, 2001), the GQ1b epitope is another attracting mimic LPS structure on *H. influenzae* in patients with MFS (Table 1). Moreover, LPSs of *Bruchella melitensis* (*B. melitensis*) were reported to have a GM1 ganglioside-like structure by Watanabe *et al.* (2005).

Oligosaccharides of LPSs or other molecules that mimic non-ganglioside structures

Potential pathogenic roles exist for other known and unknown glycolipids enriched in the Schwann cell, myelin, and axonal membranes, such as sulfated glucuronyl paragloboside (SGPG) (Yuki *et al.*, 1996), galactocerebroside, LM1, and sulfatides (Yu *et al.*, 2006). These nonganglioside-like structures also have microbial glycan mimics and could be important in the pathogenesis of neuropathy in some patients with GBS.

Galactocerebroside (GalCer) epitope, a major glycolipid in nervous tissue, is present in *M. pneumoniae*, the infection that precedes 5% or 6% of GBS cases (Jacobs *et al.*, 1998). At present, the carbohydrate structures in *M. pneumoniae* are not clearly understood, but there is a possible involvement of a GalCer-like epitope (Susuki *et al.*, 2004).

Table 2. Proposed OS of LPSs Structures Expressed by GBS- and MFS-associated Bacteria

Ganglioside mimic	LOSs class[a]	Structure
GM1a	A	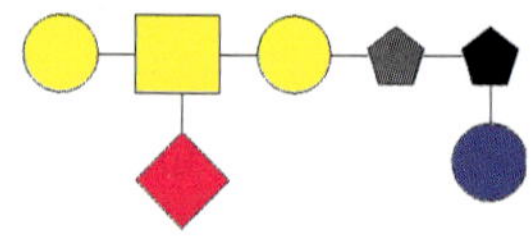
	C	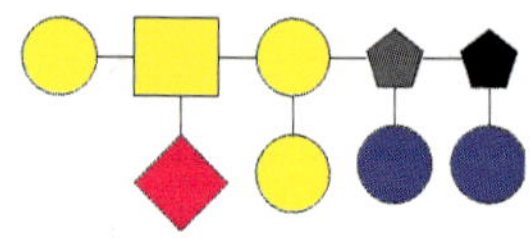
GM1b	A and B	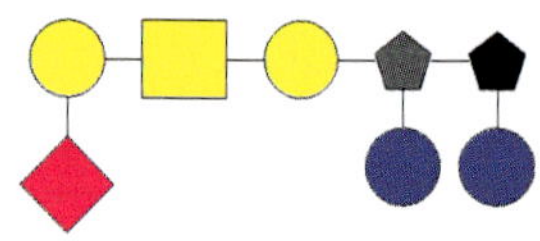
GM2	A	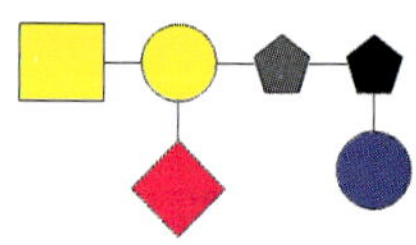
	B	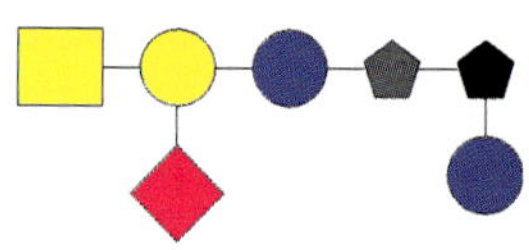
GD1a	A	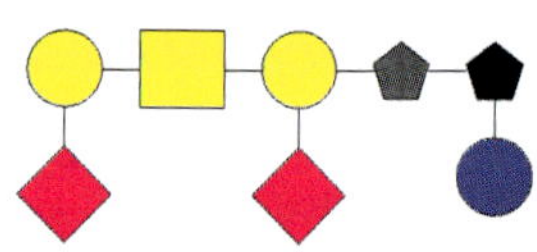
GD1c	A and B	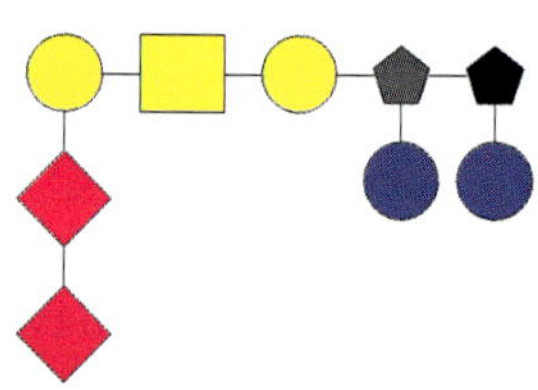
GD2	B	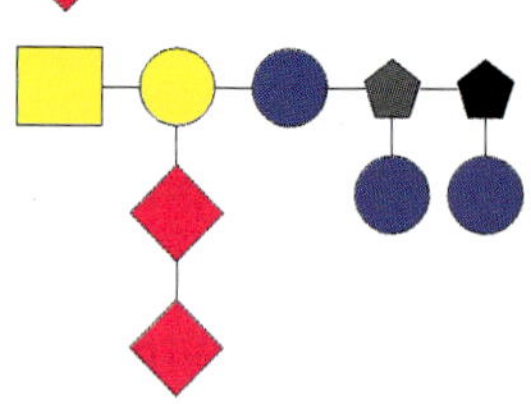

(*Continued*)

Table 2. (*Continued*)

Ganglioside mimic	LOSs class[a]	Structure
GD3	B	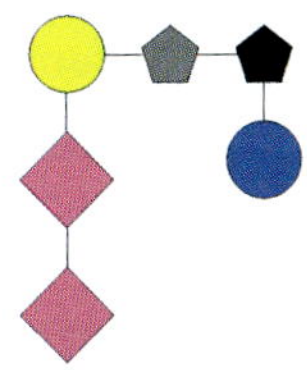
GA1	B	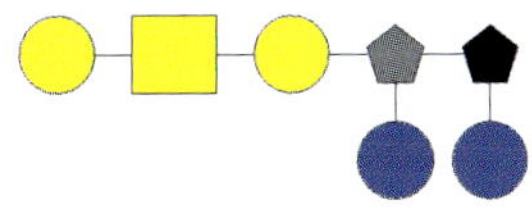
GA2	B	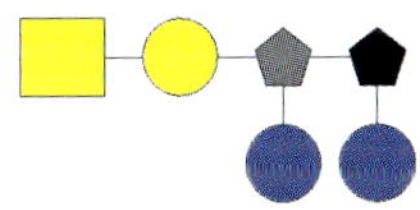
GQ1b		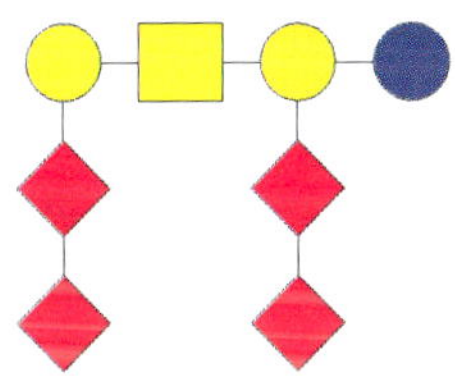
GT1a		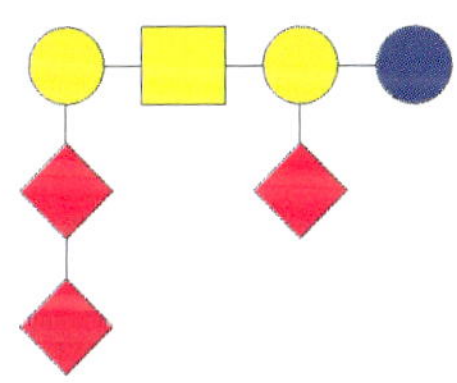

[a] The three LOS classes are based on DNA sequences of the following strains (GenBank accession number): class A — OH4384 (AF130984), OH4382 (AF167345), HS:4 (AF215659), HS:10 (AF400048), HS:19 (AF167344), HS:41 (AY044868); class B — HS:23 (AF401529), HS:36 (AF401528); class C — NCTC 11168 (AL139077), HS:1 (AY044156), HS:2 (AF400047). [Godschalk *et al.*, 2004] [Usuki *et al.*, 2005].
Key: OS = oligosaccharides; LPSs = lipopolysaccharides; LOSs = lipooligosaccharides; GBS = Guillain–Barré syndrome; MFS = Miller Fisher syndrome.

Another candidate for the pathogenesis of GBS might be CMV, which has been proved to be correlated with anti-MAG/SGPG-positive chronic polyneuropathy (Yuki *et al.*, 1998). However, until now there has been no direct evidence to identify the carbohydrate structure of the SGPG epitope in GBS-relative or MFS-relative CMV.

Chronic Inflammatory Demyelinating Polyneuropathy

Although GBS is clearly related to antecedent events, usually bacterial or viral, no such clear association is evident in CIDP. Antibodies, particularly anti-ganglioside and cross reacting with epitopes on the precipitating infective organisms appear to play important roles in GBS pathophysiology.

CIDP is characterized by progressive weakness and impaired sensory function in the legs and arms that continue to progress for longer than 4 weeks. This disorder, which is sometimes called chronic relapsing polyneuropathy, is caused by damage to the myelin sheath of the peripheral nerves (Toyka, Gold, 2003). Many monoclonal antibodies have been found in patients with CIDP with high levels, including myelin-associated glycoprotein (MAG) and SGPG (Yuki *et al.*, 1996).

Pathogenic auto-antibodies in CIDP

Unlike GBS, no glycolipid antigens have been identified as the targets of auto-antibodies in CIDP pathogenesis (Willison, Yuki, 2002). So far, many conflicting auto-antibodies have been found in patients with CIDP; the reasons might come from the different detective methods being used and the limited groups under study.

Among these antibodies have been sought to other myelin proteins, including P2 and PMP22, in CIDP (Table 3), P0 is the only protein found consistently. The potential importance of P0 as an auto-antigen is not surprising because it constitutes 50% of peripheral nervous system (PNS) myelin protein, and is illustrated by its ability to induce EAN (Zhu *et al.*, 2001).

Table 3. Antibodies to Myelin and Nerve Proteins in CIDP

Antigens related with auto-antibodies	References reporting auto-antibodies	Ig class
P0	Quarles *et al.* (1990)	IgM, IgG
	Khalili-Shirazi *et al.* (1993)	
	Meléndez-Vàsquez *et al.* (1997)	
	Yan *et al.* (2001)	
	Kwa *et al.* (2001)	
	Allen *et al.* (2005)	
P2	Hughes *et al.* (1984)	IgM, IgG
	Khalili-Shirazi *et al.* (1993)	
PMP22	Gabriel *et al.* (2000)	
	Ritz *et al.* (2000)	
	Kwa *et al.* (2001)	
$\alpha + \beta$-tublin	Manfredini *et al.* (1995)	IgM, IgG
β-tublin	Connolly *et al.* (1993)	IgM, IgG
	van Schaik *et al.* (1995)	
	Connolly *et al.* (1997)	
GD3	Usuki *et al.* (2005)	IgG
Connexin32	Kwa *et al.* (2001)	
SGPG	Inuzuka *et al.* (1998)	IgM
LM1	Fredman *et al.* (1991)	

Key: CIDP = chronic inflammatory demyelinating polyneuropathy.

In 1996, elevated IgM anti-SGPG antibody titer was found and proposed as a diagnostic marker for a subgroup of patients with CIDP with or without para-protein (Yuki *et al.*, 1996). Recently, Usuki *et al.* (2005) reported the IgG antibody titer to GD3 was remarkably elevated (titer, 1:10 000), indicating maximal activity to another tetrasaccharide epitope (–NeuAcα2–8NeuAcα2–3Galβ1–4Glc–). However, the exact mimicry mechanism should be looking for, such as mimicry structure identification, in CIDP pathogenesis.

Conclusion

In conclusion, CIDP and GBS with its variant, MFS, are autoimmune neuropathies related to the auto-antibodies. Increased auto-antibody titers in GBS are thought to be a result of the production of antibodies to a bacterial carbohydrate-containing surface antigen that cross-reacts with the myelin sheath, axons of nerve cells or non-ganglioside components, but not clear in CIDP, resulting in demyelination and axonal degeneration.

Analysis of the mimic structures of these crucial proteins (antigens) should be helpful in understanding the induction of the anti-ganglioside and anti-nonganglioside immune response that leads to GBS and related autoimmune neuropathies, CIDP.

Acknowledgments

This study was supported by grants from the Swedish Medical Research Council (K2007-62X-13133-09-3), the Swedish National Board of Health and Welfare, and the Swedish Medical Association. We acknowledge the invaluable assistance provided by Prof. Zhu at the Department of Neurobiology, Care Sciences and Society, in advising and editing the manuscript.

References

Allen D, Giannopoulos K, Gray I, Gregson N, Makowska A, Pritchard J, Hughes RA. (2005) Antibodies to peripheral nerve myelin proteins in chronic inflammatory demyelinating polyradiculoneuropathy. *J Peripher Nerv Syst* **10**: 174–180.

Ang CW, Jacobs BC, Brandenburg AH, Laman JD, van der Meché FG, Osterhaus AD, van Doorn PA. (2000) Cross-reactive antibodies against GM2 and CMV-infected fibroblasts in Guillain–Barré syndrome. *Neurology* **54**: 1453–1458.

Ang CW, Laman JD, Willison HJ, Wagner ER, Endtz HP, de Klerk MA, Tio-Gillen AP, van den Braak N, Jacobs BC, van Doorn PA. (2002) Structure of *Campylobacter jejuni* lipo-polysaccharides determines antiganglioside specificity and clinical features of Guillain–Barré and Miller Fisher patients. *Infect Immun* **70**: 1202–1208.

Ang CW, Jacobs BC, Laman JD. (2004) The Guillain–Barré syndrome: A true case of molecular mimicry. *Trends Immunol* **25**: 61–66.

Aspinall GO, McDonald AG, Raju TS, Pang H, Moran AP, Penner JLL. (1993) Chemical structure of the core region of *Campylobacter jejuni* serotype O:2 lipopolysaccharides. *Eur J Biochem* **213**: 1029–1037.

Aspinall GO, McDonald AG, Pang H, Kurjanczk LA, Penner JL. (1994) Lipopolysaccharides of *Campylobacter jejuni* serotype O:19: Structures of core oligosaccharide regions from the serostrain and two bacterial isolates from patients with the Guillain-Barré syndrome. *Biochemistry* **33**: 241–249.

Connolly A, Pestronk A, Trotter J, Feldman E, Cornblath D, Olney R (1993) High titre selective serum anti-B-tubulin antibodies in chronic inflammatory demyelinating polyneuropathy. *Neurology* **43**: 557–562.

Connolly AM, Pestronk A. (1997) Anti-tubulin autoantibodies in acquired demyelinating polyneuropathies. *J Infect Dis* **176**(Suppl 2): S157–159.

Fredman P, Vedeler CA, Nyland H, Aarli JA, Svennerholm L. (1991) Antibodies in sera from patients with inflammatory demyelinating polyradiculoneuropathy react with ganglioside LM1 and sulphatide of peripheral nerve myelin. *J Neurol* **238**: 75–79.

Gabriel C, Gregson N, Hughes R (2000) Anti-PMP22 antibodies in patients with inflammatory neuropathy. *J Neuroimmunol* **104**: 139–146.

Godschalk PC, Heikema AP, Gilbert M, Komagamine T, Ang CW, Glerum J, Brochu D, Li J, Yuki N, Jacobs BC, van Belkum A, Endtz HP. (2004) The crucial role of *Campylobacter jejuni* genes in anti-ganglioside antibody induction in Guillain–Barré syndrome. *J Clin Invest* **114**: 1659–1665.

Godschalk PC, Kuijf ML, Li J, St Michael F, Ang CW, Jacobs BC, Karwaski MF, Brochu D, Moterassed A, Endtz HP, van Belkum A, Gilbert M. (2007) Structural characterization of *Campylobacter jejuni* lipooligosaccharide outer cores associated with Guillain–Barré and Miller Fisher syndromes. *Infect Immun* **75**: 1245–1254.

Goodyear CS, O'Hanlon GM, Plomp JJ, Wagner ER, Morrison I, Veitch J, Cochrane L, Bullens RW, Molenaar PC, Conner J, Willison HJ. (1999) Monoclonal antibodies raised against Guillain–Barré syndrome-associated *Campylobacter jejuni* lipopolysaccharides react with neuronal gangliosides and paralyze muscle-nerve preparations. *J Clin Invest* **104**: 697–708.

Guerry P, Szymanski CM, Prendergast MM, Hickey TE, Ewing CP, Pattarini DL, Moran AP. (2002) Phase variation of *Campylobacter jejuni* 81–176 lipopolysaccharide affects ganglioside mimicry and invasiveness *in vitro*. *Infect Immun* **70**: 787–793.

Halstead SK, Morrison I, O'Hanlon GM, Humphreys PD, Goodfellow JA, Plomp JJ, Willison HJ. (2005) Anti-disialosyl antibodies mediate selective neuronal or Schwann cell injury at mouse neuromuscular junctions. *Glia* **52**: 177–189.

Houliston RS, Koga M, Li J, Jarrell HC, Richards JC, Vitiazeva V, Schweda EK, Yuki N, Gilbert M. (2007) A *Haemophilus influenzae* strain associated with Fisher syndrome expresses a novel disialylated ganglioside mimic. *Biochemistry* **46**: 8164–8171.

Hughes R, Gray I, Gregson N. (1984) Immune response to myelin antigens in Guillain–Barré syndrome. *J Neuroimmunol* **6**: 303–312.

Jacobs BC, Rothbarth PH, van der Meché FG, Herbrink P, Schmitz PI, de Klerk MA, van Doorn PA. (1998) The spectrum of antecedent infections in Guillain–Barré syndrome: A case-control study. *Neurology* **51**: 1110–1115.

Khalili-Shirazi A, Atkinson P, Gregson N, Hughes R. (1993) Antibody responses to P0 and P2 myelin proteins in Guillain–Barré syndrome and chronic idiopathic demyelinating polyradiculoneuropathy. *J Neuroimmunol* **46**: 245–251.

Koga M, Yuki N, Tai T, Hirata K. (2001) Miller Fisher syndrome and *Haemophilus influenzae* infection. *Neurology* **57**: 686–691.

Koga M, Gilbert M, Li J, Koike S, Takahashi M, Furukawa K, Hirata K, Yuki K. (2005a). Antecedent infections in Fisher syndrome: A common pathogenesis of molecular mimicry. *Neurology* **64**: 1605–1611.

Koga M, Koike S, Hirata K, Yuki N. (2005b) Ambiguous value of *Haemophilus influenzae* isolation in Guillain–Barré and Fisher syndromes. *J Neurol Neurosurg Psych* **76**: 1736–1738.

Kuwabara S, Bostock H, Ogawara K, Sung JY, Kanai K, Mori M, Hattori T, Burke D. (2003) The refractory period of transmission is impaired in axonal Guillain–Barré syndrome. *Muscle Nerve* **28**: 683–689.

Kuwabara S, Ogawara K, Misawa S, Koga M, Mori M, Hiraga A, Kanesaka T, Hattori T, Yuki N. (2004) Does *Campylobacter jejuni* infection elicit "demyelinating" Guillain–Barré syndrome? *Neurology* **63**: 529–533.

Kwa M, van Schaik I, Brand A, Baas F, Vermeulen M. (2001) Investigation of serum response to PMP22, connexin 32 and P0 in inflammatory neuropathies. *J Neuroimmunol* **116**: 220–225.

Lee G, Jeong Y, Wirguin I, Hays AP, Willison HJ, Latov N. (2004) Induction of human IgM and IgG anti-GM1 antibodies in transgenic mice in response to lipopolysaccharides from *Campylobacter jejuni*. *J Neuroimmunol* **146**: 63–75.

Manfredini E, Nobile-Orazio E, Allaria S, Scarlato G. (1995) Anti-alpha and beta-tubulin IgM antibodies in dysimmune neuropthies. *J Neurol Sci* **133**: 79–84.

Meléndez-Vàsquez C, Redford J, Choudhary P, Gray I, Maitland P, Gregson N, Smith K, Hughes R. (1997) Immunological investigation of chronic inflammatory demyelinating polyradiculoneuropathy. *J Neuroimmunol* **73**: 124–134.

Moran AP, Prendergast MM, Appelmelk BJ. (1996) Molecular mimicry of host structures by bacterial lipopolysaccharides and its contribution to disease. *FEMS Immunol Med Microbiol* **16**: 105–115.

Moran AP, Annuk H, Prendergast MM. (2005) Antibodies induced by ganglioside-mimicking *Campylobacter jejuni* lipopolysaccharides recognize epitopes at the nodes of Ranvier. *J Neuroimmunol* **165**: 179–185.

Mori M, Kuwabara S, Miyake M, Noda M, Kuroki H, Kanno H, Ogawara K, Hattori T. (2000) *Haemophilus influenzae* infection and Guillain–Barré syndrome. *Brain* **123**: 2171–2178.

Neisser A, Bernheimer H, Berger T, Moran AP, Schwerer B. (1997) Serum antibodies against gangliosides and *Campylobacter jejuni* lipopolysaccharides in Miller Fisher syndrome. *Infect Immun* **65**: 4038–4042.

Prendergast MM, Moran AP. (2000) Lipopolysaccharides in the development of the Guillain–Barré syndrome and Miller Fisher syndrome forms of acute inflammatory peripheral neuropathies. *J Endotoxin Res* **6**: 341–359.

Quarles R, Ilyas A, Willison H. (1990) Antibodies to ganglioside and myelin proteins in Guillain–Barré syndrome. *Ann Neurol* **27**: S48–52.

Ritter G, Fortunato SR, Cohen L, Noguchi Y, Bernard EM, Stockert E, Old LJ. (1996) Induction of antibodies reactive with GM2 ganglioside after immunization with lipopolysaccharides from *Campylobacter jejuni*. *Int J Cancer* **66**: 184–190.

Ritz M, Lechner-Scott J, Scott R, Fuhr P, Malik N, Erne B, Taylor V, Suter U, Schaeren-Wiemers N, Steck A. (2000) Characterization of autoantibodies to peripheral myelin protein 22 in patients with hereditary and acquired neuropathies. *J Neuroimmunol* **104**: 155–163.

Salloway S, Mermel LA, Seamans M, Aspinall GO, Nam Shin JE, Kurjanczyk LA, Penner JL. (1996) Miller Fisher syndrome associated with *Campylobacter jejuni* bearing lipopolysaccharide molecules that mimic human ganglioside GD3. *Infect Immun* **64**: 2945–2949.

Steininger C, Seiser A, Gueler N, Puchhammer-Stöckl E, Aberle SW, Stanek G, Popow-Kraupp T. (2007) Primary cytomegalovirus infection in patients with Guillain–Barré syndrome. *J Neuroimmunol* **183**: 214–219.

Susuki K, Odaka M, Mori M, Hirata K, Yuki N. (2004) Acute motor axonal neuropathy after Mycoplasma infection: Evidence of molecular mimicry. *Neurology* **62**: 949–956.

Takahashi M, Koga M, Yokoyama K, Yuki N. (2005) Epidemiology of *Campylobacter jejuni* isolated from patients with Guillain–Barré and Fisher syndromes in Japan. *J Clin Microbiol* **43**: 335–339.

Toyka KV, Gold R. (2003) The pathogenesis of CIDP: Rationale for treatment with immunomodulatory agents. *Neurology* **60**(8 Suppl 3): S2–7.

Usuki S, Sanchez J, Ariga T, Utsunomiya I, Taguchi K, Rivner MH, Yu RK. (2005) AIDP and CIDP having specific antibodies to the carbohydrate epitope (–NeuAcα2–8NeuAα2–3Galβ1–4Glc–) of gangliosides. *J Neurol Sci* **232**: 37–44.

Usuki S, Thompson SA, Rivner MH, Taguchi K, Shibata K, Ariga T, Yu RK. (2006) Molecular mimicry: Sensitization of Lewis rats with *Campylobacter jejuni* lipopolysaccharides induces formation of antibody toward GD3 ganglioside. *J Neurosci Res* **83**: 274–284.

Van Schaik I, Vermeulen M, van Doorn P, Brand A. (1995) Anti-beta tubulin antibodies have no diagnostic value in patients with chronic inflammatory demyelinating polyneuropathy. *J Neurol* **242**: 599–603.

Visser LH, van der Meché FG, Meulstee J, Rothbarth PP, Jacobs BC, Schmitz PI, van Doorn PA. (1996) Cytomegalovirus infection and Guillain–Barré syndrome: The clinical, electrophysiologic, and prognostic features. Dutch Guillain–Barré Study Group. *Neurology* **47**: 668–673.

Watanabe K, Kim S, Nishiguchi M, Suzuki H, Watarai M. (2005) *Brucella melitensis* infecton associated with Guillain–Barré syndrome through molecular mimicry of host structures. *FEMS Immunol Med Microbiol* **45**: 121–127.

Willison HJ, Yuki N. (2002) Peripheral neuropathies and anti-glycolipid antibodies. *Brain* **125**: 2591–2625.

Willison HJ. (2005) The immunobiology of Guillain–Barré syndromes. *J Peripher Nerv Syst* **10**: 94–112.

Yan W, Archelos J, Hartung H, Pollard J. (2001) P0 protein is a target antigen in chronic inflammatory demyelinating polyradiculoneuropathy. *Ann Neurol* **50**: 286–292.

Yu RK, Usuki S, Ariga T. (2006) Ganglioside molecular mimicry and its pathological roles in Guillain-Barré syndrome and related diseases. *Infect Immun* **74**: 6517–6527.

Yuki N, Taki T, Takahashi M, Saito K, Tai T, Miyatake T, Handa S. (1994) Penner's serotype 4 of *Campylobacter jejuni* has a lipopolysaccharide

that bears a GM1 ganglioside epitope as well as one that bears a GD1 a epitope. *Infect Immun* **62**: 2101–2103.
Yuki N, Tagawa Y, Handa S. (1996) Autoantibodies to peripheral nerve glycosphingolipids SPG, SLPG, and SGPG in Guillain–Barré syndrome and chronic inflammatory demyelinating polyneuropathy. *J Neuroimmunol* **70**: 1–6.
Yuki N, Yamamoto T, Hirata K. (1988) Correlation between cytomegalovirus infection and IgM anti-MAG/SGPG antibody-associated neuropathy. *Ann Neurol* **44**: 408–410.
Yuki N, Koga M, Hirata K. (2000) Is *Campylobacter* lipo-polysaccharide bearing a GD3 epitope essential for the pathogenesis of Guillain–Barré syndrome? *Acta Neurol Scand* **102**: 132–134.
Zhu J, Pelidou S, Deretzi G, Levi M, Mix E, van der Meide P, Winblad B, Zou L. (2001) P0 glycoprotein peptides 56–71 and 180–199 dose-dependently induce acute and chronic experimental autoimmune neuritis in Lewis rats associated with epitope spreading. *J Neuroimmunol* **114**: 99–106.

Chapter 17

Fresh Water Pearls of Wisdom on Protein Crystallization

Jan Sedzik
Editor

When crystallization experiments are undertaken, there is usually not much to do except miserable waiting. In the following, we have compiled some very insightful comments on crystallization.

Remember that crystal is sometimes the best answer.
Depressed crystal grower

If I don't get crystals, how can anyone expect me to get a Ph.D?
Ph.D. student

Less is more, although one protein crystal is often not enough,
two are often too many.
Infidel

Be kind to other crystal growers, even if they are unkind people
(they probably need it the most).
Jose and Isabella

Good judgment on crystallization comes from experience, and
a lot that comes from bad judgment.
Unknown crystal grower

Protein crystals are much too serious to be taken seriously.
GlobalSpec Inc.

The life of protein crystals is short, and offers nothing but sadness.
Fumiko Hayashi, poet

Remember that great love and great achievements involve great crystals.
Uncle don Pedro

The protein crystal starts from the nucleus.
1st *Principle of Crystal Science*

Nothing (not even crystallization) happens in a vacuum.
2nd *Principle of Crystal Science*

Optimists (crystal growers) seem less likely to die of heart disease or stroke than pessimists.
http://news.bbc.co.uk/2/hi/health/4759524.stm

A crystal grower is a fellow dedicated to making and preserving life of protein crystals when there is no crystal or even hope of doing so.
World Crystal Growth Association, Code of Ethics

When someone asks you a question on crystallization, and you don't want to answer, smile and ask, "Why do you want to know?"
Mother Nature

Never laugh at anyone's dream crystal. People who don't have dreams usually don't have protein crystals.
Code of Crystal Ethics

Never think about the future: if you worry about what may not happen with protein crystallization tomorrow and it doesn't happen, you have worried in vain. But even if it does happen, you have to worry twice.

Very experienced crystal grower

Protein crystals live *in clausura* and die in cold, silence and isolation.

Unknown monk

Ubi crystal, ibi trinitas.

Where there is crystal, there is trinity.

That is, a lover protein, a beloved crystal, and a spirit of love precipitate.

Vatican radio station

One of the key rules every good crystal grower knows is to choose the volume of crystallization drops before pH. Once the volume is there, the nucleation will rapidly follow.

Trisha Lorain, CEO

Crystallizing of proteins by the traditional way required much manual work and cheap labor. But today, we have expensive and automatic high-throughput machines. Isn't that great, how many post-docs lost their jobs?

Market oriented fanatic

Crystallization isn't about getting precipitates, it's about causing it.

Retired professor

Whatever happens in politics, crystallization will endure.

Crystal fanatic

You can have all the proteins in the world, but if you don't have the right technology (*Crystallomic*™ *of course*) to turn proteins into crystals, then all that stuff you own is worthless.
Financial advisor, Monaco

According to happiness research, successful protein crystallization has a much bigger effect on average happiness than a typical crystal grower's average income itself (£50 000 in UK, or 15 000 Skr in Sweden).
CNNp radio

In survey after survey involving huge groups of crystal growers, only significant correlations between happiness and concentration of precipitant are reported.
BBS radio

Protein crystallization is a peculiar phenomenon. It really isn't worth anything unless you do something with it.
Willi Blumm, anti-emperialist

The crystallizing of membrane proteins is for those who are not afraid of the future.
Founder of the Crystal Movement

Crystallization of membrane proteins is really simple, but we insist on making it very complicated.
Confucius, China, 479 B.C.

Crystallization of proteins is a procedure which guarantees that we will always get precipitates instead of the crystals which we deserved.
CEO, Global Crystallization Limited

Opulence (*luxuriousness*) is an ingredient of protein crystal.
Markus Langes-Swarovski

If life exists after death (*crystal of course*), what are the procedures to reanimate such crystals?

Q&A

People think there must be some magic beyond the classical crystallization — precipitates, precipitants, phase separation, pre-filled plates, robots — but there is no magical panacea, only persistence.

Howard Zinn

A bigger protein crystal can give you a massive confidence boost and a better chance of getting a grant.

Nathanael Hickmann

Russian crystal growers may have specific genes that allow them to drink more vodka and to get better crystals.

Andrei Belyakov

All that talk about 'crystal-clear' vodka being a less dangerous drink for (a crystal grower's) health is just a crystal myth.

http://english.pravda.ru/society/stories/94882-0

Life is short so make the most of it by introducing the new protein crystallization enhancement products that have been tested and sold globally.

Glen Gates

Getting large, voluminous crystals is an experience like no other and is very satisfying for crystal growers.

Fernne Magdalena, Crystallife Corp.

Trust in crystallization means permanent control of the crystallization conditions.

Joseph Stallin

I discovered that crystallization is not an emotion but an action. It is not a feeling; it is something you do.
Crystal Small-Heather

Crystalline dysfunction (CD) sometimes is referred to as *amorphous precipitation*, which is very embarrassing and a source of stress leading to heart attack. Among recent expensive advances suppressing CD is high throughput crystallization.
Jennifer Yeager

The Internet model of a crystal growth accepts only high throughput and expensive crystallization trials but without rationality.
Amnesty Crystallization International

Crystallomic is all about giving you information that will help you enjoy and live your crystal life to the fullest.
Crystal astrologist

Good crystallization practice (GCP) does not necessarily ensure one gets the right crystal, but it reduces the chances of getting the wrong ones.
Surveillance expert

If you have a problem getting or maintaining a crystal, you are not alone. In fact, more than half of all crystal growers over forty have such difficulties. This condition is called *crystalline dysfunction* (CD), but there's a safe, effective and easy method of treatment: high throughput crystallization.
CSO, Viagra Ltd.

Be engaged at least nine months before you get crystal.
Crystal grandma

Marry a man/woman you love to crystallize with. As you get older, his/her crystallization skills will be as important as any other.
Anonymous single crystal grower

Crystallization itself is a misery and nobody can tell what can be of it.
Mary Joo, financial crook

A protein crystal is not an object or a trophy; a crystal is living matter, with feelings, but without a heart that beats, without a soul that is hungry for love; crystal life is short, with a mind that aspires to have its place in science.
Aloha, Crystal Playboy

Just because venture capitalists don't invest in protein crystallization doesn't mean it's not worth investing in.
Venture capitalist, Bill Mills

If no one invests in early-stage crystallization, there will be no late-stage crystals later on.
Venture capitalist, Christopher Mira-de-belli

Scientists find extraterrestrial genes in the DNA of crystal growers. Will they crystallize in heaven?
www.agoracosmopolitan.com/home/Frontpage/2007/01/08/01288.html

Scientists discover mutated gene responsible for baldness (of crystal growers).
http://english.pravda.ru/science/tech/29-12-2006/86207-baldness-0

Nowadays high throughput crystallization is becoming a more acceptable form of sexual activity for crystal growers.
Sexuologist, Granada, Spain

Bananas are high in B vitamins that help calm the nervous system of the crystal grower after failure in crystallization.
Internet dietetician

(One in Seven) Crystal growers of New York live in households facing hunger or food insecurity!
http://www.nyccah.org/hungerinnyc.html

Every person who has seen crystals will show affection for them, but it may be as tiny as an atom.
Benjamin Crook, comedian

It is not a disgrace to get no crystals, but it is a disgrace to have no proteins for crystallization; not failure, but a lack of aim is the sin.
Father Crystal (about life)

Blessed are the crystal growers, for their stuff of low mosaicity.
Crystallomix Science Monitor

Crystallization success can come as a consequence of some truly absurd situations. For instance, it would be enough for a crystal grower to provide proof that his snoring volume is equivalent to that of a racecar engine during experiments that resulted in crystals.
IPR administrator

Getting crystals can be as enjoyable as getting thinner.
Seller of Anatrim

Being a prosperous crystal grower is not necessarily a bad thing.
Alfonzo Ben Crystal

In ten years crystallization could be a billion-dollar industry.
Venture capitalist (optimist)

In ten months crystallization could be a zero-dollar industry.
Crystal accountant (pessimist)

To be successful, crystal growers must work hard and live celibate lives.
Crystal Paris Hilton

A genetically pure crystal grower can live up to 120–150 years.
http://english.pravda.ru/russia/kremlin/16-08-2006/83981-longevity-0

Crystal growers get fat in summer due to air conditioning, eating unhealthy food (*containing purins and acrylamide*), lack of physical exercise when operating high throughput crystallization robots, and psychological stress from writing grant applications; all these factors lead, indeed, to obesity.
http://english.pravda.ru/science/health/28-07-2006/83619-obesity-0

Why are crystals better than socks?
Venture capitalists, Q&A

Crystal growers in the USA, Japan and Sweden are poor as compared to their counterparts in other rich countries. Many elderly crystal growers (*over 60*) continue to work but they receive lower wages and fewer grants, as opposed to their younger, inexperienced colleagues in Russia or China.
http://english.pravda.ru/world/americas/20-07-2006/83428-USA_Japan-0

Those who have a crystal need not precipitate.
Angel from heaven

Venture capitalists seem to follow a simple path *taken from crystallization.* They always start trading below the current bid price (*below precipitation line*) and sell a few ticks above the current offer price (*above precipitation line*).
Conclusion: To be rich you must avoid "precipitation slot"!
Annals of Financial Secrets, 2006, vol. 13, p.13

Your crystallization ability or lack thereof is embedded in your genes!
http://english.pravda.ru/news/science/31-05-2006/81290-genes-0

Why do rising commodity prices of crystals intensify international tensions?
Dr. Marc Faber, Q&A

To get better crystals, Pamela Anderson is urging protein crystal growers to go vegetarian and to get married.
http://funreports.com/fun/25-05-2006/1375-Pamela_Anderson-0

Want twin-crystals? Eat Indian dairy products.
Daily News & Analysis, India, May 19, 2006

After the loss of a crystal it can take several years to regain your previous level of well-being. So if you are a grumpy crystal grower you are always going to be a grumpy crystal grower.
Professor Ronald MacDonald

Research suggests that richer crystal growers do tend to be happier than poor ones, but once they have lab space, tenure, an HTC robotic system, protein to crystallize, home, food and clothes, then extra money does not seem to make them much happier or productive.
Research in Nucleation, Port de Crystal, Patagonia

We who flourish in protein crystallization have a moral responsibility to help others who have the desire to succeed but just need an opportunity and venture capital.
Preachers of globalization, Davos, Switzerland

To be a successful protein crystal grower you must stay informed. *Crystallomic* covers many issues you won't find elsewhere.
Crystallomic™ team

It's more like the Japanese aren't interested in news on global crystallization in general.
BBSj, Kansai

Despite its importance as the ultimate gatekeeper of scientific publication and funding (*on crystallization*), peer review is known to engender bias, incompetence, excessive expense, ineffectiveness, and corruption. A surfeit of publications has documented the deficiencies of this system.
David Kaplan, 1995 A.C.

In an age of universal lying and cheating, getting protein crystals is indeed a revolutionary act.
Crystal Bolszewik

What is the similarity between a *protein crystallographer* and a *protein crystal grower*? They are both said to be intelligent, but no-one can prove this.
Internet hacker

The *redemptorist crystal grower* is a member of the cult *Crystal Research* founded in Sweden. This is a feudal, quasi-clerical and pseudo-religious congregation engaged in

the preaching and writing about the basis and supremacy of high throughput crystallization among infidels and venture capitalists.
Crystal Intelligence Agency

Sakurajima radish and crystallization of proteins are dreams of crystal growers.
www.jsps-sto.com/websites/jsps-stocom/filbank/ newsletter%206_06.pdf

One of the most important features of crystallization is its "*inherited ability of protection*" from being copied by others. This is easily done by: (1) distorting laboratory notes, (2) destroying the crystal, or (3) applying for a patent.
IPR attorney

Once you get a microcrystal, make only one change at a time and test the crystallization after each change. It sounds so obvious that it should not need to be said.
Trial-and-error fanatic

Now is the time to stop crystallizing, and begin to understand.
Orthodox Crystal Grower

The meager meadows won't do the protein crystals.
Knut Capusta-Sauer, venture capitalist

There is nothing glorious about what our ancestors call *protein crystal*; it is simply a succession of mistakes, ignorance, lack of funding and intolerance.
Historian

"CRYSTAL" is simply a last-gasp attempt by conservatives and orthodox physical-chemists to keep humanity in ignorance

and obscurantism, through the fear of science and new
low-throughput technological advances.
Politician, Crystal Party

Revere protein crystals, love mother liquor.
Crystal growers, Meiji era

Crystallomic™ is a method to lead you to eternal happiness,
but it is only you who do it, and only you who want it. You
make the effort.
UPPs Precipitate Cartel

It's true: (Crystal?) Churchgoers are wealthier.
The Christian Science Monitor
www.csmonitor.com/2005/1114/p15s02-cogn.html

Knowing the crystal structure is the only virtue that we must
extract from acquired data or to get a degree.
Lecture notes, graduated student

In ignorance of crystallization we hurt only venture capitalists.
CrystalClear AB

Striving to make money from protein crystallization is not a sin.
Business Angel, Heavenly Temple, China

Japanese crystallization is called a *"juncture of crystal souls"*
in Buddhism. Not only does the pipette do the job, the fingers,
hands or chopsticks come into play too. Passionate Japanese (males)
invented a variation on *hara-kiri* in which swords are employed
but only in situations where there is no crystal.
NHK TV

The act of making one sitting or hanging drop puts 29 facial and finger muscles in motion. In other words, crystallizing in the long run can be used as an effective exercise to prevent wrinkles, depression, and heart attack, and is a good alternative to jogging.

Medical adviser

A quick crystallization check will burn about 2 to 3 kcalories, whereas the French method of crystallizing (openmouthed and supported by one glass of champagne) will obliterate more than 5 kcalories.

Obesity fighter

Sensitivity of the crystal grower's eye is 200 times higher than that of the fingers or SONY CCD cameras.

Optician

Glossary

In the following, we define many of the words from the field of crystallization used in this book.

Amorphous — Having no definable shape, non-crystalline, having no structure at the molecular level.

Artificial antibodies — Chromatographic matrices prepared by molecular imprinting. The gel antibodies contain cavities which have a shape complimentary to the shape of the antigen, which along with bonds between the antigen and the "wall of cavity" creates an extremely high selectivity.

Atomic Mass — The mass of an atom, molecule.

Atomic Mass Unit — See *Dalton*.

Atomic Number — The number of protons in the nucleus of each of the atoms of an element. This is a characteristic of that element.

Batch crystallization — Crystallization on the macro scale.

Crystal — Two-dimensional or three-dimensional arrays of molecules, as visually pleasant objects. Examples are: diamond, precious stones, salt, sugar, water, virus, or protein crystals.

Crystallization — A physicochemical process resulting in crystals.

Crystallization cocktail — An aqueous medium in which crystallization may appear; sometimes the composition of such a medium is difficult to determine.

Crystallomic — How to find crystallization conditions using minimum amounts of protein and labor (*low throughput crystallization*). The evaluation of crystallization trials is to assign a sign and a value of the second virial coefficient. (The word crystallomic was coined in 2002 by Dr. Jan Sedzik.)

Crystalomics™ — How to put proteins both old and new to work rapidly and efficiently in the harsh conditions typical of most practical industrial processes and in therapeutic interventions of un-precedented power and benefit.

Crystallomics — Production of highly purified proteins samples and diffraction quality crystals.
http://bioinfo-core.jcsg.org/bic/links/crystallomics.htm

Dalton (Da, D) — Unit of Atomic Mass, equal to 1/12th the mass of one atom of carbon-12 (1.6604×10^{-24} g).

Dish — A stack of wells.

Genomics — The study of genes and their function.

Hanging drop — A method of producing protein crystals on the micro scale. The drop "hangs" over a reservoir and equilibrates with it; during this process crystals may appear.

Ion — An atom, or group of atoms, that has acquired a net electric charge by the gain or loss of one or more electrons.

Mother liquor — See *crystallization cocktail.*

Native conditions — *In vitro*: natural pH, ionic strength similar to human blood values and temperatures from 4°C to 36°C.

Plate — A stack of dishes.

Proteomic — Analysis of cells and tissue; identifying and understanding biological processes and proteins expressed in cells as well as their modifications; direct analysis of the proteolytically digested proteins using liquid separation techniques and tandem mass spectrometry/database searching.

Proteomics — The analysis of complete complements of proteins. Proteomics includes not only the identification and quantification of proteins, but also the determination of their localization, modifications, interactions, activities, and ultimately, their function. Initially encompassing just two-dimensional gel electrophoresis for protein separation and identification, proteomics now refers to any procedure that characterizes large sets of proteins localized in the genome.

Proteonomics — Expression systems that can rapidly produce high levels of recombinant proteins. The baculovirus expression technology makes this system of choice in the emerging field of proteonomics, where rapid production and high yields of biologically active complex proteins are essential in the discovery of new drug targets, vaccines, and biotherapeutics.

Post-genomics — All topic areas associated with obtaining higher biological meaning and function out of raw sequence data.

Screen — Several dishes or plates, comprising random or rationally designed set for crystallization trials.

Silica — Silicon dioxide, SiO_2, a white or colorless crystalline compound, which occurs abundantly in nature as quartz, sand, flint, agate, and many other minerals. When it occurs as a dust and is inhaled, it can cause the lung disease silicosis.

Sitting drop — A method of producing protein crystals on the micro scale. The drop "sits" over a reservoir and equilibrates with it; during this process crystals appear.

Structurally based drug design — Finding an efficient drug based on knowledge of the three-dimensional structure of an infectious agent (protein, virus).

Synchrotron source — A specially built X-ray source of extremely high intensity.

Well — A single device where crystallization can be performed by hanging or sitting drop.

X-ray — Non-nuclear radiation generated by extremely high temperatures or the bombardment of heavy metals, such as tungsten, by kilovolt electrons in X-ray machines. X-rays have wavelengths between 10^{-11} to 10^{-08} m (0.1 to 100 Å), placing them between ultraviolet light and gamma-rays.

X-rays are used in medical diagnostics and to measure the separation of atoms and crystalline planes within crystals.

Ångström (Å) — A unit of length equal to 10^{-10} m. It is used to describe crystalline unit cell dimensions and the wavelengths of light and X-rays. It is more properly termed 0.1 nm (nanometer), but its long-term and widespread usage has made "Ångström" an acceptable unit of measurement. Swedish physicist Anders Jonas Ångström (1814–1874) lived in Uppsala, Sweden.

Index